Bioeconomy and Sustainability

Dirk Lanzerath • Ulrich Schurr •
Christina Pinsdorf • Mandy Stake
Editors

Bioeconomy and Sustainability

Perspectives from Natural and Social Sciences, Economics and Ethics

Springer

Editors
Dirk Lanzerath
DRZE
University of Bonn
Bonn, Nordrhein-Westfalen, Germany

Ulrich Schurr
IBG-2
Forschungszentrum Jülich
Jülich, Nordrhein-Westfalen, Germany

Christina Pinsdorf
Institute of Science and Ethics (IWE)
University of Bonn
Bonn, Germany

Mandy Stake
INM-8
Forschungszentrum Jülich
Jülich, Nordrhein-Westfalen, Germany

ISBN 978-3-030-87401-8 ISBN 978-3-030-87402-5 (eBook)
https://doi.org/10.1007/978-3-030-87402-5

This Springer imprint is published by the registered company Springer Nature Switzerland AG.
The registered company address is: Gewerbestrasse 11, 6330 Cham, Switzerland

Contents

Part I

Introduction and Overview

Introduction: Bioeconomy and Sustainability

1

Dirk Lanzerath and Ulrich Schurr

Abstract

The concept of "bioeconomy" encompasses a broad interdisciplinary field. It includes a variety of biotechnical and economic applications for more sustainable production of goods and mapping production processes more closely to nature. From an ethical point of view, bioeconomy is on the one hand promoted and embraced as a new sustainable concept for economic activity, but on the other hand, it is criticized as an ongoing form of domination and exploitation of nature. When discussing beyond-natural science aspects of "bioeconomy", one must consider that there are many definitions of bioeconomy and that the concept is by far not static. In debates on "bioeconomy", the concept's dynamic and multi-faceted definition often leads to heated discussions. Therefore, it is useful to look at the development of bioeconomy concepts and how they are embedded in wider developments. Considering the recent scientific and public debates, bioeconomy must *also put up with critical views*. How sustainable are the new biotechnological uses of natural resources? How high are the risks of labeling fraud and consumer deception? Recent years have seen this dilemma at the example of energy crops: based on the good basic idea of using plants as regenerative energy providers, this practice quickly reaches its ecological limits because energy crops are immensely space-consuming, create new monocultures, and displace biodiversity to a large extent.

D. Lanzerath (✉)
German Reference Center for Ethics in the Life Sciences (DRZE), University of Bonn, Bonn, Germany
e-mail: lanzerath@drze.de

U. Schurr
Forschungszentrum Jülich, Institute of Bio- and Geosciences, Jülich, Germany
e-mail: u.schurr@fz-juelich.de

D. Lanzerath et al. (eds.), *Bioeconomy and Sustainability*,
https://doi.org/10.1007/978-3-030-87402-5_1

Keywords

Concept of bioeconomy · Bioethics · Sustainability · Value of nature

The topic of "bioeconomy" encompasses a broad interdisciplinary field with numerous biotechnical and economic applications that aim at a more sustainable production within a national economy and that map processes more closely to nature. From an ethical point of view, bioeconomy is on the one hand promoted and embraced as a new sustainable concept of economic activity, but on the other hand also criticized as an ongoing form of domination and exploitation of nature. In order to discuss this conflict, the German Reference Centre for Ethics in the Life Sciences and the Institute of Science and Ethics at the University of Bonn, together with the Institute of Bio- and Geosciences at the Forschungszentrum Jülich, invited researchers and representatives from different disciplines such as biology, agriculture, economics, ethics, and law to Bonn and Jülich. During an intensive week of discussion, the participants contributed their diverse views on bioeconomy and the sister concept of circular economy. A special focus was also on the comparison of how German and Dutch societies in Europe deal with these issues. These study days have been supported by the German Federal Ministry of Education and Research.

When discussing beyond-natural science aspects of "bioeconomy", one must consider that there are many definitions of bioeconomy and that the concept is by far not static. In the current discourse on "bioeconomy", this diversity and the dynamic nature of the concept often leads to heated discussions. Therefore, it is useful to look at the development of different concepts of bioeconomy and how they are embedded in wider developments. One starting point of bioeconomy was the increasing availability of knowledge in the life sciences that triggered the domain of biotechnology. Here, biological systems are used and (often) engineered to obtain systems (including modified organisms), which are (better) suited to fulfill a specific task in the process from raw material to product. This approach is fully compatible with a often still implemented linear "value chain concept" from raw material to products – a perspective underlying today's fossil-based economy. However, building an economy based on depleting resources is inherently non-sustainable – irrespective of the resources being based on recently formed or fossil biomass (coal, oil, gas): today's ongoing geological processes of forming new fossil resources are significantly slower than their use. Therefore, bioeconomy per se is not necessarily sustainable.

The combination of the insight that sustainable systems are generally circular and that biological systems are organized in circularity, from processes within the cell up to ecosystem level, has lead to the concept of a sustainable bioeconomy in consent with ideas on circular economy. However, the circular setup of a process in itself is not sufficient to fulfil the criteria of sustainability. This can only be the case if the rates of the individual steps transforming one material/chemical into a next material/ chemical, are within "healthy" boundaries. This links sustainable bioeconomy and circular economy to the concept of planetary boundaries. However, planetary boundaries are only approached in the overall global and regional context. Sustainable use of resources also needs to fulfil regional and local criteria since the circularity of matter fluxes often happens in a certain region. Overconsumption of resources in one region, while being within the planetary boundaries, can still significantly damage a local ecosystem and a local economy. While this indicates that bioeconomy and the underlying technologies are not intrinsically sustainable, these ideas gain increasing attention in the wide bioeconomy community with the target to design and implement a sustainable bio-based economy. Against this background, bioeconomy must be seen as part of a wider approach of a sustainable, circular economy – in hybrid with non-biology-based parts of the economy, such as renewable energy systems.

The tight links between economics and biological sciences and ecology respectively become obvious from the reciprocal adaptation of concepts. Both domains often use approaches developed and extended by the other ones, e.g. concepts like competition, efficiency, or energy and matter flow modes are successfully used to describe and quantify economies and ecosystems. The competition for the best-suited biological and economic systems develops due to the principle "survival of the fittest," and evolution is a deeply economy-inspired concept. Therefore, it is not surprising that humans optimize and adapt plants and animals to their needs and design "biological resources."

Designing "new" biological systems gained significant momentum from discoveries on the material/chemical nature of the basis of genetics being DNA and reaches back into the early twentieth century. However, it is important to keep in mind that adaptation of plants and animals to the need of humans happened throughout the history of mankind; massively accelerated in medical and agricultural dimensions with the start of agriculture and selection of the plants and animals more suited to human needs. While this was initially empirical, with Mendel's rules being groundbreaking to systematic analysis and crossing of lines with different genetic setups, breeding became a scientific discipline to modify plants and animals for the needs of humans. On the one end, this boosted the availability of food and thus helped to overcome regularly experienced famines. To the other end, this also generated the wide diversity of pets living with humans today. Here, we can also see extremely problematic developments, e.g., dogs that can hardly breathe anymore, or even more problematic in-race-theories and their devastating consequences. This sequence clearly indicates the need for ethical debates being aligned with the development of technologies: scientific concepts and technologies – not only in the bio–/life sciences – can generate ethically beneficial as well as extremely problematic developments depending on the application, but as these applications are human innovations, they are subject to ethical norms.

If one takes an ethically inspired look at the compound term "bio-eco-nomy" with regard to the original Greek meanings of its components *bios*, *oikos* and *nomos*, then one encounters very traditional and semantically highly demanding terms.

Living entities and ways of being of life forms *(bios)* are designated in a way that distinguishes them from artifacts and technical products due to their capabilities of self-organization and reproduction. Their aliveness results from natural intra-organismic and extra-organismic circulatory systems as well as interactions of organisms with their environments in the form of other biological units and with the non-living environment. It is precisely the capacities of living systems, namely *reproduction* and *regeneration*, that are the basic prerequisites for introducing the principle of "sustainability" as a form of sparing use of resources in economics. This basic economic idea of not consuming more than what can grow back, regenerate, and be made available again in the future brings us to the second part of the concept, the *oikos*.

It is no coincidence that *oikos* is a formative semantic component in the designations of areas of human life or scientific disciplines that are perceived today as rather opposed to each other, such as *ecology* and *economy*. Many people feel that the economy is essentially detrimental to ecology. Yet both fields are about the oikos: the well-managed house, the successful household, once in nature, once in the business. Nevertheless, ecological households are initially natural systems, economic households in principle artificial ones. In the *anthropocene*, however, *natural households are becoming increasingly artificial,* while bioeconomy – as a new form of economic activity – attempts to *use or reproduce natural cycles* in order to make sustainable economic activity possible. The concept of sustainability tries to advocate economic forms compatible with nature in supporting the ability of life cycles to regenerate and reproduce. Ideally, this should lead to the preservation of natural and economic potentials for future human generations, but it also aims at the conservation of nature *as* nature, which should not only be regarded as a cheap resource. An economically appreciative view of nature is also reflected in the idea of ecosystem services. Without *ecosystem services*, human life on Earth would not be possible. Nevertheless, there is a considerable semantic and practical risk of reducing its appreciation to economic values only.

This leads us to the third conceptual component, *nomos*. Typically, nomos denotes *law* and is intended to enlighten the regularities of the relationships under investigation. It is about understanding the cycles in nature and in the economy. In its original meaning, however, nomos is also connected with the verb *nemein*, which means "to allocate" or "to divide up." A *successful allocation and division* of economic goods cannot take place in modern times without normative criteria of sustainability and justice. This not only challenges societies to ethically rethink the ways in which we manage our households, but also requires the initial involvement of natural science as well as that of societies with its stakeholders to initiate broad debates on bioeconomy. Thus, the concept of bioeconomy not only points out how we can learn from nature and use natural resources more intelligently and resource-efficiently in our living environment via modern biotechnologies, but it also requires thorough ethical reflection, not only regarding newly arising technical challenges, but also demands on ethics and law. The sustainable use of resources and strategies for the *fair distribution of scarce goods* does not come about by itself. Instead, providing good reasons for an intergenerationally fair distribution of resources and justifying the use of nature are core challenges that have to be addressed in bioeconomic approaches, which is also where their great opportunities lie if they are understood in an interdisciplinary way.

At the same time, bioeconomy must *also put up with critical views*, as it cannot be ruled out that new biotechnological processes only appear to be more natural and sustainable when they indeed describe only new, powerful forms to dominate nature. How sustainable are the new biotechnological uses of natural resources? How high are the risks of labeling fraud and consumer deception? Recent years have seen these risks at the example of energy crops, which has revealed that the good basic idea of this regenerative energy use quickly reaches its ecological limits because energy crops are immensely space-consuming, create new monocultures, and displace biodiversity to a large extent.

Sustainability has an inherently anthropocentric element, as the target is to maintain resources and opportunities for future (human) generations. However, the claim to protect nature and biodiversity exceeds the use potential, focusing on respect for other creatures and for nature as such. This sets additional boundaries for the development of future economies and specifically (the development of) bioeconomy, as the goal of conservation sets limits for the maximization of economic output – even if it is sustainable. These debates also need to include unintentional impact. In the past, with the selection of crop properties and the development of agronomic practices, humans also impacted soil microbe evolution, without even knowing that these microbes exist. Today, the fact that the expansion of land use by humans causes the mass extinction of species which we experience today and that the push into hitherto protected ecosystems can trigger pandemics for humans, animals, and crops is a clear indication that there are limits beyond what is technically feasible. However, this does not ask for a shutdown of all technological developments and applications, but rather for a proper discussion of risks and benefits of applications based on scientific findings.

It is, therefore, necessary to look very closely in order to form *constructive judgments* about new technologies and their possible uses. This can only be done in an *interdisciplinary way* between the life sciences, the economic and social sciences, and the normative disciplines of ethics and law.

In the technical-scientific section, the focus is on implementation and innovation in bioeconomy. Katrin Beer describes the complex interaction of political dimensions specifically in the context of the bioenergy sector in Germany. As a typical "wicked problem", bioenergy requires technological solutions as well as societal discussion. Leonard Prochaska and Daniel Schiller have analyzed biotechnology and bioeconomy innovation landscapes in Germany with a special focus on public innovation strategies. The nature of innovation processes in bioeconomy and economy and society in general is presented by Max Mittenzwei. Ralf Pude and colleagues illustrate practical innovations of sustainable resources – from plants to products. Lioudmila Chatalova uses models that are well established in ecology research. Based on predator–prey models, she argues that in a sustainable economy, seeking to align its economic interests with the carrying capacity of the environment, sufficiency presents a critical element of the negative-feedback mechanism that regulates resource consumption. Florian Fiebelkorn, Jacqueline Dupont, and Patrik Lammers discuss the various dimensions of alternatives to meat consumption as a central opportunity to reduce the environmental footprint of human nutrition. The discourse culture in media on bioeconomy and differences in coverage of stakeholder groups have been analyzed by Sandra Venghaus, Sophia Dieken, and Maria Belka indicating a significant difference between the techno-oriented and the behavioral perspective on bioeconomy. Jan Börner's contribution focuses on the need for social aspects of transition beyond the focus on pure technological progress to achieve the needed reduction of land use change for a sustainable future.

The normative contributions of the volume deal in particular with the different approaches to view the bioeconomic debate ethically and discuss regulatory proposals. The article by Robin Siebert, Christian Herzig, and Marc Birringer compares in a European perspective how two neighboring societies, the Dutch and the German, deal with regulations in bioeconomic fields of application, especially in the debate on the use of genome editing techniques. Maximilian Kardung describes in his article whether bioeconomy actually promotes the human pursuit of happiness against the background of the three pillars of sustainability: economic development, social development, and environmental protection. Justus Wesseler and Maximilian Kardung describe the enormous economic significance of sustainable management.

Bart Gremmen as well as Marion Stahl discuss the multifaceted ethical debate on how to deal with the bioeconomy in society in their respective articles. The latter article describes how elaborated environmental-ethical reflections can be seen as an informative basis for developing guidelines for action and implementation strategies in the context of politics, economy, and society, which allow bioeconomic strategies to be value-based and sustainable. Jana Schoop's article elaborates how Hans Jonas' ethics of responsibility helps to concretize the term of responsibility, because the current ecological situation does not solely require an economic transformation, but a fundamental philosophical reflection of the relationship between humans and nature. To specify the concept of sustainability, it is shown that sustainability, as an ethical principle, can be derived from Jonas' ethics. From the assessment of bioeconomy as a resilience-based concept, Sebastian Lenze concludes in his article that a system or entity has to fulfil three criteria in order to continue functioning in crisis situations: First, it must promote a second-order concept of resilience, that is, one that takes other contexts into account; second, the interpretation of the resilience concept used must be clarified – i.e., whether it holds a conservative-reactive, an adaptive or a transformational understanding; and, third, it has to contain a clearly defined normative dimension. Since it is precisely the sector of food production that is largely affected by bioeconomic processes, some of the contributions in this volume deal in particular with the normative issues of this area of bioeconomy. In her article, Birgit Beck emphasizes that food ethics can provide a differentiated assessment of strategies and technologies applied in bioeconomy by means of scrutinizing respective current theoretical and practical issues, for instance, those involving novel food technologies. She draws a rough sketch of food ethics in terms of a comprehensive theory of the good life and discusses central arguments concerning a paradigmatic example of technical solutions for moral problems in the context of bioeconomy, namely in vitro meat. Patrick Hohlwegler raises the question of whether the implementation of a bioeconomy will lead to moral conflicts regarding food security in the face of climate change and population growth. He argues that due to climate change and due to the anticipated population growth, moral conflicts will arise very likely regarding food security by implementing a bioeconomy based on biofuels. This leads to the assumption that our societies need to live and consume far more sustainably since our consumption patterns are the main driver of climate change. In her contribution, Mandy Stake discusses the overlapping main challenges and conflicts in the bioeconomy debate that were recognized during the immersion workshop and shows how they cluster around the key topics of sustainability, economical efficiency, human self-understanding and our role in nature. She illustrates how closely the concept of bioeconomy is bound up with the ideas of human self-understanding, for it is the human life form that decides how it intends to act, which economic system it prefers, and which relationship it adopts to nature and its organisms. These background assumptions influence how sustainable and just a society can be. In her philosophical contribution, Christina Pinsdorf analyses persisting challenges underlying the concept of bioeconomy from a more theory-driven philosophical perspective and from a rather application-oriented ethical point of view. In doing so, she reveals tensions concerning the relations between economy and man as well as between economy and nature and points to bioeconomic promises and disillusions.

Although this volume also contains introductory and comprehensive contributions on bioeconomic discourses, it is not intended as an introduction to bioeconomic discussions,[1] but rather as a volume discussing central elements of current discourse on bioeconomic research and application scenarios based on concrete examples.

Acknowledgments The joint editors would especially like to thank the German Federal Ministry of Education and Research (BMBF) for providing the funds for the workshop and the publication, as well as the staff at the German Reference Centre for Ethics in the Life Sciences (DRZE) and of the Institute of Science and Ethics (IWE) at the University of Bonn for their organizational support, namely Dorothee Güth, and editorial assistance in publishing this volume.

We would like to thank our co-editors Christina Pinsdorf and Mandy Stake for the excellent organization before and during the workshop week and additionally for the editorial preparation of the manuscripts for the provision of this book. This step was especially supported by Jana Schoop, who is also to be expressed thanks here. The entire editorial team is also grateful to all participants and contributors to the workshop week for intensive and stimulating discussions, namely (in alphabetical order) Birgit Beck, Katrin Beer, Jan Börner, Lioudmila Chatalova, Florian Fiebelkorn, Bart Gremmen, Patrick Hohlwegler, Maximilian Kardung, Julian Kinderlerer, Sebastian Lenze, Max Mittenzwei, Patricia Osseweijer, Leonard Prochaska, Ralf Pude, Jana Schoop, Robin Siebert, Marion Stahl, Dieter Sturma, Sandra Venghaus and Justus Wesseler. We also acknowledge the support of Maria Belka, Marc Birringer, Sophia Dieken, Jacqueline Dupont, Christian Herzig, Thorsten Kraska, Patrik Lammers, Daniel Schiller, and Christian Wever for their written contributions to this book.

[1]For general introduction, we propose to have a look at Pietzsch, J. (2020) Bioeconomy for Beginners Springer, ISBN 978-3-662-60390-1 or the German Version Pietzsch, J (2017) Bioökonomie für Einsteiger, Springer Verlag, ISBN 978-3-662-53763-3.

Part II

Energy and Land Use

"Global Shifting Agriculture" and Bioeconomy: Challenges for the Sustainable Use of Global Land Resources

2

Jan Börner

Abstract

For decades, introductory lectures in agricultural sciences begin by confronting students with the historical and projected future development of global population numbers. Depending on scenario assumptions, a world in 2100 may have to feed between 7 and 17 billion people. Lecturers use these demographic projections mainly to convince students that they made the right career choice: agricultural sciences will have to play a major role in developing technologies that boost primary sector productivity; allegedly the first-best strategy for the provision of sufficient food at affordable prices, while minimizing global cropland expansion.

This essay does not deny the need to develop crop varieties that produce higher and more reliable crop yields. It will argue, however, that technological innovation in agriculture is not enough to enable transformation towards a globally sustainable bioeconomy. This view is supported by the academic debate around the Sustainable Development Goals (SDGs), which highlights numerous synergies, but also tradeoffs between the multidimensional global agenda for 2030. We proceed in three steps: First, we revisit the theoretical foundations of the idea put forward by Nobel laureate Norman Borlaug, that productivity increases in agriculture reduce the demand for new farmland. Second, we synthesize recent empirical research supporting the view that Borlaug's hypothesis is a necessary, but not a sufficient condition for a sustainable global bioeconomy. And third, we highlight potential ingredients of a science and policy strategy that provides the necessary social and environmental safeguards for more sustainable innovation in agriculture.

J. Börner (✉)
Institute for Food and Resource Economics (ILR) and Center for Development Research (ZEF), University of Bonn, Bonn, Germany
e-mail: jborner@uni-bonn.de

D. Lanzerath et al. (eds.), *Bioeconomy and Sustainability*,
https://doi.org/10.1007/978-3-030-87402-5_2

Keywords

Productivity in agriculture · Sustainable global bioeconomy · Sustainable cropland use

1 Background

For decades, introductory lectures in agricultural sciences begin by confronting students with the historical and projected future development of global population numbers. Depending on scenario assumptions, a world in 2100 may have to feed between 7 and 17 billion people. Lecturers use these demographic projections mainly to convince students that they made the right career choice: agricultural sciences will have to play a major role in developing technologies that boost primary sector productivity; allegedly the first-best strategy for the provision of sufficient food at affordable prices, while minimizing global cropland expansion.

This essay does not deny the need to develop crop varieties that produce higher and more reliable crop yields. It will argue, however, that technological innovation in agriculture is not enough to enable transformation towards a globally sustainable bioeconomy.[1] This view is supported by the academic debate around the Sustainable Development Goals (SDGs), which highlights numerous synergies, but also tradeoffs between the multidimensional global agenda for 2030.[2] We proceed in three steps: First, we revisit the theoretical foundations of the idea put forward by Nobel laureate Norman Borlaug, that productivity increases in agriculture reduce the demand for new farmland. Second, we synthesize recent empirical research supporting the view that Borlaug's hypothesis is a necessary, but not a sufficient condition for a sustainable global bioeconomy. And third, we highlight potential ingredients of a science and policy strategy that provides the necessary social and environmental safeguards for more sustainable innovation in agriculture.

2 Agricultural Technology Change and "Global Shifting Agriculture"

Borlaug's intuition goes as follows: if new agricultural technologies boost per hectare crop productivity, more can be produced on the same or even a smaller amount of land. Higher crop yields increase the supply on agricultural output markets, where prices drop and thus reduce the incentives for cropland expansion. Already in the nineteenth century, British economist William Jevons challenged the

[1]We define bioeconomy, inspired by the German Bioeconomy Council, as the production and utilization of biological resources (including knowledge) to provide products, processes, and services across sectors of an economy. Defining bioeconomy in this way allows for both sustainable and unsustainable transformation outcomes.

[2]Cf. Timko et al. (2018) and Biber-Freudenberger et al. (2018).

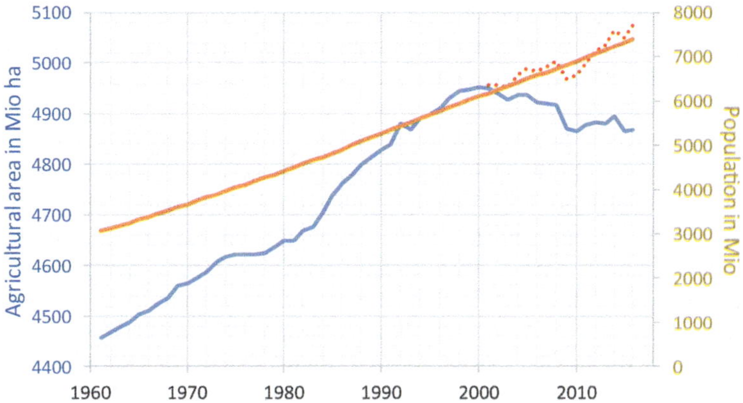

Fig. 2.1 Global growth in net-agricultural area (blue) and population (orange). The dotted red line adds cumulative net tree cover loss between 2001 and 2016 to net-agricultural area. Sources: World Bank, FAOTAT, University of Maryland

general validity of this calculus as he witnessed how technologies that enabled a more efficient use of coal triggered tremendous growth, as opposed to savings, in fossil fuel consumption during his lifetime. Clearly, fuel scarcity may have been a driver of technological innovation in the first place. Yet, over time, growth in the consumption of manufacturing goods from more fuel-efficient industries has arguably outpaced the effect of initial fuel savings on total fossil resource use. This so-called "rebound effect" is context specific and has since been observed in many other settings.[3]

In defense of Borlaug's vision, we may contend that people can only consume so much food, such that rebound effects are less likely to be mediated via food markets compared to other commodity groups due to market saturation. As we illustrate below, however, food and energy consumption patterns, especially in industrialized countries, as well as the current generation of bioeconomy strategies around the world suggest otherwise. New processing technologies increasingly allow for a variety of non-food biomass uses that could boost future demand for cropland and pastures.

Moreover, Borlaug's calculus implicitly assumes that more efficient agricultural technologies will be applied on the world's current and most productive croplands. This turns out to be a strong assumption if we consider that agricultural land markets and environmental policies in many world regions fail to effectively allocate land to its most valuable use for society.[4] As a result, we often find large tracts of agricultural land abandoned in one region, while it expands into natural ecosystems elsewhere.[5] Figure 2.1 shows that standard cropland accounting procedures grossly

[3] Cf. Herring and Roy (2007), Berbel and Mateos (2014) and Smeets et al. (2014).

[4] Cf. Deininger and Feder (2001) and Miranda et al. (2019).

[5] Cf. Schierhorn et al. (2019).

understate the actual space needed for agricultural production globally since 2001. The apparent decoupling of global population growth and agricultural area expansion after the year 2000 vanishes if we account for global net tree cover loss between 2001 and 2016.

In other words, the lion's share of global cropland and pastures is continuously used, but a considerable share shifts geographically subject to interannual dynamics of expansion and abandonment that are mediated by economic and political factors, but also through the emergence of technologies that enable agricultural production on previously unused land.[6] This global form of "shifting agriculture" has no agronomic rationale. It is not required to maintain soil productivity as in many traditional agricultural systems still practiced around the world. Instead, it leads to avoidable, often irreversible, environmental damage and not seldom provokes the displacement of traditional and indigenous populations and smallholders.[7]

Constraining the development of productivity-enhancing agricultural technologies altogether would be a poor response, nonetheless. According to Nelson Villoria, for example, additional 125 million hectares of land would have been needed to satisfy global food demand between 2001 and 2010 in the absence of technological innovation in agriculture.[8] In the context of the global food system, the Borlaug versus Jevons debate instead suggests that the opportunities of future technological change come with increasing international governance needs.[9] Failing to address these challenges may ultimately jeopardize ecosystem functions that maintain agricultural production, such as regional climate regulation, and support our ability to develop more sustainable production systems, such as genetic and species diversity.

3 Lessons from the South American Soy Boom

Land use change is an expression of societies' production and consumption patterns. The rise of the soybean economy in South America will serve us to illustrate some of the mechanisms that drive the expansion and abandonment of agricultural land. Soybeans and their derivatives are the most traded agricultural commodity worldwide. According to FAOSTAT, almost 350 million tons of soy were produced on close to 125 million hectares globally in 2018. The South American share in global soybean production has increased from roughly 25% in the 1980s to over 50% since 2010.

[6] Cf. Angelsen and Kaimowitz (2001) and Villoria (2019).

[7] Cf. Baccini et al. (2012), Obidzinski et al. (2012) and Barlow et al. (2016).

[8] Cf. Villoria (2019).

[9] Cf. Carrasco et al. (2014).

Beyond food uses, other bioeconomy sectors, such as biofuel and animal production in various world regions, were major demand side drivers of this boost in output.[10] Why did it happen in South America and what were the impacts locally?

While South American soy producers became more productive over the years, the main contribution to the observed increase in total production was a massive cropland expansion. Between 1980 and 2018, average soy yields in the region increased less than twofold, whereas harvested area increased by a factor of five to 57 Mha (FAOSTAT). Direct planting technology in combination with roundup-ready soybean varieties adapted to various South American climate zones have played a major role as push factors in bringing soy to the region's agricultural frontiers.[11] As a result, soybean production has become a major driver of the environmentally costly conversion of South America's natural and biodiverse dry and tropical forests.[12]

Potential for further agricultural expansion is considerable. According to Frey et al., Brazil's Amazon region alone holds sufficiently well-suited land to accommodate over six times more soy than the roughly 2.3 Mha planted in that region in 2014.[13] Whether or not soy can expand on these lands depends mainly on transport infrastructure investments and the effectiveness of environmental policies that legally restrict the conversion of forests to agriculture. Road and fluvial transport infrastructure improvements reduce transport costs and thus literally pave the way for farmers and processing industries to unlock the agricultural potential of remote forest zones. The effect of soy production on deforestation may not always be direct, however. Research has repeatedly produced evidence for indirect land use change, where soy producers rent or buy extensively used pastures, while cattle production expands elsewhere via both legal and illegal deforestation.[14] While most South American countries have formulated environmental policies to control illegal land cover change, lack of implementation capacities or fluctuations in political priorities limit their effectiveness.[15]

Indirect land use change can happen at local, regional, and global scale and represents one of the mechanisms through which "global shifting agriculture" occurs. While cause-effect relationships are chronically hard to establish, the environmental impacts of displacing agricultural land across the globe are sizeable. Schierhorn et al., for example, estimate that cropland abandonment in the former Soviet Union led to greenhouse gas emission savings of over 7 Gt (approximately the amount emitted by the USA in 1 year) between 1991 and 2011.[16] From the perspective of global agricultural land accounting, the abandonment roughly offset

---

[10] Cf. Bruckner et al. (2019) and Pendrill et al. (2019).

[11] Cf. Grau et al. (2005) and Trigo et al. (2009).

[12] Cf. Gasparri et al. (2013).

[13] Cf. Frey et al. (2018).

[14] Cf. Richards et al. (2014) and Gasparri and Le Waroux (2015).

[15] Cf. Nolte et al. (2017).

[16] Cf. Schierhorn et al. (2019).

the increase in cropland and pastures in South America over the same period. According to Schierhorn et al., however, emissions from land cover change in South America exceeded savings by factor four, while net biodiversity loss was not assessed.

Key lessons from the South American soy boom and the broader debate on indirect land use change can be summarized as follows: Despite the obvious economic benefits of trade in agricultural commodities, increasingly complex global value chains allow for shifts in consumption and production patterns to more effectively propagate and thus amplify regional patterns of land abandonment and expansion. Profitable technological innovations can reinforce processes of land expansion at agricultural frontiers, especially where public and private infrastructure investments improve access to land and when land use regulations and property rights are poorly enforced.

As such, the environmental and social costs of "global shifting agriculture" can be considered a collective externality of the trade system that links consumers and producers of agricultural and forestry commodities across the globe. As usual, liability and responsibility for these costs are difficult to establish, but in the absence of internationally negotiated and locally enforced land use regulations, productivity-enhancing agricultural technologies are part of the problem.

4 Bioeconomy and Global Land Resources

In line with our definition of bioeconomy, "global shifting agriculture" is inherently a bioeconomic phenomenon. As more and more countries around the world develop strategies to promote their bioeconomies, an intriguing question is whether future generations will look back at bioeconomic transformation as a Borlaugian symphony or a Jevonsian cacophony. Answers are so far speculative by nature and thus often rely on modeling studies. Escobar et al., for example, simulated alternative scenarios of policy support to increase the reliance on biomass as opposed to fossil fuels for the production of bioplastics.[17] Assuming the current state of biomass conversion technologies, they find that bio-based plastics will only pay off in terms of carbon emission savings after more than 20 years due to indirect land use change. Earlier, Hertel et al. had demonstrated similar limitations of attempts to promote the use of bio-based fuels.[18]

Beyond substituting bio-based for fossil resources, visions of future bioeconomies also embrace circular economy principles and arguably land neutral technologies, such as the use of enzymatic instead of chemical conversion processes.[19] Here it can help to differentiate between more efficient biomass uses, which, despite their potential benefits, may produce rebound effects and the

[17] Cf. Escobar et al. (2018).

[18] Cf. Hertel et al. (2010).

[19] Cf. Meyer (2017).

application of bio-based principles in land independent sectors, such as medicine or the pharmaceutical industry. Growing this latter part of bioeconomy is unlikely to put additional pressure on global land resources.

A closer look at national bioeconomy strategies, however, reveals that most countries place strategic emphasis on technological innovation in the sectors that traditionally rely on the production, conversion, and consumption of biological resources.[20] To the extent that these strategies promote alternative biomass uses and refinement, they potentially push the limits of saturation that govern traditional food markets and led Borlaug to propose that land can be saved by boosting agricultural yields.

All this would not be a problem, if the world's remaining natural landscapes were protected by effective use regulations. Unfortunately, enabling policy measures for bioeconomy and voluntary sustainability labels feature more prominently in national bioeconomy strategies than binding environmental and sustainability safeguards.[21] All else equal, bio-based innovations and enabling policy support will thus most likely align to put additional pressure on global land resources.

5 Way Forward

Coordinating action towards internalizing the costs of globally shifting land use incentives may seem like an insurmountable "wicked problem."[22] Consumers blame farmers for unsustainable production practices, farmers bemoan exaggerated consumer expectations and costly regulations, and technology developers maintain that ill-designed policies and institutions prevent their innovations from unfolding their inherent sustainability potential. Most parties ignore or downplay their own contribution to the undesirable collective outcomes.

To turn this blame game into a constructive dialogue we need to tune up the conventional Borlaugian chant of agricultural sciences with the lessons of recent multidisciplinary research on global land use change dynamics. Based on what we have learned about the drivers of land use change in various regional contexts, improved science-based decision support can help us anticipate where and when incentive structures shift in favor of land expansion at agricultural frontiers. Evidence-based methods in combination with unprecedented access to remotely sensed data on land use change have also greatly enhanced our ability to measure the effectiveness of agricultural and environmental policies. Insights from these applied fields of research can and should inform not only policy design but also goal-oriented priority setting for basic research and technology development.

At the policy level, we need to push our leaders to move from global goals to collective deeds. Few of the land-related Sustainable Development Goals (SDGs), if

[20] Cf. Dietz et al. (2018).

[21] Cf. ibid.; Grossauer and Stoeglehner (2020).

[22] Cf. DeFries and Nagendra (2017).

any, can be achieved through uncoordinated action.[23] The transaction costs involved in negotiating binding multilateral treaties are a necessary price to pay, which will reduce as we make headway towards equally sharing the benefits and costs of economic prosperity among winners and losers. To do so, we need to acknowledge that there are limits to governing global trade in bio-based commodities via improved value chain transparency and voluntary sustainability labels. Certification schemes can complement, but not substitute for functioning national and subnational land use regulations.[24] This is because adverse selection mechanisms often exclude those segments of the producer spectrum, where changes in production practices are costly, but bring about the largest sustainability gains.

A "new deal" to govern global land resources must leverage the potential power of conditional compensation mechanisms, such as the Sustainable Development Mechanism (SDM) in the Paris Agreement or Reducing Emissions from Deforestation and Degradation (REDD+), which can be flexibly designed to target agricultural frontiers affected by global shifts in economic incentives for agricultural expansion. So far publicly funded programs have made substantial progress in preparing the ground for such compensation schemes to work more effectively, for example, by establishing land cover monitoring systems and rural land cadasters.[25] Funding remains a bottleneck to scale up international compensation schemes, be it for land-based climate change mitigation or biodiversity conservation. However, we should not forget that many of the perceived benefits of bio-based transformation are expected to accrue as positive externalities and thus may require policy support.[26] This potentially creates synergies between bioeconomic transformation and the protection of global land resources at least in the context of climate policy. Emission taxes or offset trading schemes provide incentives for climate-smart (including bio-based) innovation and at the same time generate revenues that can and should be used to compensate countries for additional efforts towards protecting globally valued land resources. Compensations offered so far represent only a fraction of the actual costs of safeguarding ecologically sensitive ecosystems around the world and thus cannot be expected to effectively curb global forest loss.

References

Angelsen, A., & Kaimowitz, D. (2001). *Agricultural technologies and tropical deforestation.* CIFOR, CABI Publishing.

Auld, G., Gulbrandsen, L. H., & McDermott, C. L. (2008). Certification schemes and the impacts on forests and forestry. *Annual Review of Environmental Resources, 33*, 187–211.

[23] Cf. ibid.

[24] Cf. Auld et al. (2008) and DeFries et al. (2017).

[25] Cf. Ochieng et al. (2016) and Maniatis et al. (2019).

[26] Cf. Wesseler and von Braun (2017).

Baccini, A., Goetz, S. J., Walker, W. S., Laporte, N. T., Sun, M., Sulla-Menashe, D., Hackler, J., Beck, P. S. A., Dubayah, R., & Friedl, M. A. (2012). Estimated carbon dioxide emissions from tropical deforestation improved by carbon-density maps. *Nature Climate Change, 2*, 182–185.

Barlow, J., Lennox, G. D., Ferreira, J., Berenguer, E., Lees, A. C., Mac Nally, R., Thomson, J. R., de Barros Ferraz, S. F., Louzada, J., & Oliveira, V. H. F. (2016). Anthropogenic disturbance in tropical forests can double biodiversity loss from deforestation. *Nature, 535*, 144–147.

Berbel, J., & Mateos, L. (2014). Does investment in irrigation technology necessarily generate rebound effects? A simulation analysis based on an agro-economic model. *Agricultural Systems, 128*, 25–34.

Biber-Freudenberger, L., Basukala, A. K., Bruckner, M., & Boerner, J. (2018). Sustainability performance of National bio-economies. *Sustainability, 10*(8), 2705.

Bruckner, M., Häyhä, T., Giljum, S., Maus, V., Fischer, G., Tramberend, S., & Börner, J. (2019). Quantifying the global cropland footprint of the European Union's non-food bioeconomy. *Environmental Research Letters, 14*, 45011.

Carrasco, L. R., Larrosa, C., Milner-Gulland, E. J., & Edwards, D. P. (2014). A double-edged sword for tropical forests. *Science, 346*, 38.

DeFries, R., & Nagendra, H. (2017). Ecosystem management as a wicked problem. *Science, 356*, 265–270.

DeFries, R. S., Fanzo, J., Mondal, P., Remans, R., & Wood, S. A. (2017). Is voluntary certification of tropical agricultural commodities achieving sustainability goals for small-scale producers? A review of the evidence. *Environmental Research Letters, 12*, 33001.

Deininger, K., & Feder, G. (2001). Land institutions and land markets. In B. L. Gardener & G. C. Rausser (Eds.), *Handbook of agricultural economics. Volume 1A. Agricultural production* (1st ed., pp. 287–331), Elsevier.

Dietz, T., Boerner, J., Foerster, J. J., & von Braun, J. (2018). Governance of the bioeconomy: A global comparative study of National Bioeconomy Strategies. *Sustainability, 10*(8), 3190.

Escobar, N., Haddad, S., Boerner, J., & Britz, W. (2018). Land use mediated GHG emissions and Spillovers from increased consumption of bioplastics. *Environmental Research Letters, 13*, 125005.

Frey, G., West, T., Hickler, T., Rausch, L., Gibbs, H., & Börner, J. (2018). Simulated impacts of soy and infrastructure expansion in the Brazilian Amazon: A maximum entropy approach. *Forests, 9*, 600.

Gasparri, N. I., Grau, H. R., & Angonese, J. G. (2013). Linkages between soybean and Neotropical deforestation: Coupling and transient decoupling dynamics in a multi-decadal analysis. *Global Environmental Change, 23*, 1605–1614.

Gasparri, N. I., & Le Waroux, Y. P. (2015). The coupling of south American soybean and cattle production Frontiers. New challenges for conservation policy and land change science. *Conservation Letters, 8*, 290–298.

Grau, H. R., Aide, T. M., & Gasparri, N. I. (2005). Globalization and soybean expansion into semiarid ecosystems of Argentina. *Ambio, 34*, 265–266.

Grossauer, F., & Stoeglehner, G. (2020). Bioeconomy—Spatial requirements for sustainable development. *Sustainability, 12*(5), 1877.

Herring, H., & Roy, R. (2007). Technological innovation, energy efficient design and the rebound effect. *Technovation, 27*, 194–203.

Hertel, T. W., Golub, A. A., Jones, A. D., O'Hare, M., Plevin, R. J., & Kammen, D. M. (2010). Effects of US maize ethanol on global land use and greenhouse gas emissions: Estimating market-mediated responses. *Bioscience, 60*, 223–231.

Maniatis, D., Scriven, J., Jonckheere, I., Laughlin, J., & Todd, K. (2019). Toward REDD+ implementation. *Annual Review of Environmental Resources, 44*, 373–398.

Meyer, R. (2017). Bioeconomy strategies. Contexts, visions, guiding implementation principles and resulting debates. *Sustainability, 9*(6), 1031.

Miranda, J., Boerner, J., Kalkuhl, M., & Soares-Filho, B. (2019). Land speculation and conservation policy leakage in Brazil. *Environmental Research Letters, 14*, 45006.

Nolte, C., Le Polain de Waroux, Y., Munger, J., Reis, T. N. P., & Lambin, E. F. (2017). Conditions influencing the adoption of effective anti-deforestation policies in South America's commodity Frontiers. *Global Environmental Change, 43*, 1–14.

Obidzinski, K., Andriani, R., Komarudin, H., & Andrianto, A. (2012). Environmental and social impacts of oil palm plantations and their implications for biofuel production in Indonesia. *Ecology and Society, 17*(1), 25.

Ochieng, R. M., Visseren-Hamakers, I. J., Arts, B., Brockhaus, M., & Herold, M. (2016). Institutional effectiveness of REDD+ MRV: Countries Progress in implementing technical guidelines and good governance requirements. *Environmental Science & Policy, 61*, 42–52.

Pendrill, F., Persson, U. M., Godar, J., Kastner, T., Moran, D., Schmidt, S., & Wood, R. (2019). Agricultural and forestry trade drives large share of tropical deforestation emissions. *Global Environmental Change, 56*, 1–10.

Richards, P. D., Walker, R. T., & Arima, E. Y. (2014). Spatially complex land change. The indirect effect of Brazil's agricultural sector on land use in Amazonia. *Global Environmental Change, 29*, 1–9.

Schierhorn, F., Kastner, T., Kuemmerle, T., Meyfroidt, P., Kurganova, I., Prishchepov, A. V., Erb, K.-H., Houghton, R. A., & Müller, D. (2019). Large greenhouse gas savings due to changes in the post-soviet food systems. *Environmental Research Letters, 14*, 65009.

Smeets, E., Tabeau, A., van Berkum, S., Moorad, J., van Meijl, H., & Woltjer, G. (2014). The impact of the rebound effect of the use of first generation biofuels in the EU on greenhouse gas emissions. A critical review. *Renewable & Sustainable Energy Reviews, 38*, 393–403.

Timko, J., Le Billon, P., Zerriffi, H., Honey-Rosés, J., de la Roche, I., Gaston, C., Sunderland, T. C. H., & Kozak, R. A. (2018). A policy nexus approach to forests and the SDGs: Tradeoffs and synergies. *Current Opinion in Environmental Sustainability, 34*, 7–12.

Trigo, E., Cap, E., Malach, V., & Villarreal, F. (2009). *The Case of Zero-Tillage Technology in Argentina.* IFPRI Discussion Paper 915. Edited by International Food Policy Research Institute.

Villoria, N. (2019). Consequences of agricultural Total factor productivity growth for the sustainability of global farming: Accounting for direct and indirect land use effects. *Environmental Research Letters, 14*(12), 125002.

Wesseler, J., & von Braun, J. (2017). Measuring the bioeconomy. Economics and policies. *Annual Review of Resource Economics, 9*(1), 275–298.

Sustainable Resources: From Plants to Products

3

Ralf Pude, Christian Wever, and Thorsten Kraska

Abstract

The demand for renewable raw materials is constantly increasing. However, the cultivation of these crops is in direct competition with the areas used for food production. In addition, the area of arable land is continuously decreasing due to new settlement and devastated areas. Altogether it will become more and more important to cultivate low-input biomass plants of the second generation, which in an ideal case can be cultivated on marginal areas like industrial wasteland. This envisages the current bio-economic strategy of Germany by the Federal Ministry of Education and Research. A wider range of the so-called "ecosystem services" can be provided easily by perennial renewable biomass plants, which are introduced in this chapter.

Keywords

Sustainable resources · Perennial plants · Cascade utilization

1 Introduction

The research and development of renewable raw materials has been increased throughout the past 20 years in Germany. Area under cultivation for renewable resources ceased in 2020 at 2.58 million ha, which accounts for approximately 20% of the arable land. The land distribution for renewable resource production is mainly attributed to the production of the first generation energy crops (maize, rapeseed) with roughly 90% (2.3 million ha) followed by industrial crops at 9% (234,000 ha)

R. Pude (✉) · C. Wever · T. Kraska
INRES-Renewable Resources, University of Bonn, Rheinbach, Germany
e-mail: r.pude@uni-bonn.de; cwever@uni-bonn.de; kraska@uni-bonn.de

D. Lanzerath et al. (eds.), *Bioeconomy and Sustainability*,
https://doi.org/10.1007/978-3-030-87402-5_3

according to recent data.[1] In addition to the direct land use competition for food production, further limitations on the arable lands are imposed by settlement construction and industrial wasteland. It is thus becoming increasingly relevant to evaluate, domesticate, and cultivate the third generation low-input biomass plants that can thrive on marginal areas such as industrial wasteland,[2] as it is addressed by the current bioeconomy strategy of Germany.[3]

Moreover, perennial biomass plants provided several ecosystem services. Perennial crop cultivation could increase organic matter content in soil, prevent soil erosion, provide shelter areas for beneficial insects and small mammals (especially during winter), or provide sustenance for insects due to a long flowering period. Additionally, CO_2 is fixed in above ground biomass as well as below ground root systems. For these reasons, the European Union (EU) has included Miscanthus (*Miscanthus x giganteus*) and cup plant (*Silphium perfoliatum*) in the greening measures as of 01.01.2018.[4] This gives farmers an additional incentive to start growing such extensive low-input perennial crops. Furthermore, the acreage of bee-friendly perennial crops like cup plant is currently increasing.[5] In particular, the silphia with currently almost 3500 ha should be mentioned here, as it can be a maize substitute for biogas plants.[6] Miscanthus accounts for 4600 ha, mostly for solid fuel and bedding material for animals like horses. But it could also be used for building materials. Despite the stagnating cultivation, the utilization of renewable raw materials, both plants and residues, is increasing.[7] However, the cultivation of these crops is always in competition with land for food production. At the same time, arable land is continuously declining due to new settlement areas. This makes it all the more important to cultivate resource efficient low-input crops that can ideally be grown on marginal land or even industrial wasteland. These goals are included in Germany's current bioeconomy strategy. The question therefore arises as to how the cultivation of renewable raw materials in Germany should develop in terms of land use in the future. Here, perennial crops are of particular relevance, as they may bind considerable amounts of CO_2. This insight is also currently being considered in the climate protection debate resulting in a scientific conference held in 2019 in Sweden. Scientists met to discuss how new perennial grains like perennialized wheat or rice could foster the transformation from annual to perennial agriculture.[8] Regarding the CO_2-binding, specific plants show outstanding capabilities. Miscanthus, for example, may bind 25–30 t above and 2–4 t below ground CO_2 per hectare and year.[9] In

[1] Cf. Fachagentur Nachwachsende Rohstoffe e. V.

[2] Cf. Wever et al. (2020).

[3] Cf. Federal Ministry of Education and Research (2015).

[4] Cf. Emmerling and Pude (2017).

[5] Cf. Wever et al. (2019).

[6] Cf. Gansberger et al. (2015) and Wever et al. (2019).

[7] Cf. Fachagentur Nachwachsende Rohstoffe e. V.

[8] Cf. Lund University (2019) and Wever et al. (2020).

[9] Cf. Lewandowski et al. (1995).

Fig. 3.1 Presentation of the possibility of using perennial renewable resources; Villa Hammerschmidt, Bonn 2018

case of using CO_2-binding plants directly for energetic applications, whether for combustion, as biogas, or as fuel, the ecological advantages may result in a zero gain at best. Therefore, cascade utilization of raw materials as proposed by Thorsten Kraska et al.[10] is a much more sustainable approach. Numerous promising possibilities may furthermore be implemented in novel value chains, which changes the economic perspectives for farmers (Fig. 3.1).

2 Utilization as Substrate

A selected example for a large market is Germany's peat industry with an annual raw material requirement of 9.5 million m^3 of peat. Due to the fact that peat mining permits in Germany will expire completely within a few years, the demand for alternative substrate feedstock is increasing. In particular Miscanthus, cup plant as well as Virginia fanpetals (*Sida hermaphrodita*) seem to be promising candidates, especially due to the combination of sufficient quantities through high biomass yields and their defined material properties. Research at the University of Bonn has shown that Miscanthus can be used as a renewable growing medium not only for tomatoes, cucumbers, but also for strawberries in soilless cultivation systems.[11]

[10]Cf. Kraska et al. (2015).

[11]Cf. Kraska and Pude (2019) and Nguyen et al. (2021).

Fig. 3.2 Miscanthus as growing substrate for strawberries at Campus Klein-Altendorf; Bonn University

Plant management, especially irrigation and nutrient regimes, have to be adapted to the new substrate feedstock though (Fig. 3.2).[12]

3 Utilization as Building Material

Over several years, another large-scale market for renewable raw materials is the building material industry for alternatives to sand, which is becoming scarcer worldwide. The development of a lightweight concrete based on chopped Miscanthus with cement was tested more than 10 years ago.[13] The use of Miscanthus leads to high thermal insulation properties due to its high parenchyma content. The suitable building material strength of Miscanthus' biomass is caused by the high silica content of the leaves. However, the Miscanthus-based building material is still not authorized because it could not be tested under given conditions requested by existing DIN standards. In the end, the idea has migrated to Switzerland, where CO_2-storing and long-lasting products are promoted by the government through the so-called "Klimarappen" (climate centime). Recently, valid building material tests

[12] Cf. Kraska et al. (2018).

[13] Cf. Pude et al. (2005) and Moll et al. (2020).

Fig. 3.3 Binding agent-free building boards made of Miscanthus

Fig. 3.4 House based on Miscanthus lightweight concrete, Luxemburg 2019

are recognized within the EU, so now further houses can be built. In addition, research projects at the University of Bonn are focusing on the development of a high-performance insulating plaster made of Miscanthus, cup plant, and Jerusalem artichoke (*Helianthus tuberosum*)[14] as well as on binder-free building boards, which are formed under high pressure conditions (Figs. 3.3 and 3.4).[15]

[14] Cf. Pude and Petry (2016) and Schulte et al. (2021).

[15] Cf. Moll et al. (2018).

4 Utilization as Packaging Material

As a result of the packaging law amendment introduced 2019 in Germany, many companies are looking for environmentally friendly packaging materials, which can be produced successfully on an industrial scale. In addition, the market is still constantly growing due to the increasing online trade. In 2020, 21.4 million tons of paper were produced in Germany, with 12.4 million tons of paper, cardboard and board for packaging purposes accounting for the largest share. A large proportion of the paper produced was waste paper, which accounts for more than 79.3%. In addition, 1.2 million tons of new wood pulp was produced in Germany. This corresponds to a share of around 10%. The remainder is imported.[16]

Therefore, there is a great demand for alternative raw materials for the paper industry, which will likely increase in future. Although large quantities of waste paper are already mixed in, fresh fibers are still required to produce tear-resistant cardboards. These fibers are usually derived from the fast-growing trees like eucalyptus, which have to be chemically decomposed. This proportion of fibers could be replaced by solely mechanically disintegrated fibers from renewable raw materials. This has already been achieved with fibers from grass-like raw materials. The REWE Group has already successfully tested packaging trays with 25–50% grass fiber for the market. However, since hay was almost exclusively used as animal feed during the drought years of 2018 and 2019, the fast-growing renewable raw materials are also of particular importance here. In this context, the feasibility of mechanically processed plants such as cup plant, Virginia fanpetal, and Miscanthus has been shown recently (Fig. 3.5).[17]

The markets ideas described above for material utilization of perennial biomass plants are already being implemented by some companies. Several products have been developed to marketability. The field lab "Campus Klein-Altendorf" of the University of Bonn itself is transferring these ideas into practice: 5000 m^2 greenhouse area is supplied by heat of a biomass heating plant using shredded apple trees after 15–20 years of use in orchards.[18] Plant trials (e.g., tomatoes, cucumbers, strawberries, bell peppers, ornamentals) in the greenhouses can be conducted under practical conditions using Miscanthus as stand-alone substrate. After cultivation of tomatoes, the substrates are used in a cascading manner as a solid fuel for heating.[19] The remaining up to 15 m long tomato stems have been successfully processed into packaging material. The field lab "Campus Klein-Altendorf" has already received several awards for its innovative approaches. Already in 2013, it was selected as a "place of progress" and is a location of "KlimaExpo.NRW" ('Climate Expo NRW'). In order to accelerate the transfer of research results to companies and society, the "Transfer Center enaCom" was founded at the University

[16] Cf. Verband Deutscher Papierfabriken e. V. (2020).

[17] Frase et al. (2018) and Höller et al. (2021).

[18] Winzer et al., 2017.

[19] Cf. Kraska et al., 2018.

Fig. 3.5 Juice package with 25% mechanically processed grass fibers and packaging of apple mint tea with 40% grass fiber content

of Bonn. Many activities of higher education institutions, companies, and municipalities from the Rhein-Sieg district are already bundled in the "bio innovation park Rheinland e. V.".[20] This network for bioeconomy and green technologies was founded in 2015 and is now extended to the entire Rhineland and northern Rhineland-Palatinate.

References

Emmerling, C., & Pude, R. (2017). Introducing Miscanthus to the greening measures of the EU common agricultural policy. *GCB Bioenergy: Bioproducts for a Sustainable Bioeconomy, 9*(2), 274–279.

Fachagentur Nachwachsende Rohstoffe e. V. (FNR). (n.d.). *Nachwachsende Rohstoffe. Anbau.* Retrieved March 29, 2021, from, https://www.fnr.de/nachwachsende-rohstoffe/anbau.

Federal Ministry of Education and Research. (2015). *Bioeconomy in Germany. Opportunities for bio-based and sustainable future*. Retrieved August 14, 2019, from, https://www.bmbf.de/upload_filestore/pub/Biooekonomie_in_Deutschland_Eng.pdf

Frase, N., Höller, M., Völkering, G., & Pude, R. (2018). Prozessentwicklung in der Aufbereitung von Nachwachsenden Rohstoffen als Papierrohstoff für die Verpackungsindustrie. In *Mitteilungen der Gesellschaft für Pflanzenbauwissenschaften: From Big Data to Smart Farming* (Vol. 61, 30 ed., pp. 187–188). Tagung der Gesellschaft für Pflanzenbauwissenschaften e. V. Göttingen.

[20] Cf. www.bio-innovation.net [29.03.2020].

Gansberger, M., Montgomery, L. F. R., & Liebhard, P. (2015). Botanical characterization, crop management and potential of Silphium perpoliatum L. as a renewable resource for biogas production. A review. *Industrial Crops and Products, 63*, 362–372.

Höller, M., Lunze, A., Deutschle, A. L., Strücker, A., Frase, N., Pestsova, E., Spiess, A. C., Westhoff, P., & Pude, R. (2021). Meadow Hay, *Sida hermaphrodita* (L.) Rusby and *Silphium perfoliatum* L. as potential non-wood raw materials for the Pulp and paper industry. *Industrial Crops and Products, 167*.

Kraska, T., Kleinschmidt, B., Weinand, J., & Pude, R. (2018). Cascading use of Miscanthus as growing substrate in soilless cultivation of vegetables (tomatoes, cucumbers) and subsequent direct combustion. *Scientia Horticulturae, 235*, 205–213.

Kraska, T., & Pude, R. (2019). Miscanthus als Pflanzsubstrat für Erdbeeren—eine nachhaltige Alternative? *Obstbau, 11*, 570–574.

Kraska, T., Winzer, F., Witzel, C. P., Finger, R., & Pude, R. (2015, July 13). Cascade Utilization of Miscanthus. In *Biobased Future. Mitteilungsblatt über Biomasse für Energie und Industrie in einer nachhaltigen Wirtschaft*. No. 4. Retrieved March 23, 2021, from, http://www.a3ps.at/site/sites/default/files/newsletter/2015/no14/Biobased%20FutureJuli2015.pdf

Lewandowski, I., Kicherer, A., & Vonier, P. (1995). CO_2-balance for the cultivation and combustion of Miscanthus. *Biomass and Bioenergy, 8*(2), 81–90.

Lund University. (2019, May 6–10). *Abstracts for "Is the Future of Agriculture Perennial?"* Lund. Retrieved March 29, 2021, from, https://landinstitute.org/wp-content/uploads/2019/12/Abstract-book_-perennialagriculture.pdf

Moll, L., Kraska, T., Völkering, G., & Pude, R. (2018). Miscanthus für die nachhaltige baustoffliche Nutzung am Beispiel bindemittelfreier Faserplatten. In *Mitteilungen der Gesellschaft für Pflanzenbauwissenschaften: From Big Data to Smart Farming* (Vol. 61, 30 ed., 189–190). Tagung der Gesellschaft für Pflanzenbauwissenschaften e. V. Göttingen.

Moll, L., Wever, C., Völkering, G., & Pude, R. (2020). Increase of Miscanthus cultivation with new roles in materials production—A review. *Agronomy, 10*(2), 308.

Nguyen, V. T. H., Elfers, J., Kühn, H., Kraska, T., & Pude, R. (2021). Different Miscanthus genotypes as growing Media in Soilless Tomato Cultivation and its subsequent use for combustion. *Acta Horticulturae, 1305*, 301–308.

Pude, R., & Petry, M. (2016). *Verfahren zur Herstellung eines Dämmstoffes* (DE 10 2015 003 373 A1 2016.09.22). Offenlegungsschrift vom 22.09.2016, Deutsches Patent- und Markenamt. Retrieved March 29, 2021, from, https://patentimages.storage.googleapis.com/78/74/5b/df9474578e54c2/DE102015003373A1.pdf

Pude, R., Treseler, C.-H., Trettin, R., & Noga, G. (2005). Suitability of Miscanthus genotypes for lightweight concrete. *Die Bodenkultur, 56*(1), 61–69.

Schulte, M., Lewandowski, I., Pude, R., & Wagner, M. (2021). Comparative life cycle assessment of bio-based insulation materials: Environmental and economic performances. In *GCB Bioenergy: Bioproducts for a Sustainable Bioeconomy* [forthcoming].

Verband Deutscher Papierfabriken e. V. (VDP). (2020): *Kennzahlen Zellstoff- und Papierfabriken in Deutschland*. Statistiken vom Verband der Deutschen Papierfabriken e.V. Bonn. Retrieved March 29, 2021, from, https://www.vdp-online.de/fileadmin/0002-VDP/07_Dateien/1_Statistik/2020/PresseZahlen_2020___D__.pdf

Wever, C., Höller, M., Becker, L., Biertümpfel, A., Köhler, J., Van Inghelandt, D., Westhoff, P., Pude, R., & Pestsova, E. (2019). Towards high-biomass yielding bioenergy crop Silphium perfoliatum L.: Phenotypic and genotypic evaluation of five cultivated populations. *Biomass and Bioenergy, 124*, 102–113.

Wever, C., Van Tassel, D. L., & Pude, R. (2020). Third-generation biomass crops in the new era of De novo domestication. *Agronomy, 10*(9), 1322.

Winzer, F., Kraska, T., Elsenberger, C., Kötter, T., & Pude, R. (2017). Biomass from fruit trees for combined energy and food production. *Biomass and Bioenergy, 107*, 279–286.

Part III

Nutrition and Food Ethics

Food as a Moral Problem

4

Birgit Beck

Abstract

Bioeconomy is hailed as holding great potential for innovative and effective solutions for global problems regarding sustainability, environmental conservation, and food security. It has, however, also been criticised in view of its conceptual preconditions, unreflective use of technological fixes, and potentially adverse outcomes. Food ethics can provide a differentiated assessment of strategies and technologies applied in bioeconomy by means of scrutinising respective current theoretical and practical issues, for instance, those involving novel food technologies. The present article will (1) draw a rough sketch of food ethics in terms of a comprehensive theory of the good life, (2) analyse food as a moral problem, and (3) discuss some arguments concerning a paradigmatic example of technical solutions for moral problems in the context of bioeconomy, namely in vitro meat.

Keywords

Food ethics · In vitro meat · Bioeconomy

1 Introduction

Bioeconomy is commonly assumed to hold great potential for providing innovative and effective solutions for pressing global problems regarding sustainability, environmental conservation, mitigation of climate change, public health, and food security. It has, however, also been fiercely criticised for relying on a 'totalitarian

B. Beck (✉)
TU Berlin, Institute of History and Philosophy of Science, Technology, and Literature, Berlin, Germany
e-mail: birgit.beck@tu-berlin.de

D. Lanzerath et al. (eds.), *Bioeconomy and Sustainability*,
https://doi.org/10.1007/978-3-030-87402-5_4

approach'.[1] The 'Umwertung alles Lebendigen zum Rohstoff »Biomasse«' is seen as 'folgerichtiger letzter Schritt auf einem verhängnisvollen Weg, den die Vertreter der Bioökonomie als Rettung vor jenen Problemen ausgeben, die größtenteils gerade durch die Ausrichtung auf kurzfristige Renditeziele erzeugt wurden.'[2] Bioeconomy is further criticised for implying an 'unreflektierten Einsatz von Technologien mit kaum beherrschbaren Folgewirkungen'.[3] To frame this objection somewhat differently, bioeconomy is regarded as a 'solutionist'[4] practice which misleadingly assumes individual, social, and global structural (moral) problems to be solvable by technological fixes and shortcuts, thereby disregarding more adequate scientific, political, and ethical solutions based on 'den Prinzipien der Vorsorge, der Verantwortung, der Generationengerechtigkeit sowie der Biodiversität'.[5] A paradigmatic novel food technology from the field of cellular agriculture,[6] which can be subsumed under the heading of bioeconomy, is exemplified by the production of in vitro meat (IVM), i.e., meat from animal stem cells grown in a laboratory. The aim of producing such 'clean meat' is to provide an efficient and convenient way of reducing environmental damage, public health problems, and animal suffering without the need to alter widespread food patterns. Many authors in the debate on IVM regard this technique as a promising solution for intricate global problems. From the perspective of food ethics, however, the conceptual as well as normative presumptions of this paradigmatic instance of bioeconomy can be questioned.

Food ethics is a currently emerging philosophical discipline. The topic of food has long been of little interest for professional philosophy and ethics, at least in the German-speaking academic community. Neither theories of the good life nor normative and applied ethics have been much concerned with this issue to date. The reasons for this lack of consideration might be found in the prosaicness of the topic and in a kind of intellectual reservation: philosophical reflection on an alleged profane everyday practice like eating is presumed to be undemanding and suspected to be based on inadequate hedonistic or—even worse—paternalistic grounds. On the other hand, quite contrary to this estimation, it has been suggested that the lack of philosophical consideration of food is rather due to the complexity of the issue. This estimation appears highly plausible if we bear in mind that food ethics represents a substantial ethical account reflecting the anthropological and ethical conditions of leading a good life in an ever more technologised and globalised world by

[1]Cf. Gottwald and Krätzer (2014).

[2]Ibid., pp. 8 f., 'reevaluation of all sentient life as a commodity of 'biomass', 'the logical last step on a disastrous way regarded by adherents of bioeconomy as salvation from the problems that have for the most part been generated by that very orientation towards the goal of short-term yield' (own translation).

[3]Ibid., p. 10, 'unreflective use of technologies with hardly any controllable consequences' (own translation).

[4]Cf. Morozov (2013).

[5]Gottwald and Krätzer (2014), p. 160, "the principles of precaution, responsibility, justice between generations, and biodiversity" (own translation).

[6]Cf. Mattick (2018).

addressing various and interrelated issues of food production, consumption, and appreciation as well as their ethical, societal, and global implications.

In the following, I will (1) draw a rough sketch of food ethics in terms of a comprehensive theory of the good life, (2) analyse food as a moral problem by exploring the example of meat-eating, and (3) scrutinise arguments regarding the above-mentioned example of technical solutions for moral problems in the context of bioeconomy, namely in vitro meat.

2 Food Ethics: A Social and Philosophical Desideratum

2.1 Disregard of Food in Everyday Life

Ludwig Feuerbach's witticism that *we are what we eat* is frequently cited, but seldom taken seriously, considering the general low level of appreciation for food concerns in our everyday lives. Questions of diet are still commonly taken to be matters of socio-economic status and wealth.[7] Despite an abundance of popular cookery shows and books, further social, educational, philosophical, ethical, and eudaimonic aspects of eating are, with the exception of individual and public health, for the most part ignored. The consequences of this unreflective practice concern not only overstrained health care systems, but also environmentally and otherwise morally problematic structures in mass-scale production. The disregardful political and individual stance on our daily foodstuffs fosters global economic injustice, a shortage of natural resources, and ecological crises, resulting in food and water insecurity, a struggle for resources, depletion of a living environment, and foreseeable increased global migration.[8] Even when these correlations are recognised and publicly discussed,[9] the mere option of reconsidering and changing personal consumption habits—along with one's respective lifestyle, moral integrity, and self-understanding—is mostly associated with (decreed) abstinence and a suspected loss of autonomy and well-being, instead of with a valuable gain in moral maturity and personal quality of life.[10] Food choices as well as their related cultural and societal practices are still regarded as expressions of individual freedom in which a liberal

[7] The interrelation between (affordable) diet and socio-economic status has been of high importance throughout human history; cf. Hirschfelder (2005).

[8] Matthias Kaiser and Anne Algers identify 'five grand challenges connected to food': 'population growth', 'climate change', 'access to natural resources', 'global health issues', and 'the global markets' (Kaiser & Algers, 2016, p. 3).

[9] Adam Shriver notes that '[a]mong the public, and especially younger generations, food has become substantially more political than in past decades' (Shriver, 2020, p. 41).

[10] Mickey Gjerris assumes that reluctance to question our lifestyles and moral self-images leads to 'willed blindness', which, in turn, keeps us from realising that we 'already know that we are on thin ice, morally speaking' (Gjerris, 2015, p. 527).

society must not interfere—regardless of their potential ethical, social, and global impact.[11]

2.2 Disregard of Food in Philosophical Theory

Likewise, a thoughtful appreciation of food matters has long been of little interest for professional philosophy and—somewhat astonishingly—for ethics, at least as far as the German-speaking academic community is concerned.[12] With few exceptions,[13] neither theories of the good life[14] nor normative and applied ethics[15] have dealt much with this issue to date. While several new branches of applied ethics, for example, machine ethics,[16] ethics of artificial intelligence[17] or robot ethics,[18] are currently springing up like mushrooms, only few philosophers have taken up the task of establishing a particular discipline of food ethics. The reasons for this lack of consideration may be found in the prosaicness of the issue, as has been suggested elsewhere,[19] as well as in some kind of intellectual reservation: philosophical reflection on an alleged profane everyday practice like eating is frequently presumed to be conceptually undemanding[20] and suspected to be based on inept hedonistic

[11]Shriver takes this liberal stance to be a 'myth' and states that 'the choices we make when purchasing food have many effects on others, and can no longer [be] seen purely as a 'personal choice' (Shriver, 2020, p. 41).

[12]Paul B. Thompson, who is especially referring to the North American context, points out that there are two versions of 'food ethics': a philosophical one and 'an international social movement aimed at reforming the global food system' (Thompson, 2016, p. 61). According to his estimation, both are concerned with 'the overarching goal set' of 'right conduct, social justice and sustainability' (ibid., p. 62). However, the philosophical discipline of food ethics should not be conflated with the social-political movement because '[p]hilosophers are, of course, less confident that we know what right conduct, social justice and sustainability mean', whereas the 'popular conception of food ethics tends to presume that the overarching goal set is well-enough articulated to assess alternative policies and decision options' (ibid., p. 62) while 'there is little evidence that the practitioners of food ethics in the popular sense regard reflective and deliberative inquiry into philosophical matters to be a particularly worthwhile activity' (ibid., p. 69).

[13]Cf. e.g., Lemke (2007), Mohrs (2009, 2013), Dell'Agli (2009a), Gottwald and Boergen (2013), and Voget-Kleschin et al. (2014). See also the online journal *Epikur. Journal für Gastrosophie*, which was initiated in 2009: https://www.epikur-journal.at/ [25.03.2020]. The international journal *Food Ethics* was established in 2016. Further international contributions to food ethics can be found in, e.g., Mepham (1996), Telfer (1996), Singer and Mason (2007), Gottwald et al. (2010), Pojman and Pojman (2011), Kaplan (2012a), Sandler (2015), Thompson (2015), and Barnill et al. (2018).

[14]Cf. Hoesch et al. (2013).

[15]Cf. Düwell et al. (2011) and Stoecker et al. (2011).

[16]Cf. Misselhorn (2018).

[17]Cf. Bauberger et al. (2021).

[18]Cf. Loh (2019).

[19]Cf. Kaplan (2012b), pp. 1 f.

[20]A highly esteemed colleague of mine once stated quite distinctly: 'The professional philosopher reasons about reason. The hobby philosopher can reason about eating'.

or—even worse—perfectionist and paternalistic grounds.[21] Nevertheless, it has also been suggested, quite contrary to this estimation, that the lack of philosophical consideration of food is not to be blamed on its mundaneness or academic triviality, but rather on the complexity of the issue:

> [P]erhaps the real reason why relatively few philosophers analyze food is because it is too difficult. Food is vexing. It is not even clear what it is. [. . .] The subject quickly becomes tied up in countless empirical and practical matters that frustrate attempts to think about its essential properties. It is very difficult to disentangle food from its web of production, distribution, and consumption. Or when it is considered in its various use and meaning contexts, it is too often stripped of its unique food qualities and instead seen as, for example, *any* conceptualized object, social good, or part of nature. It is much easier to treat food as a mere case study of applied ethics than to analyze it as something that poses unique philosophical challenges.[22]

According to David Kaplan, the philosophical challenges mentioned comprise questions of food metaphysics, food epistemology, food aesthetics, food technology, food politics, food and identity, and food ethics, in particular.[23]

> The role of philosophy is to cut through the morass of contingent facts and conceptual muddle to tackle the most basic questions about food: What is it exactly? What should we eat? How do we know it is safe? How should food be distributed? What is good food? These are simple yet difficult questions because they involve philosophical questions about metaphysics, epistemology, ethics, politics, and aesthetics. [. . .] Once we have a clear understanding of philosophy's unique role, we will all be in a better position to engage in dialogue aimed at improving our knowledge, practices, and laws. We should also gain a renewed appreciation for the scope and relevance of the discipline of philosophy itself.[24]

This outlook sounds quite ambitious. However, I am convinced that Kaplan is right and that it is indeed a valuable project to establish a philosophy of food which includes a substantially rich and methodologically sound engagement with food matters from various perspectives and on multiple levels of conceptual and normative reflection. Contrary to Kaplan's conception, I would propose understanding food ethics not as an isolated further sub-discipline of applied ethics, but as an encompassing conceptual frame for the philosophical scrutiny of food issues in terms of a theory of the good life. Understood in such a broad way, (even applied) ethics comprises much more than just providing specific guidelines for political or legal regulation—which is sometimes misleadingly considered to be the only legitimate task of professional ethicists.[25] Moreover, ethics in general cannot be conducted separately from related philosophical disciplines like ontology, epistemology, or philosophical anthropology. Furthermore, particularly applied ethics

---

[21] Cf. Dieterle (2020).

[22] Kaplan (2012b), p. 2.

[23] Cf. ibid., pp. 3–18.

[24] Ibid., p. 2.

[25] Cf. Bayertz (2011) and Beck (2019a).

such as bioethics or ethics of technology has to engage with empirical subject matters relevant to the problems with which it is concerned. Therefore, food ethics has to take into consideration a whole spectrum of theoretical background knowledge and empirical insights in order to provide for suitable analyses.

Kaplan points out that '[p]hilosophers have a long but scattered history of analyzing food' and that 'the scholarship on food has real pedigree'.[26] Although this estimation may sound a bit exaggerated,[27] famous examples come to mind easily: Epicurus stated, for example, 'Anfang und Wurzel alles Guten ist die Freude des Magens'.[28] This does not mean that a theory of food ethics is a modern variant of a 'philosophy for swine'—an allegation that has been brought up against hedonistic ethics from antiquity on.[29] It should rather be conceptualised as a substantial ethical account which—by means of addressing various and interrelated issues of food—reflects on the anthropological and ethical conditions of leading a good life in a technological and globalised world. To this effect, a theory of food ethics is to be considered as a 'concrete utopia'[30] for a better (global) society. In contrast to the transhumanism-inspired utopian ideas of enhancing human morality[31] currently being discussed, food ethics is concerned with actual causes for and adequate solutions to global problems. The core of a theory of food ethics in terms of a theory of the good life, however, consists in the assumption that—contrary to public misgivings—a reflective stance on questions of diet and consumption in the broader sense and a resulting change in respective habits by no means diminishes personal autonomy and well-being, but can rather contribute substantially to leading a good and flourishing life.[32] Therefore, the fundamental task of a theory of food ethics

[26] Kaplan (2012a), p. 1.

[27] Hub Zwart, however, states that 'although the fact of this particular branch of ethics receiving a new name ('food ethics') rightly stresses its discontinuity with the past, some legacies and points of continuity can be indicated as well' (Zwart, 2000, p. 114).

[28] Epikur (1988), p. 94, "the beginning and root of all good is the pleasure of the stomach" (own translation).

[29] One reason for the continuing misestimation of hedonism appears to rest upon a misunderstanding of the very concept (cf. Horn, 2014, pp. 95 ff.). Hedonism—in classical terms—does not recommend debauchery and exuberance. Quite contrarily, hedonists like Epicurus have always advocated moderate pleasures and temperance. Another reason for the misestimation might ironically be based on an attempt to counter the allegation. In his famous defence of utilitarianism, John Stuart Mill hastened to declare that there are higher and lower pleasures and that it is 'better to be a human being dissatisfied than a pig satisfied' (Mill, 1957, p. 12). Steve Sapontzis's comment on this manoeuvre is that 'Mill's 'incalculably higher qualities' of pleasure are more successful at saving utilitarianism's dignity than its principle' (Sapontzis, 2014, p. 6).

[30] Cf. Mohrs (2003).

[31] Cf. Persson and Savulescu (2012), Beck (2015), and Giubilini and Savulescu (2018).

[32] Cf. Gjerris (2015), p. 528: 'What if acknowledging what we know about the unsustainability of our current life-style would enable us to see new opportunities. What if changing the way we live today turns out *not to be a sacrifice* of our own interests, but a way of discovering new life paths that could be described as 'good lives'?' (Emphasis added).

consists in reconsidering the relation between morality and ethics[33] and in demonstrating that practical philosophy in the classical terms of an art of living[34] is far from being obsolete or naively anachronistic in a modern, technological, and globalised world.

2.3 The Scope of Food Ethics

The field of research on food ethics is philosophically fruitful and demands interdisciplinary as well as transdisciplinary cooperation.[35] In the context of the present article, the subsequent passage can only provide a very rough and incomplete sketch of the scope and methodological orientation of a theory of food ethics.

From the perspective of normative ethics, for example, a theory of food ethics has to address questions regarding social and global justice,[36] synchronic and diachronic responsibility,[37] autonomy and paternalism,[38] possible moral overdemandingness,[39] and supererogatory action.[40] From a metaethical perspective, consideration has to be made of which (probably hybrid) ethical theory is suitable for substantiating a theory of food ethics and which criteria of adequacy have to be met in order to justify such a theory. To make an educated guess, I would propose that virtue ethics in combination with a pragmatist account provide a good starting point. There are obvious junctions to particular disciplines of applied ethics such as environmental ethics,[41] climate ethics,[42] medical ethics,[43] consumer ethics,[44] bioethics, and ethics of technology.[45] The latter include topics like cellular agriculture, meat alternatives, especially in vitro meat, and genetically modified food.[46] Since food ethics raises awareness that every instance of consumer behaviour inevitably implies a political

[33] Cf. Bayertz (2006) and Halbig (2013).

[34] Cf. Horn (2014).

[35] Cf. Kaiser and Algers (2016), p. 6.

[36] Cf. Young (2011).

[37] Cf. Bayertz (1995), Birnbacher (1988), Heidbrink (2003, 2007), Heidbrink et al. (2017), and Meyer (2018).

[38] Cf. Kühler and Nossek (2014).

[39] Cf. Chappell (2009) and Van Ackeren and Kühler (2016).

[40] Cf. Wessels (2002) and Raters (2017).

[41] Cf. Ott (2014) and Coeckelbergh (2015).

[42] Cf. Roser and Seidel (2015).

[43] Cf. Holland (2015).

[44] Cf. Heidbrink et al. (2011), Beck (2018), and McMullen and Halteman (2019).

[45] Cf. Thompson and Hannah (2008).

[46] Cf. Hopkins and Dacey (2008), Mattick (2018), Beck (2019b), Bolaños and Schäffl (2019), Bartkowski and Baum (2019), Dürnberger (2019), Nuffield Council on Bioethics (2019), and Shriver (2020).

action,[47] a further obvious relation exists to theories in political philosophy.[48] Questions traditionally raised in philosophy of mind and cognition, for example, those regarding (self-)awareness and sentience, play a role insofar as the conceptual and empirical investigation of mental and cognitive capacities of non-human animals has an effect on the argumentative basis of prominent theories in animal ethics.[49] An emerging topic that has not been extensively discussed in animal ethics is the ethical evaluation of entomophagy, i.e., the consumption of edible insects.[50] Another emerging topic which relates to animal ethics and will foreseeably demand greater attention is that of plant ethics.[51] From a philosophical anthropological perspective, an enquiry has to be made into the extent of the influence of anthropocentric world views and particular accounts of human self-understanding on attitudes towards diet and the very conception of food.[52] This includes the influence of particular language games on the formation and perpetuation of cultural customs and conventions.[53] As for interdisciplinary aspects, food ethics has to take into account empirical social research,[54] happiness studies,[55] and moral psychology,[56] for example. To conclude, as for any justification of ethical theory and moral practice, careful reflection on the underlying understanding of philosophy and the requirement of consistency of belief and action are also indispensable in the context of food ethics.[57] In light of the complexity and multidimensionality involved, the philosophical challenge lies in the consistent and coherent justification of an adequate theory of food ethics. On the basis of relevant systematic and historical philosophical literature[58] and under consideration of related empirical disciplines, a theory of food ethics can be considered as the culminating point for urgent social and global problems—as well as a desideratum of a timely theory of the good life.

[47] Cf. Lemke (2012), Lamla (2013), and Hahn (2017).

[48] Cf. Mulgan (2011) and Cohen (2019).

[49] Cf. Perler and Wild (2005), Wild (2008), Beauchamp and Frey (2011), Benz-Schwarzburg (2012) Schmitz (2014), and Ach and Borchers (2018).

[50] Cf. Shockley and Dossey (2014), Fischer (2016), Waltner-Toews and Houle (2017), and Lammers et al. (2019).

[51] Cf. Pouteau (2014) and Kallhoff et al. (2018).

[52] Cf. Ingensiep (2007), Forstmann et al. (2012), and Diamond (2012).

[53] Cf. Dell'Agli (2009b).

[54] Cf. Reitmeier (2013).

[55] Cf. Tiberius (2008) and Bayertz (2013).

[56] Cf. Schwartz (2004), Joy (2010), Mazar and Zhong (2010), Bray et al. (2011), Tiberius (2014), Piazza et al. (2015), Anderson and Barrett (2016), Hölker et al. (2019), and Armstrong et al. (2019).

[57] Cf. Ammann et al. (2011).

[58] Cf. Zwart (2000), Vogt (2018), Lagerlund (2018), and Grey and Garrett (2018).

3 Food as a Moral Problem

Paul B. Thompson observes that '[a]fter centuries of philosophical neglect, eating is once again an act that is rich in ethical significance'.[59] At least to those (philosophers) not engaging in 'willed blindness'[60] or simply ill-informed, it appears quite obvious that food poses a number of partly interwoven moral problems. However, in order to scrutinise which particular moral problems are raised by practices of food production, distribution, and consumption, it is worthwhile first to figure out what exactly a moral problem is and how it arises. Johann S. Ach and Arnd Pollmann remark that, despite the existence of extensive literature on the fundamental metaethical questions of why one should act morally at all[61] and what motivates us to act morally in specific circumstances, few philosophers have pondered upon the question of what makes a situation morally relevant in the first place, i.e., the conceptual presumptions governing us when we deal with a genuine moral problem.[62] All prominent ethical theories presuppose, at least implicitly, respective answers to this latter question; however, they frequently differ in the criteria they assume to be suitable for determining the moral relevance or irrelevance of certain situations, decisions, or actions.[63] From a Kantian position, for example, there will be different moral problems than from a utilitarian one, not least because the former includes duties to oneself, whereas the latter denies that there are such duties.[64] Although the answer might appear self-evident on the face of it, the basic question in the present context is under which conceptual presumptions there is reasonable justification for moral problems in regard to eating.[65] After a brief explanation of the concept and structure of moral problems, I will present a particular example in order to illuminate this point.

3.1 The Structure of Moral Problems

According to Ach and Pollmann, a problem generally implies a situation of uncertainty, for example, how to act or what to think.[66] To this effect, two basic sorts of problems can be differentiated: *practical* problems and *theoretical* ones. Roughly speaking, practical problems are ones of acting, whereas theoretical problems are ones of thinking, i.e., referring to beliefs, concepts, and theories. Thus, it follows that

[59] Thompson (2016), p. 73.

[60] Gjerris (2015).

[61] Cf. Bayertz (2006).

[62] Cf. Ach and Pollmann (2017), pp. 39 f.

[63] Cf. ibid., p. 40.

[64] Cf. ibid.

[65] To state the question differently: How can Shriver (2020) be shown to be right in stating that eating is not a matter of personal choice in liberal terms (any more)?

[66] Cf. Ach and Pollmann (2017), p. 42.

moral problems are practical ones (referring to the question 'What ought I to do?'); however, not every practical problem is necessarily a moral one. Although not a unanimously advanced view, it is commonly assumed that some practical problems are not morally relevant, but rather morally neutral. Such problems include technical, prudential, or aesthetic ones.[67] The authors Ach and Pollmann give the following examples. If you take your bicycle to pieces and do not know how to fix it again, you have a technical problem. Someone who wants to adopt a more healthy diet and does not know how to realise this aim has a nutritional problem.[68] In the present context, it suffices to say that only under the assumption of the existence of practical problems without moral relevance can moral problems be considered to be *specific* practical problems, namely ones that can and should be treated from the point of view of ethics—in contrast to the viewpoint of law, religion, or mere social etiquette, for example. To state that an action is morally problematic does not automatically imply that it is morally false, forbidden, or blameworthy; furthermore, moral problems are not to be equated with moral conflicts or even moral dilemmas—if one takes the latter to exist at all.[69] To speak of a moral problem only indicates that the problem is *of moral relevance*, i.e., it can and must be appropriately addressed from an ethical perspective.

To decide whether a problem is of moral relevance, however, poses a theoretical problem, namely one of (meta-)ethical theorising.[70] Therefore, moral (*practical*) problems can be differentiated from (meta-)ethical (*theoretical*) problems. One such (meta-)ethical problem consists in the question of under which circumstances a practical problem is to be regarded as a moral one. The authors call this the 'Problem-Problem'.[71] An answer to the problem-problem depends on certain meta-ethical and ethical presuppositions.

On the level of metaethics, there are, roughly speaking, two possible theoretical stances between which one has to choose: a *realist* position and a *constructivist* one.[72] According to the former, practical problems are morally problematic if and only if they are *in fact* morally problematic. Their factual being morally problematic is, in turn, granted by the (moral) fabric of the world. Thus, the realist position includes, firstly, an ontological assumption: some situations, decisions, and actions are objectively morally problematic *by virtue of moral facts*. The second assumption,

[67] Only under the assumption of morally neutral situations, decisions, and actions as well as other normative reasons rather than moral ones (e.g., prudential, legal, or religious reasons) does the question of moral overridingness actually make sense; cf. the debate in Hoffmann et al. (2017).

[68] Cf. ibid., 42 f. The nutritional problem certainly *can* be conceptualised as a moral problem if one assumes that one has a duty to oneself to care for one's own health or that an individual's healthy diet is a matter of responsibility towards public health. This conceptualisation depends on both the underlying normative ethical account and the respective concept of health.

[69] Cf. ibid., 44 f. From a strict Kantian perspective, for example, real moral dilemmas cannot exist because proper application of the categorical imperative precludes collisions of duties.

[70] Cf. ibid., p. 43.

[71] Ibid., p. 39, "problem-problem" (own translation).

[72] Cf. ibid., 46 ff.

which can (but need not necessarily) accompany the first one, is an epistemological one: the moral facts of the world are such that we can recognise them. Taking these two assumptions together, from the viewpoint of a realist position, we are able to identify a moral problem (concordantly) if we encounter one.

According to the constructivist position, moral problems are not identified when encountered, but rather conceptually constituted by virtue of ethical theories. Whether a situation, decision, or action appears morally problematic does not depend on the (moral) fabric of the world. Rather, we conceptualise practical problems as morally problematic or neutral through the particular 'lens' of our respective underlying ethical theory. The constructivist position, therefore, holds the opposite ontological and epistemological assumptions: there are no moral facts that could indicate whether a practical problem is a moral problem. The existence of moral problems and our epistemic approach to them depend solely on our respective conceptual ethical presuppositions.

Both positions have benefits and disadvantages. The realist position matches widespread assumptions of common-sense morality (as regards moral phenomenology and moral language) and allows, in principle, for a straightforward distinction between morally relevant and neutral problems. After all, many people believe that something *is* either objectively morally problematic or not. However, following John Mackie, the realist position can be criticised for presuming implausible ontological assumptions and being hardly able to explain the phenomenon of reasonable moral pluralism.[73] The latter can be explained more easily by the constructivist position. People take different practical problems to be morally relevant or neutral in light of their (even if only implicit) various ethical presumptions. However, a potential drawback of this position lies in its revisionist implications: to assume that the evaluation of a practical problem as morally relevant depends solely on its conceptual basis and not on allegedly objective moral facts might cause irritation. The estimation of an act of cruelty as morally problematic by referring to a respective theoretical framework instead of simply pointing to the fact that it *is* cruel, for example, might clash with common-sense moral intuitions.[74]

Both the realist and the constructivist positions are compatible with a possible misestimation of moral problems.[75] This misestimation can consist in *inept moralism* or *inept ignorance of moral significance.*[76]

On the one hand, it may be the case that a practical problem is *wrongly* (realist position: moral facts do not warrant the estimation) or *ill-foundedly* (constructivist position: ethical theory does not warrant the estimation) regarded as a moral problem. This kind of misestimation can be called "moralism 1". Alternatively, it may be the case that a practical problem is *rightly* (realist) or *well-foundedly* (constructivist)

[73] Cf. Mackie (1977) and Ach and Pollmann (2017), pp. 48 f.

[74] Cf. ibid., p. 53.

[75] Cf. ibid., p. 55 ff.

[76] The German expression in the contribution of Ach and Pollmann (2017), p. 58, is 'Entmoralisierung'.

regarded as a moral problem, but the degree of the moral significance of the problem is unduly exaggerated. This kind of misestimation can be called "moralism 2".

On the other hand, it may be the case that a practical problem is *wrongly* (realist) or *ill-foundedly* (constructivist) regarded as a morally *neutral* problem. This kind of misestimation can be called "ignorance of moral significance 1". Alternatively, it may be the case that a practical problem is *rightly* (realist) or *well-foundedly* (constructivist) regarded as a moral problem, but the degree of moral significance of the problem is unduly downplayed. This kind of misestimation can be called "ignorance of moral significance 2".

3.2 Example: Meat-Eating as a Moral Problem

As mentioned above, besides the commitment to a particular metaethical stance, the determination of a practical problem as a moral problem is contingent on the underlying normative ethical theory. Consider the following question on whether meat-eating is morally problematic.

From a contractualist perspective, for example, situations, decisions, or actions are morally problematic if and only if all contractual partners, viz. members of the moral community, accord with this estimation and have no reason to reject this assessment rationally. As long as the members of the moral community agree that eating meat is in their self-interest and is a matter of personal choice, they will not identify or conceptualise a moral problem in this case, but rather estimate meat-eating as a morally neutral practice. Liberalism often (tacitly) goes hand in hand with an ethical conception of contractualism, which can explain the prevailing 'myth'[77] that food choices are matters of individual negative freedom and not to be interfered with. From this perspective, the claim that eating meat is morally problematic can be rejected for being moralist 1 (eating meat poses no moral problem at all) or at least moralist 2 (eating meat may be morally problematic in principle—if the members of the moral community agree on this estimation—however, there are more salient moral problems to engage with than meat-eating).

In contrast, from a consequentialist perspective, for example, meat-eating can be considered to be morally problematic due to the negative consequences it brings about in terms of ecology, climate, animal welfare, and public health. To consider meat-eating as morally problematic does not per se specify exactly which moral problem(s) arise(s) and which solution(s) to the respective problem(s) should be implemented (I will address this question in more detail in the next paragraph). However, since liberalism also often (tacitly) goes hand in hand with consequentialism, it is no wonder that individual and public awareness of the adverse consequences of meat-eating have risen in the past decades. Accordingly, the assumption that meat-eating is morally neutral can be regarded either as an instance of ignorance of moral significance 1 (in case someone does not see a moral problem

[77] Shriver (2020), p. 41.

at all in eating meat) or as an instance of ignorance of moral significance 2 (if the adverse consequences of eating meat are recognised in principle, but unduly downplayed).

4 Technical Solutions for Moral Problems: The Example of in Vitro Meat

Assuming that eating meat is indeed considered to be morally problematic, the question arises of how to solve the particular moral problems raised by this eating habit. Meanwhile, three main kinds of moral problems concerning the production and consumption of meat are recognised: moral problems with regard to the environment, animal welfare, and public health. The most obvious solution to these problems would be simply to abstain from eating meat. However, this obvious solution faces serious cultural and traditional obstacles. Most people like eating meat. Consumption of meat can be considered to be 'normal' in two ways. Firstly, it is *statistically normal*, i.e., the majority of people in wealthy countries engage in this practice nowadays, and it is estimated that the coming decades will see a huge growth in the demand for and the consumption of meat due to the rise of middle classes in newly industrialising countries.[78] Secondly, consuming meat is *normatively normal*, i.e., it is still generally conceived to be an adequate (more or less morally neutral) way of conduct and is entangled in many ways in cultural and social practices as well as in beliefs about what constitutes a good life. Consider, for example, traditional feasts on holidays. On such occasions, serving meals comprised of meat and other animal products is still indispensable for many people. In the prevailing public opinion, this is just the way things have always been and still ought to be.

However, at the same time, many people are aware of the downsides of meat consumption, and there is an emerging debate about the issue.[79] Industrial livestock production causes considerable damage to the environment in terms of

> [e]xtensive use of arable land to feed production, deforestation to provide grazing lands, overgrazing, compaction, erosion and desertification of pastures leading to degradation of arable land, depletion of scarce water resources, eutrophication, degeneration of coral reefs and general pollution of water, air and soil caused by animal waste, hormones, antibiotics, fertilizers and pesticides spent in feed production etc. To all this can be added an extensive contribution to anthropogenic greenhouse gas emissions contributing to climate change and the enhanced stress on ecosystems globally and locally [. . .].[80]

[78] Cf. Post (2014), p. 1039.

[79] Cf. Rückert-John and Kröger (2019).

[80] Gjerris (2015), p. 523.

Moreover, the massive intake of animal protein has been suspected to threaten individual as well as public health, and, needless to say, the present-day food production processes involve massive and hardly justifiable animal suffering.[81] In light of these problems, is there any way of improving our ecological and moral footprint, so to speak, apart from a seemingly unpromising shift to comprehensive vegetarianism or veganism?[82] At first sight, what could be called a technical solution for the aforementioned moral problems which can be regarded as a paradigmatic instance of bioeconomy does indeed appear to be in sight: 'in vitro meat' (IVM).[83]

In 2013, Mark Post from Maastricht University in the Netherlands and colleagues presented the world's first lab grown hamburger at a spectacular media event in London.[84] The burger, which was prepared by a British chef and tasted by an international jury, had been created from about 20,000 bovine muscle strands grown within 3 months through stem cell technology originally developed for regenerative medical purposes. Very roughly sketched, the production of IVM functions as follows:

> Briefly, to produce artificial meat, a few stem cells taken from a limited number of animals multiply and differentiate into muscle cells (myoblasts) which then fuse (to form myotubes) and build muscle fibres. Once the muscle fibres are mature and harvested, they are assembled into a patty to create minced meat which has the shape and the visual aspect of a hamburger. Other compounds (for instance salt, seasoning, breadcrumbs, egg white powder, a binder, etc.) are added to make a firm patty with a colour similar to that of meat which can be cooked and eaten as a normal hamburger.[85]

The price of the single burger presented as a proof of concept by Post and colleagues amounted to € 250,000.[86] Since then, however, the technology has rapidly become cheaper and is considered to be a viable alternative to traditional meat production within the next few years.[87] There are still some technical difficulties to overcome in order to improve cultured meat with respect to its structure (with the aim of growing, or rather 3-D printing, thick cuts of meat), look, taste, and nutritional value.[88] Moreover, the sustainability and cost-efficiency of production still has to be ameliorated. In order to provide a truly animal-friendly alternative, efforts are being made to find a suitable culture medium without the use

[81] Cf. Post (2012), p. 298.

[82] Cf. Post (2014), p. 1039: 'Voluntarily abstaining from meat consumption is unlikely to contribute to reduction in meat demand, as the proportion of vegetarians in industrialized societies is low and has not increased over the last 35 years'. For a conclusive and witty advocacy of veganism, see Colb (2013).

[83] See also the contribution from Fiebelkorn et al. (2022) in this volume.

[84] Cf. Post (2014).

[85] Hocquette (2016), p. 169.

[86] Cf. Laestadius and Caldwell (2015), p. 2458.

[87] Cf. Mosa Meat.

[88] Cf. Hocquette (2016), p. 170.

of foetal bovine serum.[89] From the beginning of his project, Post envisioned that 'culturing of meat in laboratories and eventually in factories will transform the meat industry. This will require time, a great deal of research and development and a gradual transition in our thinking about meat'.[90] Taking this assessment for granted, what are we to think about meat, especially cultured or in vitro meat?

4.1 Ethical Arguments for and Against In Vitro Meat

In the following, I will briefly present some arguments for and against IVM which have been discussed in the relevant literature.[91] However, I will not take sides regarding the question of whether IVM provides a good answer to the aforementioned moral problems or is just an instance of 'solutionism',[92] i.e., the conceptualisation of (moral) problems in such a way that they can apparently be solved by technological shortcuts. In the words of Evgeny Morozov:

> It's not that solutions proposed are unlikely to work but that, in solving the 'problem,' solutionists twist it in such an ugly and unfamiliar way that, by the time it is 'solved,' the problem becomes something else entirely. Everyone is quick to celebrate victory, only no one remembers what the original solution sought to achieve.[93]

Instead of tackling this issue, I will question one premise that is usually regarded as self-evident, namely that meat-eating is *de facto* indispensable and that a technological fix is, therefore, the only reasonable way to go. My analysis will reveal that there are, on the one hand, no compelling reasons *for* eating meat, but that there is, on the other, no master argument *against* IVM short of perfectionist or paternalist commitments, so I will suggest a pragmatist way of dealing with the matter. My tentative conclusion will be that IVM as a means to improve our ecological and moral footprint may, in fact, turn out to be only the second best option in comparison with a shift in perspective. However, without any knock-down argument to offer from a pragmatist perspective, this suggestion is certainly open to comments and critical discussion.

4.1.1 Arguments for In Vitro Meat

Since the beginning of the debate on IVM in bioethics,[94] way before Post's proof of concept, three main kinds of arguments have been brought forward in favour of IVM, all of them related to environmental benefits, animal welfare, and public health.

---

[89] Cf. Böhm et al. (2017).

[90] Post (2012), p. 300.

[91] For a more comprehensive overview, see Beck (2019b).

[92] Cf. Morozov (2013).

[93] Ibid., p. 8.

[94] Cf. Hopkins and Dacey (2008).

From the perspective of environmental ethics, there are good arguments in favour of developing IVM as an alternative to traditional meat production. Two anticipatory life cycle analyses indicate that, compared to traditional industrial farming and meat production processes, IVM technology promises a considerable reduction of greenhouse gas emissions and of the over-exploitation of arable land and fresh water supplies.[95] In addition, it is assumed that there will be benefits for biodiversity and wildlife insofar as the land saved can be reforested as IVM 'reduces pressure for converting natural habits to agricultural land'.[96]

As for the implications for animal welfare, the intended reduction and eventual displacement of industrial factory farming and problematic slaughterhouse practices by means of IVM technology would undeniably bring about major improvement in comparison with the current circumstances. At least from the point of view of animal welfare ethics (albeit not necessarily from the viewpoint of animal rights ethics), IVM promises the benefit of a great reduction of unnecessary animal suffering.[97] Moreover, it would be possible to produce meat from rare or endangered species[98] while, at the same time, lowering the current levels of overfishing and overhunting.[99]

Last but not least, it is assumed that IVM could provide health benefits[100] by preventing the spread of epidemics like mad cow disease, swine or avian flu, and also foodborne diseases from infectious agents like *E. coli* bacteria. In addition, IVM could provide improved nutritional value as compared to traditional meat, therefore contributing to the promotion of individual as well as public health and prevention of illnesses, for example, cardiac diseases and cancer.

4.1.2 Arguments Against In Vitro Meat

Some literature has discussed possible objections to IVM. The most prominent demurs relate to a supposed lack of naturalness and authenticity, reactions of disgust, reduction of animal happiness in the world, possible anthropophagy, and disrespect for animal integrity and dignity.

Some commentators suspect that consumers might be inclined to view IVM as 'unnatural' or not being 'real' meat.[101] As to this assumption, it is questionable whether current meat production processes are much more 'natural', given the actual

[95] Cf. Tuomisto and Teixeira de Mattos (2011) and Mattick et al. (2015).

[96] Tuomisto and Teixeira de Mattos (2011), p. 6122.

[97] Cf. Pluhar (2010) and Schaefer and Savulescu (2014).

[98] Cf. Welin (2013), p. 34: 'It is not more difficult to tissue engineer meat from a red-listed and rare animal than [...] meat from a pig, a cow, or a chicken. To be able to eat, for example, sushi produced from an exotic source could be a way to sell more expensive products. It can be advertised as an ethical way of enjoying a new thrilling gastronomic experience, preferably served with a selection of exquisite wines in some fancy surroundings'.

[99] Cf. Tuomisto and Teixeira de Mattos (2011), p. 6122.

[100] Cf. Hocquette (2016), p. 171; Mattick (2018), p. 32.

[101] Cf. Hopkins and Dacey (2008), pp. 586 f.; Welin (2013), pp. 29 ff.; Dilworth and McGregor (2015), pp. 94 f.; Laestadius and Caldwell (2015), pp. 2458 f.; Marcu et al. (2015).

circumstances of industrial farming and meat processing. Moreover, IVM is, in fact, real meat cultured from animal tissue, the only difference being that it is not derived from a once living and intentionally slaughtered animal. In addition, simply equating the natural with the good implies a naturalistic fallacy. A related concern is that IVM may trigger reactions of disgust, the prominent 'yuck factor'.[102] This kind of reaction appears to express either some kind of bioconservative neophobia, as is often the case with novel biotechnologies, or, once again, a sense of unnaturalness, which can hardly be considered to be a convincing argument as long as the underlying concept of naturalness is not specified.[103]

A further objection to IVM—brought forward from a utilitarian perspective—states that reducing the need for livestock will result in fewer happy animals in the world, leading to less instantiated overall inherent value.[104] Apart from the dubious assumption that animals raised for food production, especially those on factory farms, lead what can be reasonably called happy lives, this line of argument appears to lead to Derek Parfit's famous 'repugnant conclusion'.[105] This means that as long as happiness in the world is maximised, it would be better to create more and more individuals with minimally worthwhile lives instead of fewer ones enjoying a greater amount of well-being. However, contrary to the utilitarian supposition presented, it can reasonably be assumed that what counts morally is not abstract happiness in the world, but sentient individuals and their weal and woe. Moreover, it cannot count as a convincing objection to IVM that the potential extinction of farm animals would be harmful to individual members of the respective species,[106] because there is no moral right whatsoever to be brought into existence, all the more so as a non-existing entity (a contradictio in adiecto) cannot be a subject experiencing harm or benefit.

A rather curious objection to IVM points out the possibility of obtaining human cells for the production of edible meat, which might pave the way for victimless cannibalism. It is an empirical question whether there would be widespread interest in human meat, were it actually available. However, it appears dubious that any such demand would be prevalent. G. Owen Schaefer and Julian Savulescu, who discuss this objection, assume that, due to 'a near-universal taboo in contemporary society', human IVM 'is unlikely to become a widespread craze'.[107] For one, anthropophagy usually triggers a strong yuck reaction. Moreover, it is not quite clear what would be morally wrong about eating human flesh provided that no person is killed in order to be eaten and no dead body desecrated. The easiest solution to this problem would, of course, be to ban the production of human meat by law.[108]

---

[102] Cf. Hopkins and Dacey (2008), pp. 587 f.; Hopkins (2015), p. 266.

[103] Cf. Sandin (2017).

[104] Cf. Levinstein and Sandberg (2015).

[105] Cf. Parfit (1984) and Schaefer and Savulescu (2014), p. 195.

[106] In an empirical study on the public acceptance of IVM, one participant claimed, for example, that the extinction of cattle would be 'bad for cows' (cf. Laestadius, 2015, p. 997).

[107] Schaefer and Savulescu (2014), p. 197.

[108] Cf. ibid.

Some authors object to IVM for the reason that animals, despite a reduction in their suffering, would still be objectified and instrumentalised in that they would still be seen as the mere means for human ends. This is a quite noteworthy objection. However, in order to adopt this line of argument, one has to commit oneself to a deontologically justified (abolitionist) animal rights position.[109] From this point of view, all sentient beings, non-human and human animals alike, are subjects of moral rights and inherent dignity which must not be violated. Apart from these rather demanding metaphysical and ethical assumptions, it has been pointed out that it is not quite clear why the animals which would serve as cell donors for IVM technology should be regarded as *mere* means. Provided that they could lead appropriately good lives without being killed, it is assumed that they could serve as cell donors in the course of painless standard veterinary examinations. Of course, animals are not able to give informed consent, but given the moral advance of such minimally invasive interventions in comparison with conventional food production, hypothetical consent in these cases might be warranted.[110] Obviously, discussion of this argument leads straightforwardly to a stalemate between ethical animal rights positions and animal welfare positions.

4.2 The Bedrock Premise: Eating Meat Is Indispensable

So far, we have seen that there are some prima facie good reasons for supporting the development of IVM—some scholars have, in fact, proposed regarding IVM as 'something that we may be morally required to support'[111]—as well as some rather curious and, in any case, conceptually demanding arguments against it. However, the whole discussion relies on one assumption which is hardly ever questioned in the pertinent literature, namely that the *indispensability* of meat-eating is simply *a matter of fact*. Without this prerequisite, the development of IVM as a technical solution for the above-mentioned moral problems would appear to be superfluous. However, many authors agree on the point that vegetarianism or—even worse—veganism is 'such a literally and metaphorically unappetizing lifestyle'[112] that its adoption is certainly unimaginable without severe losses in personal well-being, identity, and autonomy. Jared Piazza and colleagues provide a psychological explanation for this estimation: '[W]hen an ideology is widely endorsed, as meat eating is in most parts of the world today, the justifications supporting the ideology generally go unchallenged.'[113] As an empirical fact, this may be true; however, this does not imply straightforwardly that any conclusive justifications for eating meat will remain once they *are* challenged.

[109] Cf. Regan (2004) and Francione (2008).

[110] Cf. Schaefer and Savulescu (2014), p. 194.

[111] Hopkins and Dacey (2008), p. 595.

[112] Ibid., p. 580.

[113] Piazza et al. (2015), p. 115.

4.3 Reasons for Eating Meat: The Four Ns

Relating to the seminal work of Melanie Joy,[114] Piazza and colleagues conducted several studies on omnivores' preferred justifications for consuming meat, assuming that meat eaters are implicitly aware of moral problems regarding meat-eating, but unconsciously engage in several psychological strategies of rationalisation in order to avoid cognitive dissonance or feelings of guilt and shame. The studies revealed four principal rationalisations, which the authors termed 'the four Ns': eating meat is *natural*, *normal*, *necessary*, and *nice*.

(1) The first N (*natural*) reflects meat-eating as a part of human nature. Mother Nature—or, alternatively, God—has endowed us with carnivore capacities, therefore, being carnivores is assumed to be an essential property of humans, so to speak. Furthermore, what is deemed 'natural' in this sense is often implicitly equated with being good. (2) The second N (*normal*) appears to be based on a kind of 'normative power of the factual' argument. Humans have always eaten meat, so there is no obvious reason to change this habit. People eat meat; therefore, people should eat meat. (3) The third N (*necessary*) states that eating meat is indispensable for good health. Without meat, it is impossible to meet all the requirements for healthy nutrition. (4) The fourth N (*nice*) is based on aesthetic or hedonic grounds. People like the smell and taste of meat. Eating meat contributes to well-being and leading a good life. Do the four Ns provide good reasons for eating meat?

Even just a brief glance at the conceptual, empirical, and normative background indicates that this is not the case. The first and the second Ns (*natural* and *normal*) obviously imply a naturalistic fallacy (natural, God-given, etc. equates good) and an is-ought fallacy (something is the case; therefore, it ought to be the case). The third N (*necessary*) is empirically doubtful. Although there are competing studies, increasing medical evidence shows that a vegetarian or even vegan diet is, all things considered, healthier than an omnivore diet.[115] The fourth N (*nice*) does not appear too compelling when weighed against environmental damage and animal suffering. Taken together, the four Ns do not provide compelling reasons for eating meat in the first place.

4.4 Arguments for and against IVM Revisited

As shown above, there are good prima facie reasons for the development of IVM under the assumption that meat-eating is an indispensable part of the human diet. However, if this assumption cannot be justified conclusively, as already suggested in a very quick and superficial analysis of the four Ns, then why should IVM be developed in the first place? On the one hand, if there are no compelling arguments *for* eating meat, wouldn't it be reasonable to just refrain from eating meat and spend

[114] Cf. Joy (2010).

[115] Cf. Leitzmann (2018).

time and money on more urgent scientific projects instead of wasting resources on bioeconomic strategies like the development and mass scale production of IVM? On the other hand, eating in vitro meat could possibly be justified on the basis of aesthetic and hedonic reasons as long as it implied no harm to animals, humans, and the environment. Provided that these conditions could be met, there are no compelling arguments *against* IVM, either.

Given this situation, the question arises of whether it is wise to invest a lot of research funds and scientific effort in developing a bioeconomic technology solely for the sake of satisfying aesthetic and hedonic preferences. Should we rather try to change these very preferences? Which solution would provide us with a greater capacity to improve our environmental and moral footprint? After all, the assumption that there are no conclusive reasons for eating meat does not preclude the possibility of justifying meat consumption indirectly. The above-mentioned argument regarding negative liberty (meat-eating, like food consumption in general, is considered to belong to the private sphere with which no one should be allowed to interfere) is not an argument *for* eating meat, but rather an argument against telling people not to eat meat. This argument can certainly be applied to IVM as well. As matters stand, I think it is not self-evident that eating (in vitro) meat is covered by negative liberty, given the implications of eating (in vitro) meat in relation to harm to others and respectful relationships with others—humans and non-human animals alike. However, the argument regarding negative liberty indicates that an advocate of ethical vegetarianism or veganism is supposedly committed to some kind of perfectionism (telling people what the good is) or paternalism (telling people what *their* good is), which might explain why ethical vegetarians and vegans are oftentimes suspected of moralising (in the sense of moralising 1 or 2 explained above). Is there a way out of this argumentative stalemate?

In order to cut the Gordian knot, a change of perspective might be helpful. Perhaps compelling and overriding arguments for or against the consumption of meat or IVM are not what is needed. So far, we have accumulated a lot of facts and ethical arguments. It is widely acknowledged that eating meat under the current circumstances is morally unjustified and environmentally hazardous. Still many people find it hard or do not even consider changing their conduct. Under certain preconditions, developing IVM may be morally justifiable and environmentally beneficial, but the need for IVM arises out of a potentially misconceived assumption, namely the estimation that eating meat forms an indispensable part of leading a good human life. Refraining from eating meat, in contrast, is viewed as involving abstinence and deprivation of well-being as well as of personal autonomy. As long as this assumption remains unquestioned, IVM consequentially appears to be the only viable option for improving our environmental and moral footprint without paying too high a price (in terms of excessive moral demand). Nevertheless, the development of IVM can be regarded as a solutionist practice since the underlying moral problems (environmental damage and climate change, shortcomings regarding public health, treatment of non-human animals) are conceptualised in such a way that a technological shortcut recommends itself as the only viable option. I propose taking the bedrock premise (eating meat *is* de facto indispensable) to be the fundamental problem which is, in turn, based on a particular (implicit) concept of the good human life. Accordingly, a different outlook on the good life might lead to an alternative initial conceptualisation of the problem and on to different solutions.

However, how do we[116] come to change perspective and adopt a different view of the good life? This is nothing that could be *prescribed* in a perfectionist or paternalist manner. It is also nothing that could be achieved by appropriate reasoning alone. What we might need from a virtue ethical and pragmatist viewpoint is not more knowing-that, more sophisticated arguments and universal ethical justification, but more know-how. We might need what could—following Mark Coeckelbergh's idea of 'environmental skill'[117]—be called 'gastrosophical skill'. This implies developing adequate habits and skills regarding our evaluation of food in general and personal diet in particular to disclose unprecedented eudaemonic experiences of eating while relating to others, human and non-human animals alike, as well as our environment in morally good and ethically sound ways. In order to achieve those habits and skills, we need fresh thoughts, creativity, and inspiring role models. Melanie Joy and Cora Diamond, for example, call our attention to the strangeness of the thought that some sentient beings are considered as 'something to eat' while others are not.[118] Reconsidering this contingent categorisation might lead to a change. If (particular) animals are no longer unthinkingly considered to belong to the category 'food'—and there is no compelling reason for this conceptualisation apart from custom and convenience—there is no reason to believe that eating meat or other animal products forms an indispensable part of a good life. Quite to the contrary, what we need above all, from a pragmatist perspective, is courage to change customary practices and act in new ways. We might have to face trial and error, however; the development of skills usually needs time and a suitable social environment. There is no built-in guarantee of success. In any case, we are free to figure out our own best way of improving our moral and ecological footprints and of leading good lives. Pursuing bioeconomic strategies like the development and implementation of IVM is certainly an option, however, among effectively sustainable solutions to the above-mentioned moral problems, it might well be the second best option.

5 Conclusion

Although bioeconomy is commonly considered to hold great potential for innovative and effective solutions for global problems regarding sustainability and food security, it has been questioned in view of its conceptual preconditions, unreflective use

[116]With 'we', I refer to relatively wealthy persons in industrialised countries who can afford any dietetic choice and for whom abstaining from eating meat (and other animal products) poses no existential problem.

[117]Coeckelbergh (2015).

[118]Cf. Joy (2010) and Diamond (2012).

of technological fixes, and potentially counterproductive outcomes. As has been shown by the example of IVM, from the perspective of food ethics, novel food technologies must be scrutinised regarding their empirical, conceptual, evaluative and normative premises, and their foreseeable consequences. Food ethics is a fruitful emerging interdisciplinary field of philosophical enquiry relating to various philosophical disciplines and has to take into consideration empirical insights from the respective sciences. A substantive conception of food ethics in terms of an encompassing theory of the good life can address individual, social, and global issues regarding food production and consumption with the aim of proposing ethically justified solutions to moral problems apart from solutionist technological shortcuts from the realm of bioeconomy.

References

Ach, J. S., & Borchers, D. (2018). *Handbuch Tierethik. Grundlagen – Kontexte – Perspektiven*. J.B. Metzler.

Ach, J. S., & Pollmann, A. (2017). Moralisch problematisch – Was aus einem problem ein moralisches problem macht. In M. Hoesch & S. Laukötter (Eds.), *Natur und Erfahrung. Bausteine zu einer praktischen Philosophie der Gegenwart* (pp. 39–60). Mentis.

Ammann, C., Bleisch, B., & Goppel, A. (2011). *Müssen Ethiker moralisch sein? Essays über Philosophie und Lebensführung*. Campus Verlag.

Anderson, E. C., & Barrett, L. F. (2016). Affective beliefs influence the experience of eating meat. *Public Library of Science One, 11*(8), e0160424.

Armstrong, B., Meskin, A., & Blundell-Birtill, P. (2019). Delicious but immoral? Ethical information influences consumer expectations and experience of food. *Frontiers in Psychology, 10*, 843.

Bauberger, S., Beck, B., Burchardt, A., & Remmers, P. (2021). Ethische Fragen der Künstlichen Intelligenz. In G. Görz, U. Schmid, J. Schneeberger, & T. Braun (Eds.), *Handbuch der Künstlichen Intelligenz* (6th ed., pp. 907–934). De Gruyter.

Barnill, A., Budolfson, M., & Doggett, T. (2018). *The oxford handbook of food ethics*. Oxford University Press.

Bartkowski, B., & Baum, C. M. (2019). Dealing with rejection: An application of the exit–voice framework to genome-edited food. *Frontiers in Bioengineering and Biotechnology, 7*, 57.

Bayertz, K. (1995). *Verantwortung. Prinzip oder problem?* Wissenschaftliche Buchgesellschaft.

Bayertz, K. (2006). *Warum überhaupt moralisch sein?* C.H. Beck.

Bayertz, K. (2011). Was ist angewandte Ethik? In J. S. Ach, K. Bayertz, & L. Siep (Eds.), *Grundkurs Ethik. Band I: Grundlagen* (2nd ed., pp. 165–179). Mentis.

Bayertz, K. (2013). Empirische Antworten auf philosophische Fragen? Zum Verhältnis von philosophischer Ethik und empirischer Glücksforschung. In M. Hoesch, S. Muders, & M. Rüther (Eds.), *Glück – Werte – Sinn. Metaethische, ethische und theologische Zugänge zur Frage nach dem guten Leben* (pp. 35–47). De Gruyter.

Beauchamp, T. L., & Frey, R. G. (2011). *The Oxford handbook of animal ethics*. Oxford University Press.

Beck, B. (2015). Conceptual and practical problems of moral enhancement. *Bioethics, 29*(4), 233–240.

Beck, B. (2018). Essen und Verantwortung. Der komplizierte Ausgang des Konsumenten aus der gastrosophischen Unmündigkeit. In A. Gelfert & T. Gil (Eds.), *Philosophische Hefte* (Vol. 5). Logos Verlag.

Beck, B. (2019a). Bioethik aus Sicht einer Ethikerin. In J. Kurreck & B. Beck (Eds.), *Kursbuch Bioethik* (pp. 27–38). Universitätsverlag der TU Berlin.

Beck, B. (2019b). Ethische Argumente pro und contra in-vitro-Fleisch. In J. Rückert-John & M. Kröger (Eds.), *Fleisch. Vom Wohlstandssymbol zur Gefahr für die Zukunft* (pp. 295–321). Nomos.

Benz-Schwarzburg, J. (2012). *Verwandte im Geiste – Fremde im Recht. Sozio-kognitive Fähigkeiten bei Tieren und ihre Relevanz für Tierethik und Tierschutz*. Harald Fischer Verlag.

Birnbacher, D. (1988). *Verantwortung für zukünftige Generationen*. Reclam.

Bolaños, A., & Schäffl, I. (2019). Moderne Lebensmittelproduktion. Zelluläre Landwirtschaft/Invitro-Fleisch. In J. Kurreck & B. Beck (Eds.), *Kursbuch Bioethik* (pp. 177–187). Universitätsverlag der TU Berlin.

Böhm, I., Ferrari, A., & Woll, S. (2017). *In-vitro-Fleisch. Eine technische Vision zur Lösung der Probleme der heutigen Fleischproduktion und des Fleischkonsums?* Retrieved April, 18, 2020, from, http://www.itas.kit.edu/pub/v/2017/boua17b.pdf

Bray, J., Johns, N., & Kilburn, D. (2011). An exploratory study into the factors impeding ethical consumption. *Journal of Business Ethics, 98*, 597–608.

Chappell, T. (2009). *The problem of moral demandingness. New philosophical essays*. Palgrave Macmillan.

Coeckelbergh, M. (2015). *Environmental skill. Motivation, knowledge, and the possibility of a non-romantic environmental ethics*. Routledge.

Cohen, M. J. (2019). Let them eat promises: Global policy incoherence, unmet pledges, and misplaced priorities undercut progress on SDG 2. *Food Ethics, 4*, 175–187.

Colb, S. F. (2013). *Mind if i order the cheese burger? And other questions people ask vegans*. Lantern Books.

Dell'Agli, D. (2009a). *Essen als ob nicht. Gastrosophische Modelle*. Suhrkamp.

Dell'Agli, D. (2009b). Alles Käse? Eh Wurscht! Zur gastrosophischen Idiomatik des Deutschen. In Id (Ed.), *Essen als ob nicht. Gastrosophische Modelle*. : Suhrkamp (pp. 101–151).

Diamond, C. (2012). Fleisch essen und Menschen essen. In Id (Ed.), *Menschen, Tiere und Begriffe. Aufsätze zur Moralphilosophie*. : Suhrkamp (pp. 83–106).

Dieterle, J. M. (2020). Shifting the focus: Food choice, paternalism, and state regulation. *Food Ethics, 5*, 2.

Dilworth, T., & McGregor, A. (2015). Moral steaks? Ethical discourses of in vitro meat in academia and Australia. *Journal of Agricultural and Environmental Ethics, 28*, 85–107.

Dürnberger, C. (2019). Normative concepts of nature in the GMO protest. A qualitative content analysis of position papers criticizing green genetic engineering in Germany. *Food Ethics, 3*, 49–66.

Düwell, M., Hübenthal, C., & Werner, M. H. (2011). *Handbuch Ethik* (3rd ed.). J.B. Metzler.

Epikur (1988). *Philosophie der Freude. Briefe. Hauptlehrsätze. Spruchsammlung. Fragmente*. Translated and equipped with an epilogueby P. M. Laskowsky. : Insel Verlag.

Fiebelkorn, F., Dupont, J., & Lammers P. (2022). Acceptance of insects and in vitro meat as a sustainable meat substitute in germany: in search of the decisive nutritional-psychological factors. In D. Lanzerath, U. Schurr, C. Pinsdorf, & M. Stake (Eds.), *Bioeconomy and sustainability. Scientific, socio-economic and ethical dimensions* (pp. 77–92). Springer.

Fischer, B. (2016). Bugging the strict vegan. *Journal of Agricultural and Environmental Ethics, 29*, 255–263.

Forstmann, M., Burgmer, P., & Mussweiler, T. (2012). "The mind is willing, but the flesh is weak": The effects of mind-body dualism on health behavior. *Psychological Science, 23*(10), 1239–1245.

Francione, G. L. (2008). *Animals as persons: Essays on the abolition of animal exploitation*. Columbia University Press.

Giubilini, A., & Savulescu, J. (2018). The artificial moral advisor. The "ideal observer" meets artificial intelligence. *Philosophy & Technology, 31*, 169–188.

Gjerris, M. (2015). Willed blindness: A discussion of our moral shortcomings in relation to animals. *Journal of Agricultural and Environmental Ethics, 28*, 517–532.

Gottwald, F.-T., & Boergen, I. (2013). *Essen & Moral. Beiträge zur Ethik der Ernährung*. Metropolis.

Gottwald, F.-T., Ingensiep, H. W., & Meinhardt, M. (2010). *Food ethics*. Springer.

Gottwald, F.-T., & Krätzer, A. (2014). *Irrweg Bioökonomie. Kritik an einem totalitären Ansatz*. Suhrkamp.

Grey, J., & Garrett, A. (2018). You are what you eat, but should you eat what you are? Modern philosophical dietetics. In A. Barnhill, M. Budolfson, & T. Doggett (Eds.), *The Oxford handbook of food ethics* (pp. 773–794). Oxford University Press.

Hahn, H. (2017). Kollektivierungspflichten und ethischer Konsum. *Zeitschrift für Praktische Philosophie, 4*(1), 183–210.

Halbig, C. (2013). Tugend und Glück. In M. Hoesch, S. Muders, & M. Rüther (Eds.), *Glück – Werte – Sinn. Metaethische, ethische und theologische Zugänge zur Frage nach dem guten Leben* (pp. 143–173). De Gruyter.

Heidbrink, L. (2003). *Kritik der Verantwortung. Zu den Grenzen verantwortlichen Handelns in komplexen Kontexten*. Velbrück Wissenschaft.

Heidbrink, L. (2007). *Handeln in der Ungewissheit. Paradoxien der Verantwortung*. Kulturverlag Kadmos.

Heidbrink, L., Langbehn, C., & Loh, J. (2017). *Handbuch Verantwortung*. Springer VS.

Heidbrink, L., Schmidt, I., & Ahaus, B. (2011). *Die Verantwortung des Konsumenten. Über das Verhältnis von Markt, Moral und Konsum*. Campus Verlag.

Hirschfelder, G. (2005). *Europäische Esskultur. Geschichte der Ernährung von der Steinzeit bis heute*. Campus Verlag.

Hocquette, J.-F. (2016). Is in vitro meat the solution for the future? *Meat Science, 120*, 167–176.

Hoesch, M., Muders, S., & Rüther, M. (2013). *Glück – Werte – Sinn*. In *Metaethische, ethische und theologische Zugänge zur Frage nach dem guten Leben*. De Gruyter.

Hoffmann, M., Schmücker, R., & Wittwer, H. (2017). *Vorrang der Moral? Eine metaethische Kontroverse*. Klostermann.

Holland, S. (2015). *Public health ethics* (2nd ed.). Polity.

Hölker, S., von Meyer-Höfer, M., & Spiller, A. (2019). Inclusion of animal ethics into the consumer value-attitude system using the example of game meat consumption. *Food Ethics, 3*, 53–75.

Hopkins, P. D. (2015). Cultured meat in Western media: The disproportionate coverage of vegetarian reactions, demographic realities, and implications for cultured meat marketing. *Journal of Integrative Agriculture, 14*(2), 264–272.

Hopkins, P. D., & Dacey, A. (2008). Vegetarian meat. Could technology save animals and satisfy meat eaters? *Journal of Agricultural and Environmental Ethics, 21*, 579–596.

Horn, C. (2014). *Antike Lebenskunst. Glück und Moral von Sokrates bis zu den Neuplatonikern* (3rd ed.). C.H.Beck.

Ingensiep, H. W. (2007). "Der Mensch ist, was er isst". Natur und Kultur der Ernährung aus anthropologischer Sicht. *Essener Unikate. Berichte aus Forschung und Lehre, 30*, 52–59.

Joy, M. (2010). *Why we love dogs, eat pigs and wear cows. An introduction to carnism*. Conari Press.

Kaiser, M., & Algers, A. (2016). Food ethics: A wide field in need of dialogue. *Food Ethics, 1*, 1–7.

Kallhoff, A., Di Paola, M., & Schörgenhumer, M. (2018). *Plant ethics. Concepts and applications*. Routledge.

Kaplan, D. M. (2012a). *The philosophy of food*. University of California Press.

Kaplan, D. M. (2012b). Introduction. In *The philosophy of food* (pp. 1–23). University of California Press.

Kühler, M., & Nossek, A. (2014). *Paternalismus und Konsequentialismus*. Mentis.

Laestadius, L. I. (2015). Public perceptions of the ethics of in-vitro meat: Determining an appropriate course of action. *Journal of Agricultural and Environmental Ethics, 28*, 991–1009.

Laestadius, L. I., & Caldwell, M. A. (2015). Is the future of meat palatable? Perceptions of in vitro meat as evidenced by online news comments. *Public Health Nutrition, 18*(13), 2457–2467.

Lagerlund, H. (2018). Food ethics in the middle ages. In A. Barnhill, M. Budolfson, & T. Doggett (Eds.), *The Oxford handbook of food ethics* (pp. 759–772). Oxford University Press.
Lamla, J. (2013). *Verbraucherdemokratie. Politische Soziologie der Konsumgesellschaft.* Suhrkamp.
Lammers, P., Ullmann, L. M., & Fiebelkorn, F. (2019). Acceptance of insects as food in Germany: Is it about sensation seeking, sustainability consciousness, or food disgust? *Food Quality and Preference, 77*, 78–88.
Leitzmann, C. (2018). *Veganismus. Grundlagen, Vorteile, Risiken.* C.H. Beck.
Lemke, H. (2007). *Ethik des Essens. Eine Einführung in die Gastrosophie.* Akademie Verlag.
Lemke, H. (2012). *Politik des Essens.* In *Wovon die welt von morgen lebt.* Transcript.
Levinstein, B., & Sandberg, A. (2015). *The moral limitations of in vitro meat.* Retrieved April 18, 2020, from, http://blog.practicalethics.ox.ac.uk/2015/09/the-moral-limitations-of-in-vitro-meat/
Loh, J. (2019). *Roboterethik. Eine Einführung.* Suhrkamp.
Mackie, J. L. (1977). *Ethics: Inventing right and wrong.* Penguin Books.
Marcu, A., Gaspar, R., Rutsaert, P., Seibt, B., Fletcher, D., Verbeke, W., & Barnett, J. (2015). Analogies, metaphors, and wondering about the future: Lay sense-making around synthetic meat. *Public Understanding of Science, 24*(5), 547–562.
Mattick, C. S. (2018). Cellular agriculture: The coming revolution in food production. *Bulletin of the Atomic Scientists, 74*(1), 32–35.
Mattick, C. S., Landis, A. E., Allenby, B. R., & Genovese, N. J. (2015). Anticipatory life cycle analysis of in vitro biomass cultivation for cultured meat production in the United States. *Environmental Science and Technology, 49*, 11941–11949.
Mazar, N., & Zhong, C.-B. (2010). Do green products make us better people? *Psychological Science, 21*(4), 494–498.
McMullen, S., & Halteman, M. C. (2019). Against inefficacy objections: The real economic impact of individual consumer choices on animal agriculture. *Food Ethics, 2*, 93–110.
Mepham, B. (1996). *Food ethics.* Routledge.
Meyer, K. (2018). *Was schulden wir künftigen Generationen? Herausforderung Zukunftsethik.* Reclam.
Mill, J. S. (1957). *Utilitarianism.* Liberal Arts Press.
Misselhorn, C. (2018). *Grundfragen der Maschinenethik (2nd revised ed.).* Reclam.
Mohrs, T. (2003). *Weltbürgerlicher Kommunitarismus. Zeitgeistkonträre Anregungen zu einer konkreten Utopie.* Königshausen & Neumann.
Mohrs, T. (2009). Habe Mut, dich deines gastrosophischen Verstandes zu bedienen! Ein (kritischer) Blick auf das Postulat der gastrosophischen Mündigkeit. *Epikur. Journal für Gastrosophie*, http://www.epikur-journal.at [accessed: 25/03/2020].
Mohrs, T. (2013). Essen – Identität – Verantwortung. Konsumentenethische Reflexionen. In F.-T. Gottwald & I. Boergen (Eds.), *Essen & Moral. Beiträge zur Ethik der Ernährung* (pp. 139–150). Metropolis.
Morozov, E. (2013). *To save everything, click here: Technology, solutionism, and the urge to fix problems that don't exist.* Penguin Books.
Mosa Meat: frequently asked questions. Retrieved August 3, 2020, from, https://www.mosameat.com/faq.
Mulgan, T. (2011). *Ethics for a broken world: Imagining philosophy after catastrophe.* McGill-Queens University Press.
Nuffield Council on Bioethics. (2019). *Meat alternatives.* Bioethics briefing note. Retrived March 28, 2020, from , https://www.nuffieldbioethics.org/publications/meat-alternatives
Ott, K. (2014). *Umweltethik zur Einführung (2nd extended ed.).* Junius.
Parfit, D. (1984). *Reasons and persons.* Oxford University Press.
Perler, D., & Wild, M. (2005). *Der Geist der Tiere. Philosophische Texte zu einer aktuellen Diskussion.* Suhrkamp.

Persson, I., & Savulescu, J. (2012). *Unfit for the future. The need for moral enhancement*. Oxford University Press.

Piazza, J., Ruby, M. B., Loughnan, S., Luong, M., Kulik, J., Watkins, H. M., & Seigerman, M. (2015). Rationalizing meat consumption. The 4Ns. *Appetite, 91*, 114–128.

Pluhar, E. B. (2010). Meat and morality: Alternatives to factory farming. *Journal of Agricultural and Environmental Ethics, 23*, 455–468.

Pojman, P., & Pojman, L. P. (2011). *Food ethics*. Wadsworth Publishing.

Post, M. J. (2012). Cultured meat from stem cells: Challenges and prospects. *Meat Science, 92*(3), 297–301.

Post, M. J. (2014). Cultured beef: Medical technology to produce food. *Journal of the Science of Food and Agriculture, 94*(6), 1039–1041.

Pouteau, S. (2014). Beyond "second animals": Making sense of plant ethics. *Journal of Agricultural and Environmental Ethics, 27*, 1–25.

Raters, M.-L. (2017). Was geht uns das Elend der Welt an? Überlegungen zur Grenze zwischen Pflicht und Supererogation am Beispiel des Weltarmutsproblems. *Zeitschrift für Praktische Philosophie, 4*(2), 191–218.

Regan, T. (2004). *The case for animal rights*. University of California Press.

Reitmeier, S. (2013). *Warum wir mögen, was wir essen. Eine Studie zur Sozialisation der Ernährung*. transcript.

Roser, D., & Seidel, C. (2015). *Ethik des Klimawandels. Eine Einführung (2nd extended ed.)*. Wissenschaftliche Buchgesellschaft.

Rückert-John, J., & Kröger, M. (2019). *Fleisch. Vom Wohlstandssymbol zur Gefahr für die Zukunft*. Nomos.

Sandin, P. (2017). How to label 'natural' foods: A matter of complexity. *Food Ethics, 1*, 97–107.

Sandler, R. L. (2015). *Food ethics: The basics*. Routledge.

Sapontzis, S. F. (2014). In Defense of the pig. *Journal of Animal Ethics, 4*(1), 5–17.

Schaefer, G. O., & Savulescu, J. (2014). The ethics of producing *in vitro* meat. *Journal of Applied Philosophy, 31*(2), 188–202.

Schmitz, F. (2014). *Tierethik. Grundlagentexte*. Suhrkamp.

Schwartz, B. (2004). *The paradox of choice. Why more is less*. Ecco.

Shockley, M., & Dossey, A. T. (2014). Insects for human consumption. In J. A. Morales-Ramos, M. G. Rojas, & D. I. Shapiro-Ilan (Eds.), *Mass production of beneficial organisms. Invertebrates and entomopathogens* (pp. 617–652). Elsevier.

Shriver, A. (2020). Prioritizing the protection of welfare in gene-edited livestock. *Animal Frontiers, 10*(1), 39–44.

Singer, P., & Mason, J. (2007). *The ethics of what we eat. Why our food choices matter*. Rodale Books.

Stoecker, R., Neuhäuser, C., & Raters, M.-L. (2011). *Handbuch Angewandte Ethik*. J.B. Metzler.

Telfer, E. (1996). *Food for thought: Philosophy and food*. Routledge.

Tiberius, V. (2008). *The reflective life. Living wisely with our limits*. Oxford University Press.

Tiberius, V. (2014). *Moral psychology. A contemporary introduction*. Routledge.

Thompson, P. B. (2015). *From field to fork. Food ethics for everyone*. Oxford University Press.

Thompson, P. B. (2016). The emergence of food ethics. *Food Ethics, 1*, 61–74.

Thompson, P. B., & Hannah, W. (2008). Food and agricultural biotechnology: A summary and analysis of ethical concerns. In U. Stahl, U. E. B. Donalies, & E. Nevoigt (Eds.), *Food biotechnology* (pp. 229–264).

Tuomisto, H. L., & Teixeira de Mattos, M. J. (2011). Environmental impacts of cultured meat production. *Environmental Science and Technology, 45*(14), 6117–6123.

Van Ackeren, M., & Kühler, M. (2016). *The limits of moral obligation. Moral demandingness and ought implies can*. Routledge.

Voget-Kleschin, L., Bossert, L., & Ott, K. (2014). *Nachhaltige lebensstile: Welchen beitrag kann ein bewusster fleischkonsum zu mehr naturschutz, klimaschutz und gesundheit leisten?* Metropolis.

Vogt, K. M. (2018). Who you are is what you eat: Food in ancient thought. In A. Barnhill, M. Budolfson, & T. Doggett (Eds.), *The Oxford handbook of food ethics* (pp. 741–758). Oxford University Press.

Waltner-Toews, D., & Houle, K. (2017). Biophilia on the dinner plate: A conversation about ethics and entomophagy. *Food Ethics, 1*, 157–171.

Welin, S. (2013). Introducing the new meat. Problems and prospects. *Etikk i praksis. Nordic Journal of Applied Ethics, 7*(1), 24–37.

Wessels, U. (2002). *Die gute samariterin. Zur struktur der supererogation*. De Gruyter.

Wild, M. (2008). *Tierphilosophie zur einführung*. Junius.

Young, I. M. (2011). *Responsibility for justice*. Oxford University Press.

Zwart, H. (2000). A short history of food ethics. *Journal of Agricultural and Environmental Ethics, 12*, 113–126.

Bioeconomy and Food Security

5

Moral Conflicts Caused by Climate Change and Population Growth

Patrick Hohlwegler

Abstract

To limit global warming to well below 2 °C, as agreed upon in the Paris Agreement, a fundamental transformation of all central areas of our societies is necessary. Such a transformation includes energy supply, industry, digital infrastructure, buildings, transport, land use and our general lifestyles. The most important part of such a transformation, however, is a near-term phase-out of fossil fuels. Bioeconomy—an economy based on renewable resources and circulation—suggests a possible solution to the question of how such a transformation can be realised. Using biofuels to produce energy is expected to become a key element of every bioeconomy. Yet, using (food) crops to produce biofuels might conflict with food security—particularly in regard of climate change and population growth. In this essay, I question whether the implementation of a bioeconomy will cause moral conflicts concerning food security in regards of climate change and population growth. Based on a literature review of recent articles, I argue that due to climate change and the anticipated population growth, moral conflicts will very likely arise regarding food security by implementing a bioeconomy based on biofuels. To succeed in transforming our societies and to limit global warming to well below 2 °C, the implementation of a bioeconomy is not enough—even though the idea behind is basically right. What it takes, however, is also a change in our personal behaviour. We need to live and consume in a far more sustainable manner since our consumption patterns are the key drivers of climate change.

Keywords

Bioeconomy · Food security · Climate change

P. Hohlwegler (✉)
Philosophisches Seminar, University of Kiel, Kiel, Germany
e-mail: hohlwegler@philsem.uni-kiel.de

D. Lanzerath et al. (eds.), *Bioeconomy and Sustainability*,
https://doi.org/10.1007/978-3-030-87402-5_5

1 Introduction

A fundamental transformation of central areas of our societies, namely energy supply, industry, digital infrastructure, the building stock, transport, land use and our lifestyles—is inevitable in order to limit global warming to well below 2 °C.[1] This is particularly true of the massive consumption of fossil fuels.[2] For some years now, great hopes have been pinned on the so-called *bioeconomy* in order to successfully implement the necessary transformation within the coming decades.[3] Basically, bioeconomy is concerned with the "Aufbau einer Kreislaufwirtschaft, die im Sinne von Ressourceneffizienz und Nachhaltigkeit eine bestmögliche Verwertung sowie Mehrfachnutzung von Rohstoffen und Stoffströmen [. . .] ermöglicht".[4] A key objective of bioeconomy is to reconcile food security with the sustainable use of renewable resources for industrial purposes, while at the same time ensuring environmental protection.[5] According to a 2018 study, some 50 nations have defined bioeconomy as part of their policies.[6] However, particularly with regard to the production of bioenergy and biofuels, the question arises as to what extent the agricultural production of food could be affected.[7] Although it is stressed that the goal of food security is an essential component of the concept of bioeconomy, it is doubtful whether this goal can be achieved in the near future—especially considering the background of climate change[8], the associated impacts on agriculture[9], and a further increase in the world population, especially in sub-Saharan Africa.[10]

In this essay, I want to show that we as the present generation (a) have a moral obligation to ensure food security both *intra-* and *intergenerationally*, that we (b) have recognised and accepted this responsibility, but that we (c) do not currently fulfil this obligation, and that our current behaviour at the global level even leads us to (d) reduce, if not destroy, the conditions of possibility to achieve the goal of food security for future generations. Against this background, it is ultimately doubtful that bioeconomy, although basically pointing in the right direction, can achieve the goal of food security on a global level. In the following section, I will first explain where our moral obligation to ensure food security derives from. I will then show that we

---

[1] Cf. Falk et al. (2019).

[2] Cf. Rogelj et al. (2015).

[3] Cf. Federal Ministry of Education and Research (2014).

[4] Ibid., 5, "establishment of a circular economy which, in terms of resource efficiency and sustainability, enables the best possible recycling and multiple use of raw materials and material flows [. . .]" (own translation).

[5] Cf. European Commission (2012).

[6] Cf. Von Braun (2018).

[7] Cf. Tilman et al. (2009), Sheppard et al. (2011) and Lewandowski (2015).

[8] Cf. Intergovernmental Panel on Climate Change (2018).

[9] Cf. Intergovernmental Panel on Climate Change (2019a).

[10] Cf. United Nations, Department of Economic and Social Affairs, Population Division (2019).

have accepted this moral obligation and are working to fulfil it, but that our efforts are not yet sufficient. After that I will explain and justify my doubts that bioeconomy can achieve the objective of food security. Lastly, I shall identify some of the measures that are needed to ensure that the objective of food security can be achieved in the near future.

2 The Human Right to Adequate Food

According to Article 11, paragraph 1 of the *International Covenant on Economic, Social and Cultural Rights* (ICESCR), everyone has the right "to an adequate standard of living for himself and his family, including adequate food, clothing and housing, and to the continuous improvement of living conditions".[11] The ICESCR was adopted by the United Nations in 1966 and has been ratified by 170 states since its entry into force in 1976.[12] It can therefore be seen as a binding normative basis. The *Committee on Economic, Social and Cultural Rights* (CESCR) monitors and comments on the implementation of the rights established in the ICESCR. In 1999, the CESCR published a commentary on Article 11 of the ICESCR, further expounding the human right to an adequate standard of living. In this note, the CESCR stresses that

> the right to adequate food is indivisibly linked to the inherent dignity of the human person and is indispensable for the fulfilment of other human rights enshrined in the International Bill of Human Rights. It is also inseparable from social justice [. . .].[13]

It also follows from Article 6 of the CESCR comment that

> [t]he right to adequate food is realized when every man, woman and child, alone or in community with others, have physical and economic access at all times to adequate food or means for its procurement. The *right to adequate food* shall therefore not be interpreted in a narrow or restrictive sense which equates it with a minimum package of calories, proteins and other specific nutrients.[14]

The concept of adequacy is particularly emphasised by the CESCR and, remarkably, is linked to sustainable development in the field of food security, which should be guaranteed not only for present generations but also for future ones. Thus, Article 7 of the CESCR's comment states that

> [t]he concept of *adequacy* is particularly significant in relation to the right to food since it serves to underline a number of factors which must be taken into account in determining whether particular foods or diets that are accessible can be considered the most appropriate

[11] United Nations (1967).

[12] Cf. United Nations Human Rights Office of the High Commissioner (2020).

[13] Committee on Economic, Social and Cultural Rights (1999), Art. 4.

[14] Ibid., Art. 6.

> under given circumstances for the purposes of article 11 of the Covenant. The notion of *sustainability* is intrinsically linked to the notion of adequate food or food *security*, implying food being accessible for both present and future generations. The precise meaning of "adequacy" is to a large extent determined by prevailing social, economic, cultural, climatic, ecological and other conditions, while "sustainability" incorporates the notion of long-term availability and accessibility.[15]

The question of whether we as the present generation also have obligations to future generations is, however, a controversial ethical issue.[16] Irrespective of the discussion on this question, it follows from the CESCR's remark that such obligations exist insofar as human rights are used as an ethical basis. If we now take this as a moral yardstick and refer specifically to Article 11 of the ICESCR, this implies that we have a moral duty to end hunger in the world and to ensure that future generations have the possibility of sufficient and adequate food as well. Since human rights are egalitarian in nature and are to be applied equally to all people, the subject "we" in this essay refers to all people living at present. It remains to be noted, however, that societies of rich industrial nations should bear a special responsibility since they have the necessary (financial) resources and, furthermore, because of their economic power, are largely responsible for the current distribution of the world's resources and thus of food.[17] Nevertheless, the responsibility of rich industrial nations does not exempt other nations from their own responsibility. This is especially true with regard to the establishment of stable political conditions, which are a necessary, though not sufficient, prerequisite to the fulfilment of human rights.

However, if we look at the current situation in the world, it is clear that we are not fulfilling our moral obligation to provide adequate food for all people. According to the latest estimate of the *Food and Agriculture Organization of the United Nations* (FAO), between 720 and 811 million people were suffering from hunger in 2020, mainly in Asia and sub-Saharan Africa. In addition, more than 2.3 billion people did not have permanent access to safe and adequate food.[18] In many cases, this situation constitutes a considerable injustice,[19] which is morally reprehensible, in particular because there is de facto sufficient and adequate food for all currently living people.[20] This means that hunger in the world could be eliminated, at least at present, through a more just distribution. Adding to the moral guilt for this global injustice is the fact that our current behaviour is leading to a steady deterioration of the ecological state of the Earth, which could permanently change and even destroy the conditions for

[15] Committee on Economic, Social and Cultural Rights (1999), Art. 7.

[16] Cf. for example Ott (2004), Meyer (2012) and Caney (2018).

[17] Cf. Pogge (2005).

[18] Cf. FAO, IFAD, UNICEF, WFP and WHO (2021).

[19] An injustice and thus a moral guilt of the present generation exists when those affected do not themselves bear responsibility for the lack of adequate food, but when this lack is caused by external circumstances, especially poverty. This means that the free choice of an unbalanced diet, which can also lead to malnutrition, is not understood as injustice.

[20] Cf. Wood et al. (2018).

human life.[21] This is mainly caused by climate change and its impacts,[22] but also by the massive loss of biodiversity.[23] In this manner, our current behaviours lead to a reduction of opportunities for future generations to develop. This particularly affects the human right to adequate food since climate change will increase the intensity and frequency of extreme weather events—for example in the form of severe droughts or floods, especially in tropical regions such as sub-Saharan Africa—and thus severely impair agricultural food production. It follows, however, that not only are we failing to meet our moral obligation to our own generation, but we are also disregarding those we have to future generations.

The fact that we have recognised and accepted our moral obligation is demonstrated above all by the United Nations *Sustainable Development Goals* (SDGs).

3 The United Nations Sustainable Development Goals

In order to meet the diverse challenges facing humanity at the beginning of the twenty-first century, the United Nations has adopted the *Agenda 2030 for Sustainable Development* in 2015.[24] The overarching goal of this agenda is "to realize the human rights of all [. . .]."[25] Accordingly, among other goals, it seeks "to end poverty and hunger, in all their forms and dimensions"[26] and to take "urgent action on climate change".[27] Central to the Agenda 2030 are 17 SDGs which comprise 169 targets. Although all SDGs are "integrated and indivisible",[28] in this essay I will focus on only two SDGs that are particularly relevant for ensuring food security: SDG 2 *Zero Hunger* and SDG 13 *Climate Action*. The first and most important target of SDG 2 is to "end hunger and ensure access by all people, in particular the poor and people in vulnerable situations [. . .] to safe, nutritious and sufficient food all year round."[29] Even though the term "adequacy" is not explicitly used, it can be concluded that if SDG 2 is successfully achieved by 2030, the human right to adequate food would be fulfilled for the 2030 present generation. Thus, the above-mentioned injustice concerning the *intragenerational* moral obligation would be solved as well.

[21] Cf. Rockström et al. (2009), Steffen et al. (2015) and Steffen et al. (2018).

[22] Cf. Intergovernmental Panel on Climate Change (2018, 2019a, 2019b).

[23] Cf. Intergovernmental Science-Policy Platform on Biodiversity and Ecosystem Services (2019).

[24] Cf. United Nations (2015a).

[25] Ibid., 3.

[26] Ibid.

[27] Cf. ibid.

[28] Ibid.

[29] Ibid., 17.

Actions to achieve SDG 2 include doubling of agricultural productivity and facilitating access to various production factors, especially for disadvantaged groups such as women or indigenous people.[30] In addition, investments in various areas of agriculture, such as infrastructure and research, shall be increased and interventions in free trade, especially through agricultural subsidies, shall be ceased.

However, another, yet essential, condition to achieve SDG 2 and other SDGs is to stop climate change. This applies equally to the fulfilment of the human right to adequate food. Aware of the dangers of climate change, the United Nations has formulated SDG 13 as a further part of the Agenda 2030. According to this goal, "urgent action to combat climate change and its impacts"[31] shall be taken. The key objective of SDG 13 is to limit climate change in accordance with the 1992 United Nations *Framework Convention on Climate Change* (UNFCCC).[32] In this respect, SDG 13 is also in accordance with the 2015 Paris Agreement and its main objective of "[h]olding the increase in the global average temperature to well below 2 °C above pre-industrial levels and pursuing efforts to limit the temperature increase to 1.5 °C above pre-industrial levels, recognizing that this would significantly reduce the risks and impacts of climate change."[33]

The SDGs demonstrate the insight and declared will of the community of states to initiate and implement the necessary changes. They demonstrate that we have recognised and accepted our moral responsibility towards the securement of well-being of our own generation and towards future generations. Against the background of climate change, which has been progressing largely unabated to date, and a further increase in the world population, the question arises as to whether the SDGs can be achieved by 2030 and beyond. These doubts become particularly clear in the light of two current United Nations reports, the so-called *Emissions Gap Report* and the *World Population Prospects*, both of which are briefly presented in the following section.

4 The United Nations Reports

4.1 The Emissions Gap Report

Every year, the *United Nations Environmental Programme* (UNEP) publishes the *Emissions Gap Report*. Based on the so-called *Nationally Determined Contributions* (NDCs)—current pledges by the member states of the Paris Agreement to reduce annual greenhouse gas emissions—the gap discussed in this report addresses the amount of greenhouse gas emissions that the global community is about to overshoot

[30] Ibid.

[31] Ibid., 25.

[32] Cf. United Nations (1992).

[33] Cf. United Nations (2015b), Art. 2.

in 2030. UNEP finds that according to current NDCs there is a gap of 11 gigatons CO_2-equivalents[34] (Gt CO_2e) to the 2 °C target and a gap of 25 Gt CO_2e to the 1.5 °C target for 2030.[35] If, in addition, commitments linked to several conditions—mostly financial aid—are not met, the gap will increase by a further 2 Gt CO_2e for the 2 °C target and 3 Gt CO_2e for the 1.5 °C target. To compare: global greenhouse gas emissions in 2019 (the last available year) were 58,1 Gt CO_2e. If the 1.5 °C target of the Paris Agreement shall still be achieved, annual greenhouse gas emissions will have to be reduced by more than 50% until 2030. Such a significant decarbonisation is possible, but requires the fastest economic transformation in the history of mankind.[36] Even the 2 °C target is severely threatened in view of current efforts. Accordingly, there are considerable doubts as to whether SDG 13 and thus also other SDGs, such as SDG 2, can be met.

Based on current climate policies, global average temperature is likely to increase by 2.8 °C until 2100. Such a warming is highly likely to result in devastating consequences both for the environment and for mankind.[37]

4.2 World Population Prospects

Another report related to the issue of food security is the biennial report on the expected world population growth by the United Nations *Department of Economic and Social Affairs* (UN DESA). The last edition of the *World Population Prospects* was published in 2019.[38] It has confirmed the findings and forecasts of previous reports[39] that the world population will continue to grow and will also become older on average. According to current estimates, the world population will grow by about 2 billion to 9.7 billion people by 2050. By 2100, it is estimated that another 1.2 billion people will be added. Much of the projected growth by 2050—more than 1 billion—will take place in sub-Saharan Africa. From 2050 onwards, further growth will be concentrated almost exclusively on this region, which is already affected by recurring hunger periods.[40] If the world population develops in line with the United Nations forecast, this will very likely increase the number of people suffering from hunger in sub-Saharan Africa, thereby fundamentally jeopardising the fulfilment of SDG 2 and other SDGs.

---

[34] One gigatonne equals 10^9 tonnes. CO_2 equivalents are an artificial unit that also takes into account the effect of other greenhouse gases, such as methane (CH_4) or nitrogen oxide (NO_2), on the basis of CO_2's harmfulness to the climate.

[35] Cf. United Nations Environment Programme (2021).

[36] Cf. Falk et al. (2019).

[37] Cf. Steffen et al. (2018) and Lenton et al. (2019).

[38] Cf. United Nations, Department of Economic and Social Affairs, Population Division (2019).

[39] Cf. United Nations, Department of Economic and Social Affairs, Population Division (2015, 2017).

[40] Cf. Food and Agriculture Organization of the United Nations et al. (2021).

5 Possible Solutions

5.1 Climate Engineering

One possibility of counteracting climate change and its impacts that has been widely discussed for some years is the so-called *climate engineering* (sometimes also called *geoengineering*).[41] It is commonly defined as "the deliberate large-scale manipulation of the planetary environment [. . .]."[42] Due to the insufficient efforts to date to limit climate change, the use of climate engineering methods appears to be very likely, if not necessary, in the near future in order to achieve the 2 °C/the 1.5 °C target agreed upon in the Paris Agreement.[43]

Climate engineering can basically be divided into two categories: *Solar Radiation Management* (SRM) and *Carbon Dioxide Removal* (CDR).[44] SRM methods aim to manipulate the Earth's radiation budget by either reducing the proportion of sunlight incident or increasing the proportion of sunlight reflected by the Earth. This will counteract global warming, but not the actual cause of anthropogenic climate change, i.e. the sharp increase of atmospheric CO_2-concentration compared to the pre-industrial period. In contrast to that, CDR methods aim to reduce the CO_2-concentration in the atmosphere. In this way not only global warming would be counteracted, but also some of its impacts, e.g. ocean acidification.

SRM methods are evaluated very critically, especially from an ethical point of view, as they are associated with high environmental risks.[45] In contrast, CDR methods tend to be assessed more positively because they are (a) mostly based on natural processes, such as photosynthesis or the natural weathering of rocks, and because (b) they fight the root cause of climate change. Nevertheless, there are also various ethical reservations against CDR.[46] In principle, the best solution to climate change is a significant and immediate reduction in anthropogenic greenhouse gas emissions.[47]

Currently, the most frequently discussed CDR method is *Bioenergy with Carbon Capture and Storage* (BECCS).[48] This method combines the use of biomass to produce bioenergy (BE) with the use of Carbon Capture and Storage (CCS) technology to subsequently store the emissions generated during the production of bioenergy.[49] In addition to afforestation, the use of BECCS is assumed in most scenarios assessed by the *Intergovernmental Panel on Climate Change* (IPCC) in its

[41] Cf. The Royal Society (2009), Rickels et al. (2011) and Caldeira et al. (2013).

[42] Cf. The Royal Society (2009, p. 1).

[43] Cf. Mace et al. (2018).

[44] Cf. Rickels et al. (2011).

[45] Cf. Preston (2013).

[46] Cf. ibid.

[47] Cf. Intergovernmental Panel on Climate Change (2018).

[48] Cf. Fuss et al. (2018).

[49] Cf. Canadell and Schulze (2014).

last Assessment Report of 2013, in which global warming could be limited to below 2 °C by 2100 with a probability of more than 66%.[50] However, the area used for BECCS in these scenarios is so vast that conflicts in land use, especially in tropical regions, would be virtually unavoidable. The same applies to those scenarios in which large-scale afforestation was assumed.[51]

5.2 Sustainable Intensification of Agriculture

In view of a growing world population and a growing demand, especially for animal food, an expansion and intensification of agriculture seems almost inevitable.[52] However, an expansion of conventional agriculture, for example through extensive deforestation or the drainage of peatlands, would further intensify climate change, as additional CO_2 would be released into the atmosphere.[53] This in turn would have negative impacts on agriculture, as extreme weather events would increase. Against the background of this dilemma, the concept of sustainable intensification of agriculture has been discussed for several years now.[54] The idea behind it is to increase crop yields without increasing the ecological footprint of agricultural production and, if possible, even reducing it.[55] Essential for a sustainably intensified agriculture is that the concept of sustainability is at its core and is not just understood as an attribute to an otherwise conventional production method.[56] Accordingly, a sustainably intensified agriculture also encompasses several aspects, such as reducing food waste, adjusting subsidies for agricultural products in industrialised countries, ending the agricultural use of individual areas, using fertilisers more efficiently, using new technologies, better education, especially for women in rural regions of sub-Saharan Africa, and a strategy adapted to local conditions.[57] The use of genetically modified seeds is also not excluded in principle in the context of sustainably intensified agriculture.[58]

[50] Cf. Intergovernmental Panel on Climate Change (2014).

[51] Other CDR methods, such as *Enhanced Weathering* or *Direct Air Capture*, were not considered in the last IPCC Assessment Report and are also not discussed any further in this essay.

[52] Cf. Food and Agriculture Organization of the United Nations (2018).

[53] Cf. Ramankutty et al. (2018).

[54] Cf. Pretty and Bharucha (2014), Godfray and Garnett (2014) and Rockström et al. (2017).

[55] Cf. Sonnino et al. (2014).

[56] Cf. Rockström et al. (2017).

[57] Cf. ibid.

[58] Cf. ibid.

6 Bioeconomy

The sustainable intensification of agriculture is one way of meeting the growing demand for food in the future and at the same time counteracting the problem of hunger, especially in sub-Saharan Africa. However, this only marginally addresses the fundamental problem of climate change, as the majority of anthropogenic greenhouse gas emissions are caused by the burning of fossil fuels.[59] In order to fulfil the objectives of the Paris Agreement and to limit global warming to well below 2 °C above pre-industrial levels, a comprehensive and far-reaching transformation of all areas of life is required, especially in industrialised societies.

The concept of bioeconomy is an important part of this transformation as it aims to achieve sustainable economic growth in harmony with nature.[60] Essential for this is the renunciation of fossil fuels in favour of renewable resources. However, using plants to produce bioenergy or biofuels is controversial from an ethical point of view in regard to the human right to adequate food, as edible plant varieties such as sugar cane, corn or soya are used, too.[61] And the more these plants are used for the production of bioenergy, the smaller the proportion of agricultural land used for the production of food, which was recently only about 18%,[62] compared to 71% for the production of animal feed for livestock.[63] A further reduction of agricultural land used to produce food would very likely contribute to increasing hunger in the world—what, in turn, would be in clear contradiction to the United Nations SDGs and to our moral obligation to realise food security. Although non-edible plants, such as miscanthus or switchgrass, have been discussed as potential energy sources, too, this does not significantly change the conflict over arable land, as non-edible plants also need land to grow.[64]

Nevertheless, the topic of food security is defined as a key objective of bioeconomy.[65] In order to guarantee this, the cultivation of food should be given priority on the one hand, and food waste should be reduced on the other.[66] In principle, existing resources should be used more efficiently. However, the answer to the question of how comprehensively a changeover to renewable resources can actually be achieved by using bioeconomic methods and whether food security will in fact always be a priority remains uncertain, at least for the time being.

[59] Cf. Le Quéré et al. (2018).

[60] Cf. Federal Ministry of Education and Research (2014).

[61] Cf. Thompson (2012) and Gamborg et al. (2012).

[62] Cf. German Environment Agency (2013).

[63] Cf. ibid.

[64] Cf. Murphy et al. (2011).

[65] Cf. Federal Ministry of Education and Research (2014).

[66] Cf. ibid.

7 Discussion

Both climate change and projected population growth will have a major impact on agricultural food production in the near future. Both factors will make SDG 2 and other SDGs more difficult to achieve. Furthermore, the use of climate engineering will become almost unavoidable in the near future. According to current models, this will also have a strong impact on agricultural food production, especially in tropical regions. This will make it even more difficult to meet our moral obligation to provide adequate food for all people.

Bioeconomic processes aim at a sustainable production, in which fossil fuels are to be dispensed with. In view of climate change and its associated impacts, this is also urgently required, since a large proportion of anthropogenic greenhouse gas emissions are caused by the use of fossil fuels. To this extent, bioeconomy represents a plausible approach to limit climate change. However, there are justified doubts as to whether bioeconomy can also guarantee the goal of food security. Although currently there would be enough food available to feed all people adequately, it is realistically unlikely that there will be a change in global food distribution in the near future. For this reason, and in view of the expected population growth, an intensification and expansion of (conventional) agriculture, especially in sub-Saharan Africa, will hardly be avoidable in order to stop hunger in the world in the sense of SDG 2.

Even though there are various ways of counteracting hunger without expanding and intensifying conventional agriculture—including new technologies for managing agricultural land, more efficient use of fertilisers and more education, a reduction in food waste and an adjustment of agricultural subsidies in developed countries—it remains doubtful whether the goal of food security can actually be achieved through bioeconomy. The abandonment of fossil fuels will increase the demand for biomass. Although biomass can be obtained in part from organic waste and agricultural residues, these sources alone will hardly be able to meet the entire demand. To this extent, bioeconomy will lead to increased competition for arable land, which will result in an expansion and more intensive use of such land. This in turn will intensify climate change, which will have a negative impact on agriculture as extreme weather events will increase. However, the use of climate engineering is also likely to lead to competition for agricultural land, as very large areas of land are required both for the use of BECCS and afforestation. A portfolio solution that includes various CDR methods would also increase the competition for agricultural land, as both, BECCS and afforestation, would very likely be part of such a solution. In this respect, moral conflicts in land use as a result of climate change and population growth will hardly be avoidable in the near future.

Since we have a moral obligation to fight hunger, we will have to increase the amount of agricultural land available for food production. However, this will exacerbate climate change. In order to limit global warming it will very likely require the use of climate engineering in the near future, given our current efforts. Now, if the two methods mentioned above are used, this will limit the amount of land available to agriculture. The conflict of goals between food production and climate protection will also give rise to various moral conflicts. People in tropical regions of the world will have to decide whether they want to continue to grow food on their land or whether they want to grow special plants or trees for climate protection. They will have to decide whether they want to sell and migrate their land in the face of climate change or continue to live in their familiar surroundings under much worse conditions.

But inhabitants of industrial nations will have to ask themselves, too, whether they are willing to pay more for sustainably produced food. Whether they are willing to use their wealth and technical capabilities to combat the impacts of climate change particularly in tropical regions, which have been caused by burning large amounts of fossil fuels. If we do not fundamentally change our way of life until 2030 and fulfil our moral obligations, we will neither achieve the United Nations SDGs nor continue to live in a world as we know it today.

Even if the approach of bioeconomy is basically convincing: In order to actually be able to achieve the goal of food security and also to limit climate change, more than just a change in the economy is needed. It requires a fundamental change in our way of life. Our enormous appetite for consumption is a driving force behind climate change and global injustice, and thus also for the unfair distribution of food in the world. If industrialised societies manage to make their consumption more sustainable and conscious, there is hope that the SDGs can actually be achieved by 2030. Sustainable and conscious consumption means, for example, that a new T-shirt cannot be bought for €3.00 if it is supposed to have been produced under fair and climate-neutral conditions. Sustainable consumption also implies higher prices for animal food, especially meat, as such food is very emission-intensive to produce and thus have a negative impact on the climate. Another important aspect of sustainable and conscious consumption concerns the way we travel. Instead of travelling by plane, in many cases we could use a train. Instead of owning and using a private car, we could use a bicycle, car-sharing or even public transport. These are just a few aspects of a whole range of possibilities for tackling climate change and hunger on a small scale. Sustainable and conscious consumption does not necessarily mean renunciation. But it does mean that we are prepared to recognise the true value of things and are accordingly willing to pay for them. In the rich societies of the West, sustainable and conscious consumption would not mean existential cuts, but it would do much more justice to our moral obligation to global food security, which has already been accepted in principle.

References

Caldeira, K., Bala, G., & Cao, L. (2013). The science of geoengineering. *Annual Review of Earth and Planetary Science, 41*, 231–256.

Canadell, J. G., & Schulze, E. D. (2014). Global potential of biospheric carbon management for climate mitigation. *Nature Communications, 5*(5282).

Caney, S. (2018). Justice and future generations. *Annual Review of Political Science, 21*, 475–493.

Committee on Economic, Social and Cultural Rights (CESCR). (1999). General comment no. 12: *The right to adequate food* (Art. 11).

European Commission. (2012). *Innovating for sustainable growth: A bioeconomy for Europe*. European Commission.

Falk, J., Gaffney, O., Bhowmik, A. K., Bergmark, P., Galaz, V., Gaskell, N., Henningsson, S., Höjer, M., Jacobson, L., Jónás, T., Kåberger, T., Klingenfeld, D., Lenhard, J., Loken, B., Lundén, D., Malmodin, J., Malmqvist, T., Olausson, V., Otto, I., . . ., Shalit, T. (2019). *Exponential Roadmap 1.5. Scaling 36 Solutions to Halve Emmissions by 2030*. Retrived July 6, 2020 https://exponentialroadmap.org/wpcontent/uploads/2020/03/ExponentialRoadmap_1.5.1_216x279_08_AW_Download_Singles_Small.pdf

Federal Ministry of Education and Research, Federal Ministry of Food and Agriculture. (2014). *Bioökonomie in Deutschland. Chancen für eine biobasierte und nachhaltige Zukunft*. Federal Ministry of Education and Research, Federal Ministry of Food and Agriculture.

Food and Agriculture Organization of the United Nations (FAO). (2018). *The future of food and agriculture – Alternative pathways to 2050*.

Food and Agriculture Organization of the United Nations (FAO), International Fund for Agricultural Developement (IFAD), United Nations Childrens's Fund (UNICEF), World Food Programme (WFP) and World Health Organization (WHO). (2021): *The State of Food Security and Nutrition in the World 2021. Transforming food systems for food security, improved nutrition and affordable healthy diets for all*. Rome, FAO. doi: https://doi.org/10.4060/cb4474en

Fuss, S., Lamb, W. F., Callaghan, M. W., Hilaire, J., Creutzig, F., Amann, T., Beringer, T., de Oliveira Garcia, W., Hartmann, J., Khanna, T., Luderer, G., Nemet, G. F., Rogelj, J., Smith, P., Vicente, J. L. V., Wilcox, J., del Mar Zamora Domingues, M., & Minx, J. C. (2018). Negative emisions–part 2: Costs, potentials and side effects. *Environmental Research Letters, 13*(063002).

Gamborg, C., Millar, K., Shortall, O., & Sandøe, P. (2012). Bioenergy and land use: Framing the ethical debate. *Journal of Agricultural and Environmental Ethics, 25*(6), 909–925.

German Environment Agency. (2013). *Globale Landflächen und Biomasse nachhaltig und ressoucenschonend nutzen*. Retrieved June 30, 2020, from, https://www.umweltbundesamt.de/sites/default/files/medien/479/publikationen/globale_landflaechen_biomasse_bf_klein.pdf

Godfray, H. C. J., & Garnett, T. (2014). Food security and sustainable intensification. *Philosophical Transactions of the Royal Society B, 369*(1639).

Intergovernmental Science-Policy Platform on Biodiversity and Ecosystem Services (IPBES). (2019). Summary for policymakers of the global assessment report on biodiversity and ecosystem services of the intergovernmental science-policy platform on biodiversity and ecosystem services. Retrieved June 30, 2020, from, https://ipbes.net/sites/default/files/ipbes_7_10_add.1_en_1.pdf

Intergovernmental Panel on Climate Change (IPCC) (2014). *Climate change 2014: Synthesis report*. Contribution of Working Groups I, II and III to the Fifth Assessment Report of the Intergovernmental Panel on Climate Change. Geneva, Switzerland. Retrieved June 30, 2020, from, https://www.ipcc.ch/site/assets/uploads/2018/05/SYR_AR5_FINAL_full_wcover.pdf

Intergovernmental Panel on Climate Change (IPCC). (2018). Summary for policymakers. In *Global warming of 1.5°C. An IPCC special report on the impacts of global warming of 1.5°C above pre-industrial levels and related global greenhouse gas emission pathways, in the context of strengthening the global response to the threat of climate change, sustainable development, and efforts to eradicate poverty*. Retrieved June 30, 2020, from, https://www.ipcc.ch/site/assets/uploads/sites/2/2019/05/SR15_SPM_version_report_LR.pdf

Intergovernmental Panel on Climate Change (IPCC). (2019a). Summary for policymakers. In *Climate change and land. An IPCC special report on climate change, desertification, land degradation, sustainable land management, food security, and greenhouse gas fluxes in terrestrial ecosystems*. Retrieved June 30, 2020, from, https://www.ipcc.ch/site/assets/uploads/sites/4/2020/02/SPM_Updated-Jan20.pdf

Intergovernmental Panel on Climate Change (IPCC). (2019b). Summary for policymakers. In *IPCC special report on the ocean and cryosphere in a changing climate*. Retrieved June 30, 2020, from, https://www.ipcc.ch/site/assets/uploads/sites/3/2019/11/03_SROCC_SPM_FINAL.pdf

Lenton, T. M., Rockström, J., Gaffney, O., Rahmstorf, S., Richardson, K., Steffen, W., & Schellnhuber, H. J. (2019). Climate tipping points – Too risky to bet against. *Nature, 575*, 592–595.

Le Quéré, C., Andrew, R. M., Friedlingstein, P., Sitch, S., Hauck, J., Pongratz, J., Pickers, P. A., Korsbakken, J. I., Peters, G. P., Canadell, J. G., Arneth, A., Arora, V. K., Barbero, L., Bastos, A., Bopp, L., Chevallier, F., Chini, L. P., Ciais, P., Doney, S. C., . . . Zheng, B. (2018). Global carbon budget 2018. *Earth System Science Data, 10*(4), 2141–2194.

Lewandowski, I. (2015). Securing a sustainable biomass supply in a growing bioeconomy. *Global Food Security, 6*, 34–42.

Mace, M. J., Fyson, C. L., Schaeffer, M., & Hare, W. L. (2018). *Governing large-scale carbon dioxide removal: Are we ready?* Carnegie Climate Geoengineering Governance Initiative (C2G2).

Meyer, L. H. (2012). *Intergenerational justice*. Routledge.

Murphy, R., Woods, J., Black, M., & McManus, M. (2011). Global developments in the competition for land from biofuels. *Food Policy, 36*(1), 52–61.

Ott, K. (2004). Essential components of future ethics. In R. Döring & M. Rühs (Eds.), *Ökonomische rationalität und praktische vernunft: Gerechtigkeit, ökologische ökonomie und naturschutz. eine festschrift anlässlich des 60. geburtstags von prof. Dr. Ulrich Hampicke* (pp. 83–110). K&N.

Pogge, T. (2005). Recognized and violated by international law: The human rights of the global poor. *Leiden Journal of International Law, 18*(4), 717–745.

Preston, C. J. (2013). Ethics and geoengineering: Reviewing the moral issues raised by solar radiation management and carbon dioxide removal. *WIREs Climate Change, 4*(1), 23–37.

Pretty, J., & Bharucha, Z. P. (2014). Sustainable intensification in agricultural systems. *Annals of Botany, 114*(8), 1571–1596.

Ramankutty, N., Mehrabi, Z., Waha, K., Jarvis, L., Kremen, C., Herrero, M., & Rieseberg, L. H. (2018). Trends in agricultural land use: Implications for environmental health and food security. *Annual Review of Plant Biology, 69*, 789–815.

Rickels, W., Klepper, G., Dovern, J., Betz, G., Brachatzek, N., Cacean, S., Güssow, K., Heintzenberg, J., Hiller, S., Hoose, C., Leisner, T., Oschlies, A., Platt, U., Proelß, A., Renn, O., Schäfer, S., & Zürn, M. (2011). *Gezielte Eingriffe in das Klima?Eine Bestandsaufnahme der Debatte zu climate engineering. Sondierungsstudie für das Bundesministerium für Bildung und Forschung*. Kiel Earth Institute.

Rockström, J., Steffen, W., Noone, K., Persson, Å., Chapin, F. S., III, Lambin, E., Lenton, T. M., Scheffer, M., Folke, C., Schellnhuber, H. J., Nykvist, B., de Wit, C. A., Hughes, T., van der Leeuw, S., Rodhe, H., Sörlin, S., Snyder, P. K., Costanza, R., Svedin, U., . . . Foley, J. (2009). Planetary boundaries: Exploring the safe operating space for humanity. *Ecology and Society, 14*(22), 32.

Rockström, J., Williams, J., Daily, G., Noble, A., Matthews, N., Gordon, L., Wetterstrand, H., DeClerck, F., Shah, M., Steduto, P., de Fraiture, C., Hatibu, N., Unver, O., Bird, J., Sibanda, L., & Smith, J. (2017). Sustainable intensification of agriculture for human prosperity and global sustainability. *Ambio: A Journal of the Human Environment, 46*(1), 4–17.

Rogelj, J., Luderer, G., Pietzcker, R. C., Kriegler, E., Schaeffer, M., Krey, V., & Riahi, K. (2015). Energy system transformations for limiting end-of-century warming to below 1.5 °C. *Nature Climate Change, 5*(6), 519–527.

Sheppard, A. W., Gillespie, I., Hirsch, M., & Begley, C. (2011). Biosecurity and sustainability within the growing global bioeconomy. *Current Opinion in Environmental Sustainability, 3*(1), 4–10.

Sonnino, R., Moragues-Faus, A., & Maggio, A. (2014). Sustainable food security: An emerging research and policy agenda. *International Journal of Sociology of Agriculture and Food, 21*(1), 173–188.

Steffen, W., Richardson, K., Rockström, J., Cornell, S. E., Fetzer, I., Bennett, E. M., Biggs, R., Carpenter, S. R., de Vries, W., de Wit, C. A., Folke, C., Gerten, D., Heinke, J., Mace, G. M., Persson, L. M., Ramanathan, V., Reyers, B., & Sörlin, S. (2015). Planetary boundaries: Guiding human development on a changing planet. *Science, 347*(6223), 736–746.

Steffen, W., Rockström, J., Richardson, K., Lenton, T. M., Folke, C., Livermann, D., Summerhayes, C. P., Barnosky, A. D., Cornell, S. E., Crucifix, M., Donges, J. F., Fetzer, I., Lade, S. J., Scheffer, M., Winkelmann, R., & Schellnhuber, H. J. (2018). Trajectories of the earth system in the Anthropocene. *Proceedings of the National Academy of Sciences of the United States of America, 115*(33), 8252–8259.

The Royal Society. (2009). *Geoengineering the climate: Science, governance, and uncertainty*. The Royal Society.

Thompson, P. B. (2012). The agricultural ethics of biofuels: The food vs. fuel debate. *Agriculture, 2*(4), 339–358.

Tilman, D., Socolow, R., Foley, J. A., Hill, J., Larson, E., Lynd, L., Pacala, S., Reilly, J., Searchinger, T., Somerville, C., & Williams, R. (2009). Beneficial biofuels – The food, energy, and environment trilemma. *Science, 325*(5938), 270–271.

United Nations, Department of Economic and Social Affairs, Population Division (UN DESA). (2015). *World Population Prospects: The 2015 Revision, Key Findings and Advance Tables*. Working Paper No. ESA/P/WP.241. Retrieved July 6, 2020, from, https://population.un.org/wpp/publications/files/key_findings_wpp_2015.pdf

United Nations, Department of Economic and Social Affairs, Population Division (UN DESA). (2017). *World population prospects: The 2017 revision, key findings and advance tables*. Working Paper No. ESA/P/WP.248. Retrieved July 6, 2020, from, https://population.un.org/wpp/Publications/Files/WPP2017_KeyFindings.pdf

United Nations, Department of Economic and Social Affairs, Population Division (UN DESA). (2019). *World population prospects 2019:* Highlights (ST/ESA/SER.A/423). Retrieved July 6, 2020, from, https://population.un.org/wpp/Publications/Files/WPP2019_Highlights.pdf

United Nations Environment Programme (2021). Emissions Gap Report 2021: The Heat Is On – A World of Climate Promises Not Yet Delivered. Nairobi.

United Nations Human Rights Office of the High Commissioner. (2020). *Status of Ratification Interactive Dashboard*. Treaty: International Covenant on Economic, Social and Cultural Rights. Retrieved June 20, 2020, from, https://indicators.ohchr.org/

United Nations. (1967): *International covenant of economic, social and cultural rights* (ICESCR). Retrieved July 6, 2020, from, https://treaties.un.org/doc/Treaties/1976/01/19760103%2009-57%20PM/Ch_IV_03.pdf

United Nations. (1992). *United nations framework convention on climate change*. Retrieved July 13, 2020, from, https://unfccc.int/resource/docs/convkp/conveng.pdf

United Nations. (2015a). *Transforming our world: The 2030 agenda for sustainable development* (A/RES/70/1*). Retrieved July 6, 2020, from, https://www.un.org/ga/search/view_doc.asp?symbol=A/RES/70/1&Lang=E

United Nations. (2015b). *Paris agreement*. Retrieved July 6, 2020, from, https://unfccc.int/files/essential_background/convention/application/pdf/english_paris_agreement.pdf

Von Braun, J. (2018). Bioeconomy – The global trend and its implications for sustainability and food security. *Global Food Security, 19*, 81–83.

Wood, S. A., Smith, M. R., Fanzo, J., Remans, R., & DeFries, R. S. (2018). Trade and the equitability of global food nutrient distribution. *Nature Sustainability, 1*, 34–37.

6 Acceptance of Insects and In Vitro Meat as a Sustainable Meat Substitute in Germany: In Search of the Decisive Nutritional-Psychological Factors

Florian Fiebelkorn, Jacqueline Dupont, and Patrik Lammers

Abstract

A further intensification of industrial agriculture alone cannot be a viable solution to sustainably feed a steadily growing world population in the future. Besides technical innovations, individual eating habits must become more sustainable. The production of insects and in vitro meat offers several advantages over the production of conventional meat, such as lower CO_2 emissions and land use as well as reduced water consumption. Due to the great sustainability potential, interest in both meat alternatives has not only increased significantly in the media and science in recent years, but also large companies in the food industry, such as Nestlé and Wiesenhof, have already recognized their economic potential. Whether the two meat alternatives will prevail in Germany in the future depends—apart from technical and legal factors—strongly on the acceptance of potential consumers. It has already been shown that disgust and fear of novel foods have a negative influence on the acceptance of insects and in vitro meat as food. In addition, it has already been investigated to what extent other nutritional and environmental psychological factors, such as attitudes towards edible insects and in vitro meat, sensation seeking or sustainability consciousness, might have an influence on the acceptance of both meat alternatives. In addition to an overview of selected sustainability and health indicators of both meat alternatives, the book chapter primarily focuses on the above-mentioned environmental and nutritional-psychological factors influencing the acceptance of edible insects and in vitro meat in Germany.

F. Fiebelkorn (✉) · J. Dupont · P. Lammers
Department of Biology Didactics, School of Biology and Chemistry, University of Osnabrück, Osnabrück, Germany
e-mail: fiebelkorn@biologie.uni-osnabrueck.de; dupont@biologie.uni-osnabrueck.de

D. Lanzerath et al. (eds.), *Bioeconomy and Sustainability*,
https://doi.org/10.1007/978-3-030-87402-5_6

Keywords

Edible insects · Entomophagy · Novel foods · Meat substitutes · Cultivated meat · Consumer acceptance

1 Increase in World Population and Meat Consumption

The world population is expected to grow from 7.6 billion people at present to 9.8 billion by 2050 and to 11.2 billion by 2100.[1] An increase in meat consumption is predicted to accompany this growth.[2] Meat consumption is already very high in industrialized nations compared to developing and emerging countries and will exceed the 100 kg/capita/year mark in 2030.[3] Global per capita consumption of meat is predicted to increase from 41.3 kg in 2015 to 49.4 kg in 2050, mainly due to population growth and higher per capita income in developing and emerging countries.[4] This "nutrition transition" in many parts of the world will require global food production to increase by approximately 60% by 2050. In Germany alone, more than two million animals are slaughtered each day to cover our meat consumption and to export meat; thus, a further intensification of agriculture cannot be the only solution to satisfy our hunger for animal proteins.[5]

2 Consequences of High Meat Consumption for the Environment and Health

Industrial animal husbandry is responsible for 18% of global CO_2 emissions, with some authors estimating a share of over 50%.[6] It also contributes significantly to the loss of biodiversity.[7] Besides the consequences for biodiversity and the climate, high consumption of meat has many negative effects on human health. For example, high meat consumption is associated with an increased risk of cancer and cardiovascular disease.[8] David Tilman and Michael Clark describe this phenomenon as the "food-environment-health trilemma".[9]

[1]Cf. United Nations (2017).

[2]Cf. Food and Agriculture Organization of the United Nations and Agricultural Development Economics Division (2012).

[3]Cf. Food and Agriculture Organization of the United Nations (2003).

[4]Cf. ibid.; Food and Agriculture Organization of the United Nations and Agricultural Development Economics Division (2012).

[5]Cf. ibid. and Fiebelkorn (2017).

[6]Cf. Livestock, Environment and Development Initiative and Food and Agriculture Organization of the United Nations (2006) and Goodland and Anhang (2009).

[7]Cf. Campbell et al. (2017).

[8]Cf. Micha et al. (2013), Campbell and Campbell (2017) and Willett et al. (2019).

[9]Tilman and Clark (2014).

3 Alternatives to Meat Production and Consumption

In order to reduce the environmental impact of conventional meat production, a wide range of different measures and strategies are proposed. In industrialized countries, where protein consumption far exceeds the nutritional requirements of the World Health Organization (WHO), a reduction in the consumption of animal products would be particularly effective in reducing negative environmental impacts.[10] Improving the efficiency of animal production systems (by optimizing feed production, for example) could also reduce negative environmental impacts.[11]

Another strategy would be to switch consumption to alternative sources of protein, such as mycoproteins, microalgae or protein-rich plants such as lentils and rapeseed.[12] These alternatives to animal protein are in many cases not only more sustainable in production than conventional meat, but also contain higher-quality nutrients and minerals.[13]

In addition to these plant alternatives, insects and in vitro meat have been discussed in recent years as sustainable alternatives to conventional meat.[14] The interest in both meat alternatives has increased immensely, not only in the media, but also in science and industry, due to their considerable ecological and economic potential.[15] For example, Germany's largest poultry breeder and processor, the PHW Group, better known under the brand name Wiesenhof, has already invested millions of euros in the development of food products made from insects and in vitro meat.[16]

According to the report "How will cultured meat and meat alternatives disrupt the agricultural and food industry?" by management consultants at A. T. Kearney, in vitro meat has a particularly high potential to change the global meat market.[17] The authors of the report estimate that in 2040, up to 35% of global meat consumption will be of in vitro meat, representing a market value of US$ 630 billion out of an estimated total market value for meat of US$ 1.8 trillion.[18] According to a market report by the Barclays Investment Bank, the global market for edible insects is estimated to reach a financial volume of up to US$ 8 billion by 2030.[19]

From a purely economic point of view, it is therefore not surprising that the world's largest food producer, Nestlé, in its study on the future "Wie is(s)t Deutschland 2030?",[20] also saw the consumption of insects and in vitro meat as a

[10] Cf. Poore and Nemecek (2018).

[11] Cf. Ooninex (2017).

[12] Cf. Nadathur et al. (2016).

[13] Cf. ibid.

[14] Cf. ibid. and Alexander et al. (2017).

[15] Cf. Zukunftsforum (2015), Verbeke (2015) and Verbeke et al. (2015).

[16] Cf. Ksienrzyk (2018).

[17] Cf. Gerhardt et al. (2019).

[18] Cf. ibid.

[19] Cf. Barclays Investment Bank (2019).

[20] Zukunftsforum (2015); "What will Germany be like/eat in 2030?" (own translation).

possible food trend to which German consumers, as members of a resource-saving and value-oriented society, can adapt by 2030 at the latest.[21] It should be kept in mind, however, that the different options for sustainable production of proteins are not mutually exclusive. Therefore, all statements about industrial production of insects and in vitro meat should be made not only in comparison with the production of conventional meat products, but also with other realistic, non-animal-alternatives.[22]

4 How Sustainable and Healthy Are Insect Foods?

Human consumption of insects—also known as entomophagy—is part of the traditional eating habits of more than 2 billion people in over 130 countries.[23] In contrast to countries like Belgium and the Netherlands, which have allowed the sale of insects as novel food already before 2018, in Germany, the sale of insects as a novel food has only been permitted since the Novel Food Regulation came into effect on 1 January 2018 (see section "Legal framework for the authorization of insects and in vitro meat as novel foods in Germany").[24] Various insect-based foods, such as muesli bars, pasta, chocolate, or insect burgers, are already available in German supermarkets.

The production of insects offers several advantages over the production of conventional meat, including the high feed conversion efficiency of insects and their ability to feed on a wide variety of food sources.[25] In addition, the production of many insect species requires much lower CO_2 emissions and water consumption per kg of food generated than that of pigs and cattle. Insects also require a much smaller area to produce the same amount of protein as conventional meat.[26] However, according to Dennis Oonincx and Imke de Boer, the energy required for insect production is comparable to the energy used in conventional livestock breeding.[27]

Nevertheless, given the large number of edible insect species—currently estimated at more than 2111 species—one should be careful about making generalizations about the sustainability of their production.[28] In order to assess the sustainability potential, specific measurements and calculations of selected sustainability indicators, such as CO_2 emissions or energy, land and water

[21] Zukunftsforum (2015).

[22] Cf. Gamborg et al. (2018).

[23] Cf. Fiebelkorn (2017).

[24] Cf. ibid.

[25] Cf. Food and Agriculture Organization of the United Nations (2013).

[26] Cf. Oonincx et al. (2010), Oonincx and de Boer (2012) and Oonincx (2017).

[27] Cf. Oonincx and de Boer (2012).

[28] Cf. Fiebelkorn (2017) and Jongema (2017).

consumption, must be determined in so-called "Life Cycle Assessments" (LCAs) for each edible insect species.[29]

For example, Oonincx and de Boer have carried out an LCA of greenhouse gas emissions, energy consumption and land use for mealworms (*Tenebrio molitor*) and compared the data with those from the production of conventional animal proteins such as cow's milk, poultry, pork, and beef.[30] The production of 1 kg of these animal proteins requires the same amount of energy, but produces more greenhouse gases and requires much more land area than the production of 1 kg of protein from mealworms.[31] Similar results were obtained in a study in Thailand, in which the production of crickets (*Acheta domesticus*) and field crickets (*Gryllus bimaculatus*) was compared with that of broilers.[32] Due to the high complexity of the investigations, detailed LCAs are only available for a few insect species, such as mealworms and crickets.[33]

In addition to the potential for sustainable production, many edible insect species have favourable nutritional values. They are rich in protein, provide sufficient amounts of essential amino acids, minerals, and vitamins and have a better ratio of saturated to unsaturated fatty acids than, for example, fish or poultry.[34] However, as with LCAs, detailed nutrient analyses are available for relatively few of the edible insect species, so the beneficial nutritional properties should not be generalized to all edible insect species.[35] Among the 12 insect species that, according to the European Food Safety Authority (EFSA), have the greatest potential to be used as food and feed in the EU are buffalo worms (*Alphitobius diaperinus*), mealworms, and crickets.[36]

5 How Sustainable and Healthy Is In Vitro Meat?

In the USA, Israel, Japan and the Netherlands, there are already several start-up companies specializing in the commercial production of in vitro meat.[37] The German PHW Group has invested in the Israeli start-up company SuperMeat, which specializes in the production of in vitro chicken meat.[38] Although information on

[29] Cf. Fiebelkorn and Kuckuck (2019).
[30] Cf. Oonincx and de Boer (2012).
[31] Cf. ibid.
[32] Cf. Halloran et al. (2017).
[33] Cf. Oonincx and de Boer (2012), Halloran et al. (2016) and Halloran et al. (2017).
[34] Cf. Belluco et al. (2013) and Rumpold and Schlüter (2013).
[35] Cf. Rumpold and Schlüter (2013) and Dossey et al. (2016).
[36] Cf. European Food Safety Authority (2015).
[37] Cf. German Bundestag (2018).
[38] Cf. ibid.

in vitro meat can already be found in the REWE online shop,[39] it is not yet sold in online shops, nor in German supermarkets or restaurants.[40]

In addition to challenges of introducing in vitro meat to markets and the technical challenges of scaling up its production, legal barriers remain: in vitro meat is not yet approved as a novel food in Germany (see section "Legal framework for the authorization of insects and in vitro meat as novel foods in Germany"). According to Mark Post, Professor of Vascular Physiology at Maastricht University, and his team, who have significantly advanced the development of in vitro meat, the cost of an in vitro meat burger has already been reduced to approximately 11 US dollars.[41] Post also predicted that by 2021, the development will have progressed enough and the price will have been reduced enough to bring in vitro meat burgers to the market.[42] Other experts estimate that in vitro meat will probably not be commercially available until 2025–2030.[43]

To what extent the production of in vitro meat is more sustainable than that of conventional meat is currently the subject of controversial debate.[44] As there are currently no industrial production facilities for in vitro meat, the LCAs available for evaluating the sustainability potential are mostly based on extrapolations from values obtained for the production of in vitro meat on a smaller scale under laboratory conditions.[45] These LCAs have shown that compared to conventional meat production, in vitro meat production requires less land area and emits less greenhouse gases.[46] However, the energy consumption for the production of in vitro meat appears to be much higher than that for beef, pork, sheep, and poultry meat.[47]

The nutrient composition of in vitro meat depends on the nutrients in the nutrient solutions of the bioreactors in which the muscle cells are grown. Detailed nutrient analyses of in vitro meat are not yet available, but they should be very similar to those of conventional fillet meat (without connective and fatty tissue).

[39] Cf. https://www.rewe.de/ernaehrung/in-vitro-fleisch/

[40] Cf. Böhm et al. (2017).

[41] Cf. ibid.

[42] Cf. Maastricht University.

[43] Cf. Office of Technology Assessment at the German Bundestag (2016).

[44] Cf. Hocquette (2016) and Post and Hocquette (2017).

[45] Cf. ibid.; Tuomisto and Teixeira de Mattos (2011) and Tuomisto et al. (2014).

[46] Cf. ibid.

[47] Cf. Tuomisto et al. (2014), Alexander et al. (2017). For a detailed presentation of bioethical arguments for and against in vitro meat, please see Beck (2022) in this volume.

6 Legal Framework for the Authorization of Insects and In Vitro Meat as Novel Foods in Germany

The European Commission, in cooperation with the EFSA, is responsible for the approval of novel foods in the EU and Germany. According to the Novel Food Regulation (EU 2015/2283), the term "novel foods" covers all foods that were not used for human consumption to any significant extent in the EU before 15 May 1997.[48] In addition, they must fall into at least one of the following categories mentioned in Article 3 of the Novel Food Regulation:

- Food with a new or specifically modified molecular structure,
- food consisting of, or isolated or produced from micro-organisms, fungi or algae,
- food consisting of, or isolated or produced from materials of mineral origin,
- food consisting of, or isolated or produced from plants or parts of plants,
- food consisting of, or derived from, animals or parts of animals,
- foodstuffs consisting of, or isolated or produced from cell or tissue cultures obtained from animals, plants, micro-organisms, fungi or algae,
- food consisting of engineered nanomaterials.

With the enactment of the amendment to the Novel Food Regulation on January 1, 2018, food from insects can be approved for the German market. In contrast, in vitro meat has not yet been approved as a novel food in Europe, which is why we can only speak of a *possible* legal framework.[49] However, it can be assumed that in vitro meat will also be approved as a novel food in the future in accordance with the Novel Food Regulation (Art. 3 para. 2 lit. a sublit. iv) and will therefore be subject to the same legal regulations as food from insects.

7 Nutritional-Psychological Factors Influencing the Acceptance of Insects and In Vitro Meat

Whether the two novel meat alternatives will prevail in Germany in the future will depend above all on consumer acceptance. Although the media coverage of insect-based food and in vitro meat has increased significantly in Europe and other Western countries, the willingness to consume the two meat alternatives is still relatively low in many European countries.[50]

Studies by Filiep Vanhonacker, Ellen Van Loo, Xavier Gellynck, and Wim Verbeke in Belgium showed that the willingness to consume food from insects has increased

[48] Cf. European Parliament and Council of the European Union (2015).

[49] Cf. German Bundestag (2019).

[50] Cf. Goodwin and Shoulders (2013), Hopkins (2015), Hartmann and Siegrist (2017), Shockley et al. (2017), Bryant and Barnett (2018) and Mancini et al. (2019).

from approximately 5% in 2013 to approximately 20% in 2015.[51] Similar results were found among consumers in Hungary, Switzerland, Poland, and the Netherlands.[52] Christina Hartmann et al. found that German residents are less likely to consume insects than residents of China are.[53] The willingness of consumers from Germany and Switzerland to consume processed insect-based products is higher than for unprocessed products.[54] Nonetheless, a study by Oliver Meixner and Leonhard Mörl von Pfalzen in Germany, Austria, and Switzerland concluded that only a quarter of the respondents were willing to eat insects.[55]

Consumer acceptance of in vitro meat has been investigated in four studies thus far, each with different results.[56] Jean-François Hocquette et al. found that a minority of respondents in France (5% to 11%) would recommend or accept eating in vitro meat instead of conventional meat.[57] In addition, they found that only a small percentage of respondents (9% to 19%) believed that in vitro meat will be accepted by consumers in the future. Despite these relatively low acceptance levels, 38% to 47% of respondents would support research on in vitro meat.[58] Peter Slade reported that only 11% of his Canadian subjects would choose an in vitro meat burger over a beef and/or veggie burger.[59] Wim Verbeke, Pierre Sans and Ellen Van Loo found that 51% of their Belgian subjects (mainly students) had never heard of in vitro meat.[60] After being informed about the technical production of in vitro meat, 23.9% were willing to try in vitro meat (and 66.7% responded that they might be willing to do so). After receiving additional information on the health and sustainability aspects of in vitro meat, 42.5% of the respondents were willing (and 51.4% perhaps willing) to try in vitro meat. Matti Wilks and Clive Phillips reported that 65.3% of their US sample would be willing to test in vitro meat.[61] Of these, 32.6% would be willing to eat it regularly and 31.5% would be willing to use it as a substitute for conventional meat. According to Christopher Bryant and Julie Barnett, the differences in the studies can probably be attributed to the different groups being used as study participants, the inconsistent description of in vitro meat and the study designs.[62] Overall, however, the results suggest that a relatively large number of consumers

[51] Cf. Vanhonacker et al. (2013) and Verbeke (2015).

[52] Cf. Schösler et al. (2012), Tan et al. (2015), Gmuer et al. (2016), Gere et al. (2017), Kostecka et al. (2017) and Schlup and Brunner (2018).

[53] Cf. Hartmann et al. (2015).

[54] Cf. ibid.; Gmuer et al. (2016).

[55] Cf. Meixner and Mörl von Pfalzen (2018).

[56] Cf. Verbeke et al. (2015), Hocquette et al. (2015), Wilks and Phillips (2017) and Slade (2018).

[57] Cf. Hocquette et al. (2015).

[58] Cf. ibid.

[59] Cf. Slade (2018).

[60] Cf. Verbeke et al. (2015).

[61] Cf. Wilks and Phillips (2017).

[62] Cf. Bryant and Barnett (2018).

would be willing to try in vitro meat, but a much smaller proportion would be willing to use it instead of traditional meat or other meat alternatives.[63]

According to the latest nutrition report of the Federal Ministry of Food and Agriculture, 31% of the German population can imagine buying insect-based food as an alternative to conventional meat in a supermarket or other grocery store. For in vitro meat—described as "meat from the test tube"—the willingness was 17%. Men (40%) were more willing to buy insect-based food than women (22%). Similar trends were observed for in vitro meat: 25% of men and 10% of women could imagine buying in vitro meat.[64] Younger subjects (14–29 years) were more likely to consider buying insect-based foods (43%) and in vitro meat (32%) than subjects over 60 years of age (insects 18%; in vitro meat 9%). Furthermore, 29% of the German population were of the opinion that the increased consumption of alternative meat types, such as in vitro meat or insect-based food, was an appropriate measure to ensure the nutrition of the growing world population. Again, men (36%) were more receptive to insect-based and in vitro meat foods than women (22%). Younger subjects aged 14–29 years, in particular, were much more likely to see the two meat alternatives as a sensible measure to feed the growing world population than subjects aged over 45 years (42% vs. 24%).[65]

The nutritional-psychological factors that play a role in the acceptance of consuming insects and in vitro meat are outlined below.[66]

1. ***Meat consumption:*** Hanna Schösler, Joop de Boer, and Jan J. Boersema showed that subjects with a lower meat consumption are more willing to accept meat substitutes such as seitan, tofu, or insects.[67] Meanwhile, subjects with a higher meat consumption showed a higher willingness to try in vitro meat.[68]
2. ***Reducing meat consumption:*** In the study by Verbeke, study participants who wanted to reduce their meat consumption showed a higher acceptance of insects as a meat substitute,[69] whereas Wilks and Phillips found that vegetarians and vegans are less willing to use in vitro meat as a meat substitute.[70]
3. ***Attitudes:*** It has already been shown that a positive attitude towards food from insects has a significantly positive effect on the willingness to consume insect-based foods.[71] For in vitro meat, it has been shown that a positive attitude of

[63] Cf. ibid.

[64] Cf. forsa Politik- und Sozialforschung GmbH (2018) and Federal Ministry of Food and Agriculture (2019).

[65] Cf. ibid.

[66] For a detailed description of selected influencing factors, please refer to the three review articles Hartmann and Siegrist (2017), Bryant and Barnett (2018) and Mancini et al. (2019).

[67] Cf. Schösler et al. (2012).

[68] Cf. Mancini and Antonioli (2019).

[69] Cf. Verbeke (2015).

[70] Cf. Wilks and Phillips (2017).

[71] Cf. Ruby et al. (2015) and La Barbera et al. (2020).

subjects towards the health, safety and nutritional aspects of in vitro meat is associated with a higher propensity to buy.[72]

4. ***Food Neophobia:*** Food neophobia is defined as a person's aversion to novel foods.[73] Many studies have documented a negative correlation between food neophobia and the willingness to consume insects and in vitro meat.[74]
5. ***Food Disgust:*** Food disgust describes a person's feeling of disgust caused by nutritional triggers. Studies have already shown that food disgust has a negative effect on both the willingness to eat insects and the willingness to consume in vitro meat.[75]

The factors that determine the acceptance of insects as food have already been investigated in several studies.[76] In particular, food neophobia and food disgust were important influencing factors.[77] Gender, attitudes towards the consumption of insects and the previous consumption of insects were also found to be significant predictors of acceptance.[78]

The predictors for the acceptance of in vitro meat as a food have also been investigated in several studies in recent years.[79] As with insects, food disgust and food neophobia have been shown to be significant factors influencing consumer willingness to eat in vitro meat.[80] Age and sex also showed an influence on acceptance with younger respondents and males being more likely to accept the consumption of in vitro meat.[81]

8 Current Research in Biology Didactics on the Acceptance of Insects and In Vitro Meat in the German Population

One of the main research priorities of the Department of Biology Didactics at the University of Osnabrück is the question of how the German population—especially the younger generation—accepts novel food products such as insects and in vitro meat. Several research projects have already addressed selected environmental and nutritional-psychological factors, such as sustainability consciousness, food neophobia,

---

[72] Cf. Gómez-Luciano et al. (2019).

[73] Cf. Pliner and Salvy (2006).

[74] Cf. Hartmann et al. (2015), Hartmann and Siegrist (2018), Lammers et al. (2019) and Wilks et al. (2019).

[75] Cf. Hartmann and Siegrist (2018), Lammers et al. (2019) and Wilks et al. (2019).

[76] Cf. Sogari et al. (2019).

[77] Cf. Hartmann et al. (2015), Verbeke (2015), Hartmann and Siegrist (2018) and Lammers et al. (2019).

[78] Cf. Hartmann et al. (2015), Ruby et al. (2015), Verbeke (2015) and Lammers et al. (2019).

[79] Cf. Hartmann and Siegrist (2017) and Bryant and Barnett (2018).

[80] Cf. Wilks et al. (2019).

[81] Cf. ibid.; Shaw and Mac Con Iomaire (2019).

and food disgust. As an example of the research activities, summaries of two master theses that have already been published are given below. Patrik Lammers' master thesis deals with the acceptance of food from insects by the German population. Jacqueline Dupont's master thesis focuses on the willingness of children and adolescents to consume insects and in vitro meat.

8.1 Case Study 1: Acceptance of Insects as Food in Germany—In Search of the Decisive Nutritional-Psychological Factors

The study by Patrik Lammers, Liza Ullmann and Florian Fiebelkorn examined the acceptance of insect-based foods among German consumers.[82] The nationwide online survey ($N = 516$; $M_{Age} = 47.07$ years, $SD = 16.06$; *female* $= 51.6\%$) attempted to determine which factors have the greatest influence on the consumption of insect burgers and buffalo worms. In addition to sociodemographic factors, meat consumption and the "classical" variables in the field of entomophagy (familiarity, previous insect consumption, food neophobia and food technology neophobia, the fear of novel food technologies), the study focused on the variables sensation seeking, sustainability consciousness and food disgust, which had not yet been considered. In total, 41.9% of the participants were willing to consume an insect burger. In contrast, only 15.9% of the participants were willing to consume buffalo worms, a main ingredient of the insect burger. Using hierarchical multiple regressions, it was shown that food disgust was the most important predictor of the acceptance of edible insects, followed by previous insect consumption, food neophobia, gender, sensation seeking, and food technology neophobia. The high influence of food disgust shows that not only the explicit disgust of insects, but also the disgust of food in general, is decisive for the consumer acceptance of insect-based products. In contrast to food disgust, sustainability consciousness was not a significant predictor of the willingness to consume insects, despite the strong sustainability awareness of the study participants.

8.2 Case Study 2: Attitudes and Acceptance of Young People Towards the Consumption of Insects and Cultured Meat in Germany

The study by Jacqueline Dupont and Florian Fiebelkorn examined the willingness of children and adolescents ($N = 718$; $M_{Age} = 13.67$ years, $SD = 2.31$; *female* $=$ 57.5%) from Germany to consume insects and in vitro meat.[83] One focus of the study was to compare attitudes towards insect and in vitro meat foods in general and in the form of a specific product, an insect or in vitro meat burger. Another focus of

[82] Cf. Lammers et al. (2019).

[83] Cf. Dupont and Fiebelkorn (2020).

the study was to analyse the influence of selected nutritional-psychological factors on the willingness of children and adolescents to consume these products. In addition to sociodemographic factors (age, gender) and meat consumption, familiarity, attitudes, food neophobia and food disgust were included in the analysis. The children and adolescents showed a significantly higher willingness to consume the in vitro meat burger, with no difference in attitude towards the two alternatives as food. Using a hierarchical multiple regression, it could be shown that the attitude towards the burger was the strongest predictor for the willingness to consume both burger alternatives. The negative influence of food neophobia was also confirmed in this study. In contrast, food disgust was not a significant predictor for the willingness to consume the two meat alternatives. This demonstrates that among children and adolescents, their attitude is the most decisive factor for the acceptance of food made from insects and in vitro meat.

References

Alexander, P., Brown, C., Arneth, A., Dias, C., Finnigan, J., Moran, D., & Rounsevell, M. D. A. (2017). Could consumption of insects, cultured meat or imitation meat reduce global agricultural land use? *Global Food Security, 15*, 22–32.

Barclays Investment Bank. (2019). *Insect protein: Bitten by the bug*. Retrieved June 25, 2020, from, https://www.investmentbank.barclays.com/our-insights/insect-protein-bitten-by-the-bug.html

Beck, B. (2022). Food as a moral problem. In D. Lanzerath, U. Schurr, C. Pinsdorf, & M. Stake (Eds.), *Bioeconomy and sustainability. Scientific, socio-economic and ethical dimensions* (pp. 33–60). Springer.

Belluco, S., Losasso, C., Maggioletti, M., Alonzi, C. C., Paoletti, M. G., & Ricci, A. (2013). Edible insects in a food safety and nutritional perspective: A critical review. *Comprehensive Reviews in Food Science and Food Safety, 12*(3), 296–313.

Böhm, I., Ferrari, A., & Woll, S. (2017). *In-Vitro-Fleisch. Eine technische Vision zur Lösung der Probleme der heutigen Fleischproduktion und des Fleischkonsums?* Institut für Technikfolgenabschätzung und Systemanalysen (ITAS) am Karlsruher Institut für Technologie.

Bryant, C., & Barnett, J. (2018). Consumer acceptance of cultured meat: A systematic review. *Meat Science, 143*, 8–17.

Campbell, B. M., Beare, D. J., Bennett, E. M., Hall-Spencer, J. M., Ingram, J. S. I., Jaramillo, F., Ortiz, R., Ramankutty, N., Sayer, J. A., & Shindell, D. (2017). Agriculture production as a major driver of the earth system exceeding planetary boundaries. *Ecology and Society, 22*(4), 8.

Campbell, T. C., & Campbell, T. M. (2017). *China study: Pflanzenbasierte Ernährung und ihre wissenschaftliche Begründung*. Verlag Systemische Medizin.

Dossey, A. T., Morales-Ramos, J. A., & Rojas, M. G. (2016). *Insects as sustainable food ingredients*. In *Production, processing and food applications*. Academic Press.

Dupont, J., & Fiebelkorn, F. (2020). Attitudes and acceptance of young people toward the consumption of insects and cultured meat in Germany. *Food Quality and Preference, 85*, 103983.

European Food Safety Authority (EFSA). (2015). Risk profile related to production and consumption of insects as food and feed. *EFSA Journal, 13*(10), 4257.

European Parliament; Council of the European Union. (2015). Regulation (EU) 2015/2283 of the European Parliament and oft the Council of 25 November 2015 on Novel Foods, Amending Regulation (EU) No 1169/2011 of the European Parliament and of the Council and Repealing Regulation (EC) No 258/97 of the European Parliament and of the Council and Commission

Regulation (EC) No 1852/2001. *Official Journal of the European Union* L327/1, 11.12.2015, 1–22.

Federal Ministry of Food and Agriculture. (2019). *Deutschland, wie es isst. Der BMEL-Ernährungsreport 2019*. Retrieved June 25, 2020, from, https://www.bmel.de/SharedDocs/Downloads/DE/Broschueren/Ernaehrungsreport2019.pdf?__blob=publicationFile&v=4

Fiebelkorn, F. (2017). Entomophagie – Insekten als Nahrungsmittel der Zukunft. *Biologie in unserer Zeit, 47*(2), 104–110.

Fiebelkorn, F., & Kuckuck, M. (2019). Insekten oder In-vitro-Fleisch – Was ist nachhaltiger? Eine Beurteilung mithilfe der Methode des "Expliziten Bewertens". *Praxis Geographie, 6*, 14–20.

Food and Agriculture Organization of the United Nations (FAO). (2003). *World agriculture: Towards 2015/2030: An FAO perspective*. Retrieved June 25, 2020, from, https://www.fao.org/3/y4252e/y4252e00.htm

Food and Agriculture Organization of the United Nations (FAO). (2013). *Edible insects: Future prospects for food and feed security*. FAO Forestry Paper No. 171. Retrieved June 25, 2020, from, https://www.fao.org/3/i3253e/i3253e.pdf

Food and Agriculture Organization of the United Nations (FAO), Agricultural Development Economics Division (ESA). (2012). *World agriculture towards 2030/2050: The 2012 revision*. ESA Working Paper No. 12–03. Retrieved June 25, 2020, from, https://www.researchgate.net/publication/270890453_World_Agriculture_Towards_20302050_The_2012_Revision

Forsa Politik- und Sozialforschung GmbH. (2018). *So will Deutschland essen. Ergebnisse einer repräsentativen Bevölkerungsbefragung*. Retrieved June 25, 2020, from, https://www.bmel.de/SharedDocs/Downloads/DE/_Ernaehrung/Forsa_Ernaehrungsreport2019-Tabellen.pdf?__blob=publicationFile&v=3

Gamborg, C., Röcklinsberg, H., & Gjerris, M. (2018). Sustainable proteins? Values related to insects in food systems. In A. Halloran, R. Flore, P. Vantomme, & N. Roos (Eds.), *Edible insects in sustainable food systems* (pp. 199–211). Springer.

Gere, A., Székely, G., Kovács, S., Kókai, Z., & Sipos, L. (2017). Readiness to adopt insects in Hungary: A case study. *Food Quality and Preference, 59*, 81–86.

Gerhardt, C., Suhlmann, G., Ziemßen, F., Donnan, D., Warschun, M., & Kühnle, H.-J. (2019). *How will cultured meat and meat alternatives disrupt the agricultural and food industry?* Retrieved August 4, 2020, from, https://www.kearney.com/documents/20152/2795757/How+Will+Cultured+Meat+and+Meat+Alternatives+Disrupt+the+Agricultural+and+Food+Industry.pdf/06ec385b-63a1-71d2-c081-51c07ab88ad1?t=1559860712714

German Bundestag. (2018). *Sachstand: In-vitro-Fleisch*. No. WD 5–3000-009/18. Retrieved June 25, 2020, from, https://www.bundestag.de/resource/blob/546674/6c7e1354dd8e7ba622588c1ed1949947/wd-5-009-18-pdf-data.pdf

German Bundestag. (2019). *Ausarbeitung: Einzelfragen zu In-vitro-Fleisch*. No. WD 5-3000-151/18. Retrieved June 25, 2020, from, https://www.bundestag.de/resource/blob/592836/5d0ea08045a3e9bafc92393495d754a2/WD-5-151-18-pdf-data.pdf

Gmuer, A., Nuessli Guth, J., Hartmann, C., & Siegrist, M. (2016). Effects of the degree of processing of insect ingredients in snacks on expected emotional experiences and willingness to eat. *Food Quality and Preference, 54*, 117–127.

Gómez-Luciano, C. A., de Aguiar, L. K., Vriesekoop, F., & Urbano, B. (2019). Consumers' willingness to purchase three alternatives to meat proteins in the United Kingdom, Spain, Brazil and the Dominican Republic. *Food Quality and Preference, 78*, 103732.

Goodland, R., & Anhang, J. (2009). Livestock and climate change. *World Watch, 22*(6), 10–19.

Goodwin, J. N., & Shoulders, C. W. (2013). The future of meat: A qualitative analysis of cultured meat media coverage. *Meat Science, 95*(3), 445–450.

Halloran, A., Hanboonsong, Y., Roos, N., & Bruun, S. (2017). Life cycle assessment of cricket farming in north-eastern Thailand. *Journal of Cleaner Production, 156*, 83–94.

Halloran, A., Roos, N., Eilenberg, J., Cerutti, A., & Bruun, S. (2016). Life cycle assessment of edible insects for food protein: A review. *Agronomy for Sustainable Development, 36*, 57.

Hartmann, C., Shi, J., Giusto, A., & Siegrist, M. (2015). The psychology of eating insects: A cross-cultural comparison between Germany and China. *Food Quality and Preference, 44*, 148–156.
Hartmann, C., & Siegrist, M. (2017). Consumer perception and behaviour regarding sustainable protein consumption: A systematic review. *Trends in Food Science & Technology, 61*, 11–25.
Hartmann, C., & Siegrist, M. (2018). Development and validation of the food disgust scale. *Food Quality and Preference, 63*, 38–50.
Hocquette, A., Lambert, C., Sinquin, C., Peterolff, L., Wagner, Z., Bonny, S. P. F., Lebert, A., & Hocquette, J.-F. (2015). Educated consumers don't believe artificial meat is the solution to the problems with the meat industry. *Journal of Integrative Agriculture, 14*(2), 273–284.
Hocquette, J.-F. (2016). Is in vitro meat the solution for the future? *Meat Science, 120*, 167–176.
Hopkins, P. D. (2015). Cultured meat in Western media: The disproportionate coverage of vegetarian reactions, demographic realities, and implications for cultured meat marketing. *Journal of Integrative Agriculture, 14*(2), 264–272.
Jongema, Y. (2017). *Worldwide list of recorded edible insects*. Retrieved June 25, 2020, from, https://www.wur.nl/upload_mm/8/a/6/0fdfc700-3929-4a74-8b69-f02fd35a1696_Worldwide list of edible insects 2017.pdf.
Kostecka, J., Konieczna, K., & Cunha, L. M. (2017). Evaluation of insect-based food acceptance by representatives of polish consumers in the context of natural resources processing retardation. *Journal of Ecological Engineering, 18*(2), 166–174.
Ksienrzyk, L. (2018). *Warum investiert der Geflügelzüchter Wiesenhof in Fake-Fleisch?* Retrieved November 21, 2019, from, https://ngin-food.com/artikel/wiesenhof-in-vitro-fleisch-vegan-beyond-meat-supermeat-startups/
La Barbera, F., Verneau, F., Nørgaard Videbæk, P., Amato, M., & Grunert, K. G. (2020). A self-report measure of attitudes toward the eating of insects: Construction and validation of the entomophagy attitude questionnaire. *Food Quality and Preference, 79*, 103757.
Lammers, P., Ullmann, L. M., & Fiebelkorn, F. (2019). Acceptance of insects as food in Germany: Is it about sensation seeking, sustainability consciousness, or food disgust? *Food Quality and Preference, 77*, 78–88.
Livestock, Environment and Development Initiative (LEAD); Food and Agriculture Organization of the United Nations (FAO). (2006). *Livestock's long shadow: Environmental issues and options*. Retrieved June 25, 2020, from, http://www.fao.org/docrep/010/a0701e/a0701e.pdf
Maastricht University (n.d.). *Cultured beef – Frequently asked questions*. Retrieved June 25, 2020, from, https://culturedbeef.mumc.maastrichtuniversity.nl/sites/intranet.mumc.maastrichtuniversity.nl/files/culturedbeef_mumc_maastrichtuniversity_nl/frequently_asked_questions.pdf
Mancini, M. C., & Antonioli, F. (2019). Exploring consumers' attitude towards cultured meat in Italy. *Meat Science, 150*, 101–110.
Mancini, S., Moruzzo, R., Riccioli, F., & Paci, G. (2019). European consumers' readiness to adopt insects as food. A review. *Food Research International, 122*, 661–678.
Meixner, O., & Mörl von Pfalzen, L. (2018). *Die akzeptanz von insekten in der ernährung. Eine studie zur vermarktung von insekten als lebensmittel aus konsumentensicht*. Springer Gabler.
Micha, R., Michas, G., Lajous, M., & Mozaffarian, D. (2013). Processing of meats and cardiovascular risk: Time to focus on preservatives. *BMC Medicine, 11*, 136.
Nadathur, S. R., Wanasundara, J. P. D., & Scanlin, L. (2016). *Sustainable protein sources*. Academic Press.
Office of Technology Assessment at the German Bundestag (TAB). (2016). *Fleisch 2.0 – unkonventionelle Proteinquellen*. Retrieved June 25, 2020, from, http://www.tab-beim-bundestag.de/de/pdf/publikationen/themenprofile/Themenkurzprofil-005.pdf
Oonincx, D. G. A. B. (2017). Environmental impact of insect production. In A. van Huis & J. K. Tomberlin (Eds.), *Insects as food and feed: From production to consumption* (pp. 79–93). Wageningen Academic Publishers.

Oonincx, D. G. A. B., & de Boer, I. J. M. (2012). Environmental impact of the production of mealworms as a protein source for humans—A life cycle assessment. *Public Library of Science One, 7*(12), e51145.
Oonincx, D. G. A. B., van Itterbeeck, J., Heetkamp, M. J. W., van den Brand, H., van Loon, J. J. A., & van Huis, A. (2010). An exploration on greenhouse gas and ammonia production by insect species suitable for animal or human consumption. *Public Library of Science One, 5*(12), e14445.
Pliner, P., & Salvy, S.-J. (2006). Food neophobia in humans. In R. Shepherd & M. Raats (Eds.), *The psychology of food choice* (pp. 75–92). CABI.
Poore, J., & Nemecek, T. (2018). Reducing food's environmental impacts through producers and consumers. *Science, 360*(6392), 987–992.
Post, M., & Hocquette, J.-F. (2017). New sources of animal proteins: Cultured meat. In P. P. Purslow (Ed.), *New aspects of meat quality: From genes to ethics* (pp. 425–441). Woodhead Publishing.
Ruby, M. B., Rozin, P., & Chan, C. (2015). Determinants of willingness to eat insects in the USA and India. *Journal of Insects as Food and Feed, 1*(3), 215–225.
Rumpold, B. A., & Schlüter, O. K. (2013). Nutritional composition and safety aspects of edible insects. *Molecular Nutrition & Food Research, 57*(5), 802–823.
Schlup, Y., & Brunner, T. (2018). Prospects for insects as food in Switzerland: A Tobit regression. *Food Quality and Preference, 64*, 37–46.
Schösler, H., de Boer, J., & Boersema, J. J. (2012). Can we cut out the meat of the dish? Constructing consumer-oriented pathways towards meat substitution. *Appetite, 58*(1), 39–47.
Shaw, E., & Mac Con Iomaire, M. (2019). A comparative analysis of the attitudes of rural and urban consumers towards cultured meat. *British Food Journal, 121*(8), 1782–1800.
Shockley, M., Allen, R. N., & Gracer, D. (2017). Product development and promotion. In A. van Huis & J. K. Tomberlin (Eds.), *Insects as food and feed: From production to consumption* (pp. 399–420). Wageningen Academic Publishers.
Slade, P. (2018). If you built it, will they eat it? Consumer preferences for plant-based and cultured meat burgers. *Appetite, 125*, 428–437.
Sogari, G., Menozzi, D., & Mora, C. (2019). The food neophobia scale and young adults' intention to eat insect products. *International Journal of Consumer Studies, 43*(1), 68–76.
Tan, H. S. G., Fischer, A. R. H., Tinchan, P., Stieger, M., Steenbekkers, L. P. A., & van Trijp, H. C. M. (2015). Insects as food: Exploring cultural exposure and individual experience as determinants of acceptance. *Food Quality and Preference, 42*, 78–89.
Tilman, D., & Clark, M. (2014). Global diets link environmental sustainability and human health. *Nature, 515*, 518–522.
Tuomisto, H. L., Ellis, M. J., & Haastrup, P. (2014). Environmental impacts of cultured meat: Alternative production scenarios. In R. Schenck & D. Huizenga (Eds.), *Proceedings of the 9th International Conference on Life Cycle Assessment in the Agri-Food Sector (LCA Food 2014)*, San Francisco, CA. Retrieved June 25, 2020, from, https://core.ac.uk/download/pdf/38629617.pdf
Tuomisto, H. L., & Teixeira de Mattos, M. J. (2011). Environmental impacts of cultured meat production. *Environmental Science & Technology, 45*(14), 6117–6123.
United Nations, Department of Economic and Social Affairs, Population Division. (2017). *World population prospects: The 2017 revision, key findings and advance tables*. Working Paper No. ESA/P/WP/248. Retrieved August 11, 2020, from, https://population.un.org/wpp/Publications/Files/WPP2017_KeyFindings.pdf
Vanhonacker, F., Van Loo, E. J., Gellynck, X., & Verbeke, W. (2013). Flemish consumer attitudes towards more sustainable food choices. *Appetite, 62*, 7–16.
Verbeke, W. (2015). Profiling consumers who are ready to adopt insects as a meat substitute in a Western society. *Food Quality and Preference, 39*, 147–155.
Verbeke, W., Sans, P., & Van Loo, E. J. (2015). Challenges and prospects for consumer acceptance of cultured meat. *Journal of Integrative Agriculture, 14*(2), 285–294.

Wilks, M., & Phillips, C. J. C. (2017). Attitudes to in vitro meat: A survey of potential consumers in the United States. *Public Library of Science One, 12*(2), e0171904.

Wilks, M., Phillips, C. J. C., Fielding, K., & Hornsey, M. J. (2019). Testing potential psychological predictors of attitudes towards cultured meat. *Appetite, 136*, 137–145.

Willett, W., Rockström, J., Loken, B., Springmann, M., Lang, T., Vermeulen, S., Garnett, T., Tilman, D., DeClerck, F., Wood, A., Jonell, M., Clark, M., Gordon, L. J., Franzo, J., Hawkes, C., Zurayk, R., Rivera, J. A., De Vries, W., Sibanda, L. M., . . . Murray, C. J. L. (2019). Food in the Anthropocene: The EAT-lancet commission on healthy diets from sustainable food systems. *The Lancet, 393*(10170), 447–492.

Zukunftsforum, N. (2015). *Wie is(s)t Deutschland 2030? Nestlé Zukunftsstudie*. Deutscher Fachverlag.

Part IV

Technology and Governance

Characteristics of Innovation in Bioeconomy

7

Max Mittenzwei

Abstract

In recent years, the research field of bioeconomy has experienced significant global growth based on an increasing number of annual publications in the last 10 years. Bioeconomy received a strong political push by European policymakers after the instalment of a "knowledge-based bioeconomy" 15 years ago. While playing an essential role in recent European Union (EU) policies, bioeconomy still lacks a coherent understanding across multiple layers and especially regarding innovation activities. Innovations undoubtedly form one of the basic building blocks of the success of the knowledge-based bioeconomy and its increasing reach, but it must nevertheless be noted that frequently their innovation is not well-understood, and misconceptions prevail. Therefore, this chapter attempts to characterize innovation in bioeconomy. Based on a theoretical discussion of different concepts and aspects of innovation and a literature review at the intersection of bioeconomy and innovation, a catalogue of criteria about what can influence innovation in bioeconomy is proposed. Thus, seven criteria categories are deduced, as well as multiple keywords assigned to each of them. The proclaimed categories are then discussed and ultimately help to identify innovation triggers for bioeconomy. Thus, the article attempts to propose a realistic foundation and theoretical assessment of innovation in bioeconomy to reinforce future discourse on the matter.

Keywords

Concept of bioeconomy and innovation · Theoretical assessment · Catalogue of criteria

M. Mittenzwei (✉)
Faculty of Economic and Social Geography, University of Greifswald, Greifswald, Germany
e-mail: max.mittenzwei@uni-greifswald.de

D. Lanzerath et al. (eds.), *Bioeconomy and Sustainability*,
https://doi.org/10.1007/978-3-030-87402-5_7

1 Introduction and Motivation

In 2004, based on the knowledge-based bioeconomy, the term "bioeconomy" found its way into the policy discussion in Europe.[1] Fourteen years later, within the framework of the Global Bioeconomy Summit (GBS) in 2018, over 700 representatives from politics, science, civil society, as well as the business sector from more than 70 countries gathered up to discuss the challenges and future of bioeconomy.[2] One may expect as an outcome of such an event a polished action plan of what exactly the next steps towards the implementation of bioeconomy need to look like. The actual result, however, tends to reduce one's optimism. The question of a universal and streamlined definition of what precisely bioeconomy means includes and implicates on a global level, remained unanswered, and all that was gained is another document that offers general recommendations and states a general urgency, without providing concreteness. Especially against the backdrop of an official European Commission document—released about a year before the summit—explicitly stating the need for a common framework and giving concrete recommendations, makes the whole event appear to be redundant. Not surprisingly, more and more authors have started to focus on the negative aspects of the recent developments in bioeconomy. It has become "a buzzword used by public institutions",[3] gets criticized "for being a weak form of ecological modernization aiming for increased exploitation of natural resources"[4] and the ongoing academic discussion "about its environmental aspects and its questionable and variegated integration of sustainability perspectives"[5] gains increased publicity. Whereby at its core, bioeconomy is not just a catchword, if some things are kept in mind.[6] First of all, the attempt to frame and define bioeconomy as a sector will not lead to satisfactory results. Various authors state the need to refer to bioeconomy as a multi-dimensional concept instead of a sharply defined sector. One of the main reasons for that is the fact that bioeconomy in itself is exceedingly fuzzy,[7] still in its infancy[8] and is, per se, nothing new.[9] These points have had a considerable influence on the predominant definition problem of bioeconomy. In general, the bioeconomy concept entails the sustainable use of renewable biomass instead of finite fossil resources for the development and production of various bio-based, value-added products, services, and energy. These work as substitutes for existing fossil fuel-based products, services, and energy and are a part of a broader societal transition to a

[1]Cf. Golembiewski et al. (2015).
[2]Cf. von Braun (2018).
[3]Vivien et al. (2019, p. 1).
[4]Bauer (2018, p. 1).
[5]Albrecht (2019, p. 3).
[6]Cf. Golembiewski et al. (2015) and Peltomaa (2018).
[7]Cf. Golembiewski et al. (2015).
[8]Cf. Ibid. and von Braun (2018).
[9]Cf. Pietzsch (2017).

low-carbon future.[10] It also promotes the Circular Economy concept as a natural fit[11] as well as the adoption of cascading, meaning to initially process biomass into high-value products, before using the residues for lower value applications until a minimum of waste remains at the end.[12] With being primarily conceptually based, we can think of bioeconomy "as a wholesale shift in the way our economies—and necessarily our societies and polities—are organized and coordinated such that they are no longer based on fossil fuels".[13] However, in inhabiting this kind of conceptual flexibility, bioeconomy can be exploited to promote different and contrasting objectives[14] and gets gutted as an irrelevant buzzword in many publications, policies, and reports. It has proven attractive to many different actors because it can mean something for everyone—it is many things to many people.[15] Its holistic approach can thus be seen as its strength on the one hand, but also as its weakness on the other: a "fetishization of everything bio-"[16] takes places, while the role of bioeconomy as a powerful meta-discourse[17] should not be underestimated.[18] In conclusion, bioeconomy has most definitely the potential to affect a fundamental change in the industry,[19] although it is not as straightforward as many researchers, politicians, and decision-makers may think.

At the same time, our economy faces a lock-in into a fossil-based and CO_2-intensive production mode,[20] which certainly is a significant hurdle for bioeconomy to overcome. Matteo de Besi and Kes McCormick[21] see the solution in a transformative change that involves long-term approaches and interactions at all levels of society. Their vision gets supported by Birch, as he sees bioeconomy as a socio-technical transition.[22] However, "[. . .] the geographical dimensions of such transitions are often ignored or overlooked in existing research"[23] but are a vital element for a successful transition. It is indeed a transformation that would change the social, technical, and material elements of specific systems.[24] For this transition, innovation is seen by various authors as one, if not the critical factor for moving

[10] Cf. van Lancker et al. (2016) and Birch (2019).

[11] Cf. Näyhä (2019).

[12] Cf. van Lancker et al. (2016).

[13] Birch (2019, p. 2).

[14] Cf. Peltomaa (2018).

[15] Cf. Staffas et al. (2013) and Vivien et al. (2019).

[16] Birch and Tyfield (2012, 3).

[17] Cf. Bauer et al. (2018).

[18] Cf. Birch (2019).

[19] Cf. Schütte (2018).

[20] Cf. Pyka (2017).

[21] Cf. de Besi and McCormick (2015).

[22] Cf. Birch (2019).

[23] Ibid., 19.

[24] Cf. ibid.

forward.[25] However, the innovation term is again used quite inflationary, even more so in the bio-economic context. Especially in some European Union (EU) policies, the combination of both terms—bioeconomy and innovation—needs to be critically reviewed.[26] The research landscape regarding innovations in a bio-economic context appears to be quite empty so far,[27] even though the above-mentioned authors mutually agreed on it being one of the building blocks of bioeconomy. Thus, the motivation for this article is to showcase what the innovation term explicitly implicates for the concept of bioeconomy and which factors can influence innovation in a bio-economic context.

2 Innovation as a Concept

The introduction shows that bio-economic innovation is, as well as bioeconomy itself, neither well defined nor understood. Thus, this article will first focus on the theoretical foundations of innovation. The general importance and relevance of the concept of innovation were emphasized repeatedly in research both in the twentieth century and at the beginning of the twenty-first century. Especially for the (long-term) competitiveness of companies and regions, it is seen as one of the main driving forces, because of the implementation of novelty and variety. Succeeding in innovation lets companies prosper; innovative countries and regions have a higher income than less innovative ones and catching up with innovation leaders means increasing a company's innovation activity.[28] In conclusion, innovation is seen as a pretty necessary factor. However, the meaning of innovation and especially how and when it occurs are not entirely clear.[29] Innovation itself is not a new phenomenon, it is arguably as old as humankind itself.[30] While we know quite well where innovation leads to, we know much less about the why and how innovation occurs. Since multiple researchers in different working fields tried to grasp innovation and customize it to fit their specific scientific area, a certain "fuzziness" around the term and its various conceptual framings can be noticed.[31] In the following, essential currents of the different types, models, and finally, levels of innovation are briefly presented in order to form a basis for the bio-economic discourse.

[25] Cf. Golembiewski et al. (2015), van Lancker et al. (2016), Dabbert et al. (2017), Bauer et al. (2018), Purkus et al. (2018), Schütte (2018) and Birch (2019).

[26] Cf. Birch and Tyfield (2012).

[27] Cf. van Lancker et al. (2016).

[28] Cf. Fagerberg et al. (2011).

[29] Cf. Fernandes Rodrigues Alves et al. (2018).

[30] Cf. Fagerberg et al. (2011).

[31] Cf. ibid.

2.1 Innovation: Models, Types, and Levels

The linear model of innovation is, without a doubt, one of the first frameworks which got developed for understanding the relation of science and technology to economy. It implies that innovation starts with basic research, followed by applied research and development, before ending with production and diffusion.[32] However, in Jan Fagerberg's opinion, innovation has little to do with this linear model. He argues that it is based on the assumption of innovation being applied science, while in reality, firms usually innovate because of a commercial need to do so.[33] Benoît Godin opposes this by saying that the model is merely a "rhetorical entity, [...] a thought figure"[34] that makes the otherwise fuzzy concept of innovation easier for administrators and agencies to grasp.[35] Besides, Schumpeter is, without a question, the most influential name when talking about innovation. He invented the "trinity" of the innovation process, resulting in the indistinction between invention (new ideas are generated), innovation (ideas are developed into processes and products), and diffusion (spreading these processes and products across markets).[36] Joseph Schumpeter therefore not only introduced innovation as a process, but also made the vital distinction between invention and innovation into two separate parts of the concept, which nowadays get mixed up quite often. The linear model of innovation arose only due to interpreters of Schumpeter's work, who anchored it into the context of the technology-push and demand-pull debate.[37] Simple models, like the differentiation into product and process, as well as physical and intangible innovations, can be found as the basis of more advanced concepts (Fig. 7.1). Often used for policy recommendations, the innovation systems perspective achieved scientific attention in recent years. It combines all essential economic, social, political, organizational, and other factors that influence the development, diffusion, and use of innovations,[38] while also stressing out linkages between these actors.[39] Thus, all innovation processes are naturally embedded in innovation systems. Further, the concepts of "Technology Innovation Management" (TIM) and "Open Innovation" (OI) tend to get highlighted quite often in recent innovation literature.[40] TIM "seeks to understand how novel technologies and innovations emerge and how they can be commercialized successfully".[41] It thus attempts to decipher the most-asked question since the days of Schumpeter. OI, on the other hand, gets mentioned as a subfield to

32 Cf. Godin (2016).

33 Cf. Fagerberg et al. (2011).

34 Godin (2016, p. 659).

35 Cf. ibid. 660.

36 Cf. Schumpeter (1939).

37 Cf. Godin (2016).

38 Cf. Purkus et al. (2018).

39 Cf. Pyka (2017).

40 Cf. Birch (2009), Golembiewski et al. (2015), van Lancker et al. (2016).

41 Golembiewski et al. (2015, p. 2).

Fig. 7.1 Innovation fields in manufacturing firms (Kirner et al. 2009)

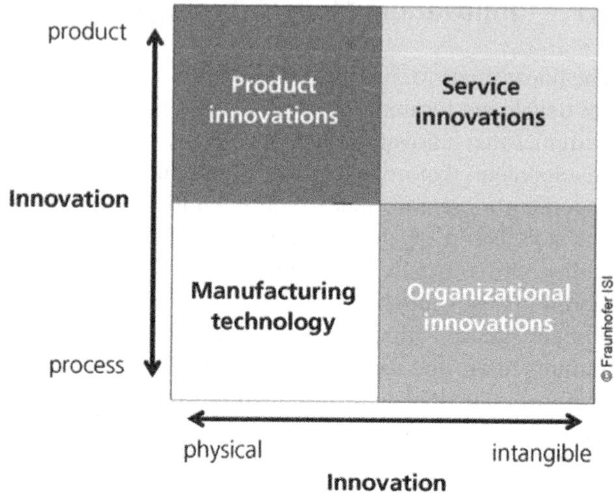

TIM that is rapidly becoming a dominant approach innovation.[42] It can be defined as "the use of purposive inflows and outflows of knowledge to accelerate internal innovation and expand the markets for external use of innovation", thus considering the "boundaries between the firm and its surrounding environment [. . .] to be more porous which allows knowledge and innovation to move more easily between the two".[43]

That leads us to one of the basic terms of the innovation vocabulary: knowledge. It provides a crucial input to innovation in that it enables actors to understand the world and make decisions that affect it.[44] Kean Birch also underlines the importance to differentiate between different types of knowledge: appropriable (restricted access) and non-appropriable (free to access),[45] as well as tacit (knowing-how) and explicit (knowing-that) knowledge.[46] These terms are essential in the further course of the article, especially for the understanding of spillovers, collaborations, and Birch's knowledge-space dynamic.

It becomes apparent that the concept of innovation can be combined with different approaches, which can be understood as a renewed indication of its adaptability but do also provide another argument for its breadth and fuzziness. Besides models, this affects types of innovation as well. Tzeng, for example, distinguishes between the following three leading schools of innovation[47] (Table 7.1):

[42] Cf. van Lancker et al. (2016).

[43] Ibid., 4.

[44] Cf. Birch (2009).

[45] Cf. Birch (2009).

[46] Cf. Nonaka and Takeuchi (1995).

[47] Cf. Tzeng (2014).

Table 7.1 The main schools of Schumpeterian innovation (Tzeng, 2014)

	Corporate capability school	Entrepreneurship school	Culture school
General perspective	Economic	Social	Cultural
Nature of innovation	Institutionalized capability	Innovation as grassroots impetuses	Innovation as deep craft
Inherent logic of innovation	Evaluate	Engage	Envision
Relationship among members	Instruction-based	Identity-based	Intergenerational

Terms like "technical innovation" and "administrative" or "management innovation" were brought forward as well, resulting in even more spin-offs, like organizational innovation.[48] The latter is defined by the Organisation for Economic Co-operation and Development (OECD) as "the implementation of a new organizational method in a firm's business practices, workplace organization or external relations".[49] It is furthermore stated that "other scholars also developed typologies for understanding organizational innovation; however, many of them are overlapped",[50] thus providing another argument for a conceptual "one size fits all"-mentality of innovation. Into the same category fall responsible innovation and social innovation. Responsible innovation includes the future-oriented organization of development and is defined as a "transparent, interactive process by which societal actors and innovators become mutually responsive to each other with a view to the (ethical) acceptability, sustainability and societal desirability of the innovation process and its marketable products".[51] Social innovation, on the other hand, emphasizes the importance of active citizenship in innovation.[52] By now, the diverse phenomenon of innovation and its redundant concept become clear.[53]

Besides the mentioned knowledge, another core term is "creative destruction", or, respectively, "incremental" versus "fundamental change". Nowadays, this dichotomy is also described as the level of innovation and, spanning back to Schumpeter, creative destruction is one of the two possibilities for change to occur. The incremental type describes small improvements along well-known trajectories, while the fundamental, or creative destruction type, which leads to structural changes, for example the emergence of new and the disappearance of old industries,[54] meaning a

[48] Cf. Fernandes Rodrigues Alves et al. (2018).

[49] Organisation for Economic Co-operation and Development/Statistical Office of the European Communities (2005, p. 177).

[50] Fernandes Rodrigues Alves et al. (2018, 3).

[51] von Schomberg (2012, 50).

[52] Cf. Pyka (2017).

[53] Cf. Kirner et al. (2009).

[54] Cf. Suroso and Azis (2015) and Pyka (2017).

"wholesale transformation of socio-technical systems".[55] By now, it has become evident that there seems to be a jungle of innovation concepts, lots of "alternative models, with their multiple feedback loops [that] look more like modern artwork or a "plate of spaghetti and meatballs" than [...] useful analytical framework[s]".[56] Bioeconomy by itself was identified as a fuzzy concept, and the innovation concept does not look much different. At a basic level, innovation is doing the old in a new way, while the idea behind bioeconomy is pretty much the same. Sadly, combining minus and minus does not automatically result in plus like in mathematics, so bio-economic innovations need to be individually reviewed.

2.2 Innovation in Bioeconomy

With the beginning of the twenty-first century, a paradigmatic shift towards a somewhat sustainable and smart economy is in the air.[57] Various authors agree on the appraisal of bioeconomy as one of the central factors for this change, which is unfortunately impaired by a fundamental uncertainty.[58] The creative destruction gets mentioned,[59] and the transformation process is believed to span over a large part of the twenty-first century.[60] This will lead to the reorganization of the whole world economic system, thus being an indispensable part of our future society.[61] The lack of systematic assessment, however, is seen as one of the hurdles for this transition to take place[62]; the diffuse nature and unclearness remain to be seen as problems that need fixing as soon as possible.[63] But how exactly does this *lack of systematic assessment* look like? A publication analysis, conducted in the database Web of Science Core Collection (WoS), with the advanced search string.

TS = (bioeconomy AND innovat) OR TS=(bioeconomy AND innovat*) OR TS=(bio-eco* AND innovat*),*

resulted in a total of 292 found publications in the research field of bio-economic innovations (Fig. 7.2).

The exponential growth of annual publications since 2014 can be seen, proving a significant interest in the topic in recent years. The reason behind that might be an increasing number of countries incorporating bioeconomy into their national

[55] Birch (2019, p. 18).

[56] Godin (2016, p. 660).

[57] Cf. Pyka (2017).

[58] Cf. ibid.

[59] Cf. Pyka (2017), Fernandes Rodrigues Alves et al. (2018), Schütte (2018) and Birch (2019).

[60] Cf. Saviotti (2017).

[61] Cf. Saviotti (2017) and Bauer et al. (2018).

[62] Cf. Bauer et al. (2018).

[63] Cf. Purkus et al. (2018).

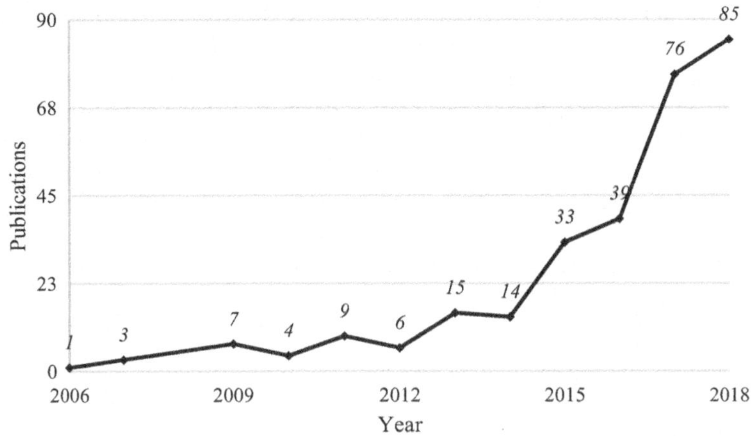

Fig. 7.2 Resulting numbers of publications in the database Web of Science

strategies and policies as well as thereby triggering scientific interest in the topic. However, of these 292 publications, only 13 include statements about bio-economic innovation factors. An explanation here could be the degree of fuzziness of both concepts. The hurdle of lacking assessment and again, the breadth of the bioeconomy concept can thus be underlined. Together with drivers that can benefit or even trigger innovations from the general innovation literature, influential factors that are found in these 13 publications can now be further looked at.

Birte Golembiewski et al. conducted a publication analysis to achieve an overview of the current research landscape dealing with bioeconomy[64] and highlight the challenges of technology and innovation management (TIM) for bioeconomy. They state the cross-sectorial character of bioeconomy and thus the need for interdisciplinary approaches. The need for broader, holistic approaches to bioeconomy can be found in other publications as well. Fredric Bauer speaks of the demand for a long-term, holistic perspective and adaptive policymaking,[65] Georg Schütte states the need for "holistic, systemic perspectives and solutions",[66] while Dries Maes and Steven van Passel reject approaches that focus on research and development alone.[67] As already briefly mentioned, knowledge is commonly seen as a core factor for innovation,[68] and this is no different in the bio-economic context.[69] Actually, in the early days of bioeconomy, it was called the "knowledge-based bioeconomy" in the European Union. Marlon Fernandes Rodrigues Alves et al. see knowledge as the

[64] Cf. Golembiewski et al. (2015).

[65] Cf. Bauer et al. (2018).

[66] Schütte (2018, p. 6).

[67] Cf. Maes and van Passel (2019).

[68] Cf. Kirner et al. (2009).

[69] Cf. Golembiewski et al. (2015).

"most important resource and thus learning as the most important process".[70] Knowledge also is deeply intertwined with location or space of origin. It can come from diverse locations and in many forms, while every spatial context is unique, knowledge entails geographical specificity.[71] Birch calls that connection the "knowledge-space dynamic".[72] He argues that innovation occurs in specific locations, where firms and other organizations have access to complementary capabilities because of their co-location and proximity to one another. Knowledge can thus leak between actors, lead to an iterative process of learning and bolster the occurrence of bio-economic innovation.[73] Birch's knowledge-space dynamic is solidly underpinned by a Schumpeterian understanding of innovation. Pyka frames it as a Neo-Schumpeterian approach: they highlight the complementary interplay in knowledge generation and diffusion processes between firms, consumers, and government institutions,[74] thus emphasizing innovation as an interactive process between multiple actors.[75] Bauer states the crucial link between university research and private sector research, therefore cross-sectoral research, while Birch also mentions the relevance of multi-scale, therefore international linkages.[76] The importance of the encompassing environment, as seen in Birch's knowledge-space model, needs to be kept in mind as well.[77] By looking at company's internal processes, factors that influence the emergence of innovation can of course also be identified there. Cheng-Hua Tzeng highlights the importance of long-time commitment to financing the development of new technologies.[78] He further argues, in the sense of the cultural innovation school, that technical innovation is not necessarily the outcome of digging information out of books or articles, but rather is a set of skills that cannot be reduced to a science.[79] Innovation in bioeconomy is seen as a "rather complex, collaborative, and multi-level process which is embedded in innovation systems",[80] and it is, in general, a good idea to "broaden one's perspective on innovation".[81] It needs to be assured, though, that there are different innovation paths. Not every firm innovates by developing new products; services can be innovative as well as the introduction of innovative manufacturing technologies or the implementation of innovative organizational concepts.[82] Jan Fagerberg et al. stated the importance of

[70] Fernandes Rodrigues Alves et al. (2018, p. 6).
[71] Cf. Birch (2009).
[72] Ibid.
[73] Cf. Birch (2009) and Birch (2012).
[74] Cf. Pyka (2017).
[75] Cf. Bauer et al. (2018).
[76] Cf. Birch (2009).
[77] Cf. Fagerberg et al. (2011).
[78] Cf. Tzeng (2014).
[79] Cf. ibid.
[80] Kirner et al. (2009, p. 1).
[81] Tzeng (2014, p. 17).
[82] Cf. Kirner et al. (2009).

the environment for innovation, which is also a major factor in the Open Innovation concept. However, Open Innovation relies heavily on trust between actors. Most collaborations are undertaken with already known partners, to reduce the risk of knowledge theft or involuntarily outgoing spillovers.[83] Of course, one could always argue that a certain openness towards new collaborations and, following that, knowledge exchange needs to be the standard case, but it is not an easy task to achieve—and definitely cannot be taken for granted. Especially with regard to the bioeconomy concept and its uncertainty, the acceptance of firms seems to be a problem and is considered a significant hindrance to innovation. Not only that, but the lack of acceptance of consumers and thus the society in general is a hurdle as well.[84] A limited consumer understanding of bioeconomy might as well reduce the market demand and the innovation capacity as a whole,[85] because apparently "a bio-economic innovation will only be successful if consumers accept it".[86] This is why authors recommend, besides Open Innovation that includes consumers and users into the innovation process, a whole portfolio of policy changes, to address all actors relevant in a given innovation system. Louise Staffas et al. argue that various national strategies and policies include innovation, but few go beyond a general recommendation.[87] The need for coherence of national and international strategies is stated widely,[88] as well as a coordinated and in-depth approach that includes entrepreneurial activities, knowledge diffusion, guidance, market formation help, resource mobilization, and the creation of legitimacy.[89] Policies especially need to account for the fact that innovation is not only taking place within R&D intensive high-tech sectors or in high-tech firms alone.[90] Bauer explains further that the transition also needs a general change of behaviour and expectations among consumers and an institutional change regarding norms, standards, and regulations.[91] He also states the need to let firms innovate at their own pace, because innovation is, as shown, nothing that can be triggered, but something that can be positively influenced. What is more, science and technology alone will not manage to solve the transition puzzle, politics need to intervene and help to initiate the change.[92] An appropriate innovation agenda, a national strategy that influences all policy areas, supports new technologies and finds new ways of financing deployment and diffusion of innovation is needed.[93]

[83] Cf. van Lancker et al. (2016).

[84] Cf. Pyka (2017).

[85] Cf. Wensing et al. (2019).

[86] Pyka (2017, p. 9).

[87] Cf. Staffas et al. (2013).

[88] Cf. Stadler and Chauvet (2018) and Schütte (2018).

[89] Cf. Purkus et al. (2018).

[90] Cf. Kirner et al. (2009).

[91] Cf. Bauer (2018).

[92] Cf. ibid.

[93] Cf. Bauer (2018) and Bauer et al. (2018).

3 Criteria for Bioeconomy Innovations

The findings of the previous chapter are now to be compiled within the framework of a criteria catalogue. Van Lancker et al. deliver a useful entry point for this. They incipiently state the importance of radically new and disruptive innovations, such as new business models, reconfigured value chains or the creation of entirely new value chains, while also considering the intricate knowledge base of various sciences. Cooperation between different actors can help develop this sophisticated knowledge, while commercialization and adoption of new bio-economic technologies and products are seen as a challenge, due to high switching costs and the locked-in state of economy. Complex and fragmented policy schemes form another challenge, as many of the new concepts are expected to comply with a number of different policy schemes and are also subject to regulation from different administrative levels. The authors conclude that "innovation processes [...] are best considered as transdisciplinary endeavours, open to relevant stakeholders, with ample room for iterativety between idea development, invention and commercialization".[94] Organizations need to "[strive] to innovate towards the bioeconomy"[95] while "leadership should embrace innovation and openness",[96] and the "organizational culture should reflex this".[97] "Available knowledge, expertise and technology need to be scrutinized, [...] relational capability and absorptive capacity need to be adequate".[98] Additionally, Tzeng emphasizes that "most important pathways include joint or cooperative ventures, contract research, consulting, informal interactions, conferences, and publications".[99] Based on the comprehensive literature work in the previous chapters, the following criteria catalogue can be established (Table 7.2).

In the following, the criteria and accompanied keywords are described in detail. Regarding *Knowledge and Awareness*, some knowledge base needs to be present. This knowledge base can consist of human capital, an experience shared inside a company, a cooperation with a research institute, or any other form that is capable of providing knowledge. The distinction between appropriable and non-appropriable knowledge is needed as well because the barriers and hurdles that need to be overcome to get inputs are important factors for the successful acquisition and should be known to the company. Besides general awareness over the recent activities in their particular working field, an idea about potential spillover effects and how knowledge flows inside, but also outside of the firm, are regarded as influential factors. Talking about the barrier between a company and the surrounding world introduces the following criteria: *openness* and *collaboration*. While the known distinction between vertical and horizontal cooperation is again more on

[94] Van Lancker et al. (2016, p. 7).

[95] Ibid.

[96] Ibid.

[97] Ibid.

[98] Ibid.

[99] Tzeng (2014, p. 6).

Table 7.2 Criteria catalogue based on literature

Number	Criteria	Keywords
1	Knowledge and awareness	Presence of knowledge base; appropriable/non-appropriable knowledge; diffusion and spillover effects
2	Openness and collaboration	Vertical/horizontal cooperation; multi-scale linkages; degree of connection; level of trust
3	Supportive environment	Proximity; supportiveness; dynamic/undynamic; suitability for innovation
4	Assisting policies and government	Coordination; holistic approach; coherency and clear understanding; funding and support; creative destruction
5	Society and consumers	Acceptance; understanding; certainty; demand for new products
6	Company management	Capability; acceptance; interactions; openness; R&D expenditures; long-term planning; demand and need
7	Feasibility	Technological, social, environmental, ecological feasibility; sufficiency and efficiency; available resources

the "beneficial-when-known-and-exploited" side of things, multi-scale linkages across more than one layer are regarded as highly potent factors for innovation. Especially when talking about the cooperation and collaborations of a company, the general rule seems to be that the more are present and used, the better, because of the unavoidable flow of knowledge. Of course, the degree and intensity of the connections and linkages are essentially important, as well as trust between the actors. Trust is an even more essential factor of the Open Innovation approach, which supports dismantling strict company boundaries about knowledge transfer and is proven to influence innovation activities. Nevertheless, not only the company itself inherits certain criteria that could potentially favour the creation of innovation, a *supportive environment* is a bolstering factor as well. Not only is proximity regarded as a big driver, because the knowledge flow can occur with a much higher frequency and also often in a face-to-face manner, but the supportiveness of the surrounding plays an essential role too. Without it, companies lose a potential partner on a political level and do also run the risk to antagonize it against them, which always creates an obstructing atmosphere. The supportiveness often influences and is directly influenced by the dynamic of a surrounding region and its actors. New ways of thinking, living, and guiding political decisions create a favourable environment that is suitable to handle innovation that may influence their daily living. *Assisting policies and government* need to have, most of all, a clear and with higher authorities coherent understanding of the target of a bioeconomy process to be able to support companies and actors at the right places and times. A holistic approach, instead of a narrow sectorial-based one to the bioeconomy can help decision-makers to receive a better outline of the term and its implications for our future but also yields a synopsis over certain connections, which otherwise would have been overseen. Funding and support can thus also reach otherwise overlooked actors and firms, and again, the aforementioned holistic view creates a bigger picture for policymakers to decide financial support on. Acknowledging the need for a transformative change and thus creative destruction of the present lock-in state can go hand in hand with open-mindedness regarding bioeconomy and innovations in general and thus is seen as another favourable factor. Not only politics and governmental activities can create a benefitting environment for innovation to occur, but also the *society and consumers* play a significant role. The importance of their acceptance and understanding of bio-economic principles has already been described, but a particular degree of certainty regarding future developments in economy but also politics supports them in making educated decisions and take on a progressive standpoint. At the consumer side, the demand for a new product or process can create an increasingly strong pull and thus urges actors to come after it, often being innovative in adapting their production systems to the new market demand. *Company management* naturally needs financial and social capability in order to be actively engaged in innovative activities. Acceptance and also knowledge about said bio-economic principles is regarded as important as allocating R&D expenditures. The significance of a certain openness, especially towards incoming and not-yet-known linkages and further towards broader ideas, developments, and implications, was again described above. Long-term planning does not favour innovative undertakings on its own, but when paired with knowledge about the need to change current economic or ecologic behaviour can become a driver for innovation. Watching the market demand closely and acting upon being aware of potential gaps may also provide companies with opportunities to establish new products. The last criterion that got deducted is *feasibility*. It can be seen as an outlier because it is assumed that innovation is not triggered simply because something is feasible or not. Rather it should be seen as a supportive criterion once an innovation is already on its way to establishment. It was shown that innovation needs implementation; if any one of the technological, social, environmental, or ecological feasibility is not given, implementation will face serious barriers along its way. The same holds for sufficiency and efficiency; innovators need to assure both for a smooth transition from the invention- to the innovation-phase. Lastly, the needed resources need to be available and adequate with a sustainable infrastructure in place.

At this point, the question about criteria specific to bioeconomy rises; the literature review did not yield any specifics, which is why the above criteria catalogue does not include any. One may think initially of sustainability as a criterion. However, sustainability is another buzzword, encompassing already existing criteria and thus would only add another unnecessary layer on top of the other two, bioeconomy and innovation. A company may undertake activities that result in innovation, but the actual reasoning behind it is often not the need or want to be more sustainable, but to be more efficient or effective, and thus it may use sustainability as a disguise. Otherwise, when a company is forced by an external entity to be more sustainable, sustainability can be seen as a trigger for innovation. Actors that use biological resources, biomass, see themselves as sustainable by definition, as their work needs to be in a sustainable manner in order to secure their livelihood for the present and future. Sustainability is promoted on many political levels, present in the policy discussion for at least 20 years, and promoted all over the world, whereas at its core, it is the simple concept of not destroying what you live on. *Sustainability* may thus be regarded as a trigger for bio-economic innovations but will not be included in the above catalogue, because of its over-usedness, buzzword-character, and unspecific approach.

4 Conclusion

Innovation plays a vital role in our modern economy and society. Bioeconomy, especially in light of the ongoing development of a "new green revolution", appears to manifest itself as an essential factor when talking about possible ways out of a fossil lock-in. With the help of a literature review and a theoretic outlook, this article highlights what factors possibly influence innovation in the context of bioeconomy. Its relevance thus lies in providing a holistic overview of the combination of two terms that are by themselves not easy to frame, thus making the first step towards a remarkable, new research area within the growing bioeconomy discourse. The importance of a shift towards this new economic principle has been stated numerous times in recent years. As this catalogue of criteria is based solely on theory, it needs to be validated with practical examples as a next step; the work on it is far from finished. However, using it as a mere guideline should provide researchers with a good foundation for their work. The article's general approach towards innovation and bioeconomy topics may also help conceiving them from another, maybe new, point of view. However, what has also become clear is the lack of criteria unique to bioeconomy in the literature. Neither the cascading nor the circular economy approach are mentioned as triggers for innovation, while they are perfect examples for innovation out of necessity and thus need to be further investigated. Then again, because bioeconomy cannot be described as a single economic sector, but rather as a concept that spans across multiple sectors, finding particular innovation criteria for it is not an easy task. Sustainability was mentioned but got disregarded because of its comprehensive approach. This means that, in the end, innovation in bioeconomy seems to be based mostly on general criteria. Thus, as a result, the conceptuality of the innovation term can be underlined as well as the broadness of bioeconomy itself which leads to the insight that further research on the topic is still needed.

References

Albrecht, M. (2019). (Re-)producing bioassemblages. Positionalities of regional bioeconomy development in Finland. *Local Environment, 24*(4), 342–357.

Bauer, F. (2018). Narratives of biorefinery innovation for the bioeconomy. Conflict, consensus or confusion? *Environmental Innovation and Societal Transitions, 28*, 96–107.

Bauer, F., Hansen, T., & Hellsmark, H. (2018). Innovation in the bioeconomy—Dynamics of biorefinery innovation networks. *Technology Analysis & Strategic Management, 30*(8), 935–947.

Birch, K. (2009). The knowledge-space dynamic in the UK bioeconomy. *Area, 41*(3), 273–284.

Birch, K. (2012). Knowledge, place, and power. Geographies of value in the bioeconomy. *New Genetics and Society, 31*(2), 183–201.

Birch, K. (2019). *Neoliberal bio-economies? The co-construction of markets and nature*. Cham: Springer International Publishing.

Birch, K., & Tyfield, D. (2012). Theorizing the bioeconomy. *Science, Technology, & Human Values, 38*(3), 299–327.

Dabbert, S., Lewandowski, I., Weiss, J., & Pyka, A. (Eds.). (2017). *Knowledge-driven developments in the bioeconomy*. Cham: Springer International Publishing.

De Besi, M., & McCormick, K. (2015). Towards a bioeconomy in Europe. National, regional and industrial strategies. *Sustainability, 7*(8), 10461–10478.

Fagerberg, J., Mowery, D. C., & Nelson, R. R. (Eds.). (2011). *The Oxford handbook of innovation*. Oxford: Oxford Univ. Press.

Fernandes Rodrigues Alves, M., Vasconcelos Ribeiro Galina, S., & Dobelin, S. (2018). Literature on organizational innovation. Past and future. *Innovation & Management Review, 15*(1), 2–19.

Godin, B. (2016). The linear model of innovation. *Science, Technology, & Human Values, 31*(6), 639–667.

Golembiewski, B., Sick, N., & Bröring, S. (2015). The emerging research landscape on bioeconomy. What has been done so far and what is essential from a technology and innovation management perspective? *Innovative Food Science & Emerging Technologies, 29*, 308–317.

Kirner, E., Kinkel, S., & Jaeger, A. (2009). Innovation paths and the innovation performance of low-technology firms—An empirical analysis of German industry. *Research Policy, 38*(3), 447–458.

Maes, D., & van Passel, S. (2019). Effective bioeconomy policies for the uptake of innovative technologies under resource constraints. *Biomass and Bioenergy, 120*, 91–106.

Näyhä, A. (2019). Transition in the Finnish Forest-based sector. Company perspectives on the bioeconomy, circular economy and sustainability. *Journal of Cleaner Production, 209*, 1294–1306.

Nonaka, I., & Takeuchi, H. (1995). *The knowledge creating company. How Japanese companies create the dynamics of innovation*. New York: Oxford Univ. Press.

Organisation for Economic Co-operation and Development (OECD); Statistical Office of the European Communities (Eurostat). (2005). *Oslo manual. Guidelines for collecting and interpreting innovation data* (3rd ed.). Retrieved June 23, 2020, from, https://www.oecd-ilibrary.org/docserver/9789264013100-en.pdf?expires=1592903529&id=id&accname=guest&checksum=EDCA5591254052A63BC784C8379C5953

Peltomaa, J. (2018). Drumming the barrels of Hope? Bioeconomy narratives in the media. *Sustainability, 10*(11), 4278.

Pietzsch, J., & Schurr, U. (Eds.). (2017). *Bioökonomie für Einsteiger*. Berlin: Springer Spektrum.

Purkus, A., Hagemann, N., Bedtke, N., & Gawel, E. (2018). Towards a sustainable innovation system for the German wood-based bioeconomy. Implications for policy design. *Journal of Cleaner Production, 172*, 3955–3968.

Pyka, A. (2017). Transformation of economic systems: The bio-economy case. In S. Dabbert, I. Lewandowski, J. Weiss, & A. Pyka (Eds.), *Knowledge-driven developments in the bioeconomy* (pp. 3–16). Cham: Springer International Publishing.

Saviotti, P. P. (2017). Structural change, knowledge and the bioeconomy. In S. Dabbert, I. Lewandowski, J. Weiss, & A. Pyka (Eds.), *Knowledge-driven developments in the bioeconomy* (pp. 17–32). Cham: Springer International Publishing.

Schumpeter, J. A. (1939). *Business cycles. A theoretical, historical and statistical analysis of the capitalist process*. New York/London: McGraw-Hill.

Schütte, G. (2018). What kind of innovation policy does the bioeconomy need? *New Biotechnology, 40*(Pt A), 82–86.

Stadler, T., & Chauvet, J.-M. (2018). New innovative ecosystems in France to develop the bioeconomy. *New Biotechnology, 40*(Pt A), 113–118.

Staffas, L., Gustavsson, M., & McCormick, K. (2013). Strategies and policies for the bioeconomy and bio-based economy. An analysis of official National Approaches. *Sustainability, 5*(6), 2751–2769.

Suroso, E., & Azis, Y. (2015). Defining mainstreams of innovation. A literature review. In *Advances in Economics, Business and Management Research* 5. Proceedings of the International Conference on Economics and Banking (pp. 387–398).

Tzeng, C.-H. (2014). A review of contemporary innovation literature. A Schumpeterian perspective. *Innovations, 11*(3), 373–394.

Van Lancker, J., Wauters, E., & van Huylenbroeck, G. (2016). Managing innovation in the bioeconomy. An open innovation perspective. *Biomass and Bioenergy, 90*, 60–69.

Von Braun, J. (2018). Bioeconomy—The global trend and its implications for sustainability and food security. *Global Food Security, 19*, 81–83.

von Schomberg, R. (2012). Prospects for technology assessment in a framework of responsible research and innovation. In *Technikfolgen abschätzen lehren* (pp. 39–61). Wiesbaden: VS Verlag für Sozialwissenschaften.

Vivien, F.-D., Nieddu, M., Befort, N., Debref, R., & Giampietro, M. (2019). The hijacking of the bioeconomy. *Ecological Economics, 159*, 189–197.

Wensing, J., Carraresi, L., & Bröring, S. (2019). Do pro-environmental values, beliefs and norms drive farmers' interest in novel practices fostering the bioeconomy? *Journal of Environmental Management, 232*, 858–867.

8 Spatial Implications of the Leitmotif Shift from Biotechnology to Bioeconomy

Leonard Prochaska and Daniel Schiller

Abstract

With the introduction of a dedicated bioeconomy strategy in 2010, the German government altered its bio-based innovation policy from an approach driven by the biotechnology towards the more holistic notion of a "bioeconomy". The shift towards the bioeconomy is motivated by an increasing urgency to develop alternatives to fossil raw materials, the mitigation of anthropogenically induced climate change, and contributions to far-reaching societal issues. The production and processing of biomasses based on principles of the biotechnology and the production of bio-based products have the potential to strengthen rural regions which are otherwise underrepresented when it comes to knowledge-based growth. In this study, spatially comparative empirical analyses are carried out, which show what parameters were decisive for the acquisition of publicly funded R&D projects in the biotechnology and bioeconomy in the period from 1995 to 2015. Using the German subsidy database "Förderkatalog" of the Federal Ministry of Education and Research, we find that regions with a lower share of knowledge-intensive industries are more often engaged in bioeconomy projects outside of the biotechnology core. Moreover, population density does not play an important role for the spatial distribution of bioeconomy projects unlike projects within the biotechnology. However, biotechnological knowledge is still the main driver for the development of the knowledge-based bioeconomy.

L. Prochaska (✉)
Faculty of Geography and Geology, University of Greifswald, Greifswald, Germany
e-mail: leonard.prochaska@uni-greifswald.de

D. Schiller
Faculty of Economic and Social Geography, University of Greifswald, Greifswald, Germany
e-mail: daniel.schiller@uni-greifswald.de

D. Lanzerath et al. (eds.), *Bioeconomy and Sustainability*,
https://doi.org/10.1007/978-3-030-87402-5_8

Keywords

Innovation policy · R&D funding · Rural regions

1 Introduction

Most recently, two studies assessed that an increasing geographical polarisation in Germany led to differently developed regions in terms of their infrastructure, living conditions and economic performance.[1] Next to a historically caused west-east divide, a clear south-north descent is one apparent finding. Although it was previously known, the most publicly communicated and alarming conclusion is the steadily growing gap between agglomerations and rural regions.[2] In order to reduce those regional disparities both the German government and the European Union (EU) implemented several policies. Classical regional policies on EU level that aim at the conversion of regional developments are the Cohesion Policy and manifold structural funds, namely, e.g., European Agricultural Fund for Rural Development (EAFRD), European Regional Development Fund (ERDF) and European Social Fund (ESF). On the national tier, Germany deploys the instrument Bund-Länder-Gemeinschaftsaufgabe "Verbesserung der regionalen Wirtschaftsstruktur" (GRW) (Joint Task "Improvement of the Regional Economic Structure") to foster regions with the intention of achieving an equivalent spatial development in economic and social terms. This goal is mainly pursued, among other things, by investing in the region's business-related infrastructure or subsidising the economic activities (e.g. expansion of production capacities) of resident small and medium enterprises (SME).

However, the underlying approaches in regional policy have gradually changed. In particular, the importance of innovation-based growth, especially in peripheral regions, has grown significantly in decades.[3] At the same time, classical innovation policy successively moved away from sectoral and geographical growth pole approaches and thus, stressed attention at the value of the spatial dimension as well as the significance of the "region" as an actor.[4] Therefore, to allocate the limited resources most efficiently as well as to take account of the varying regional endowments and capacities, one-size-fits-all strategies have been renounced in both literature and policies.[5]

One example is the emergence of the bioeconomy in innovation policy. According to the official Research Strategy on Bioeconomy in Germany (NFSB)

[1] Cf. Halle Institute for Economic Research – Member of the Leibniz Association (2019) and Federal Ministry of the Interior, Building and Community (2019).

[2] Cf. Federal Ministry of the Interior, Building and Community (2019).

[3] Cf. Organisation for Economic Co-operation and Development (2011).

[4] Cf. Koschatzky (2005).

[5] Cf. Tödtling and Trippl (2005) and Organisation for Economic Co-operation and Development (2011).

published in 2010, innovation and diffusion should not only be driven by urban centres, but should rather develop collectively in all regions. The bioeconomy strategy itself evolved out of previous biotechnology funding, merged with regional and agricultural innovation approaches and replaced them. While past calls for competitions for the systematic subsidisation of biotechnology locations took place predominantly in agglomerations (e.g. BioRegio & BioProfile), the bioeconomy concept is also applied in peripheral and rural regions. For this reason, we analysed the biotechnology and bioeconomy funding histories from 1995 to 2015 as one of the current cases for a more modern approach in Germany's innovation policy.[6] Due to the nature of bioeconomy and its focus on natural resources, whose production is often tied to less developed and rural regions, the potential for the so-called left behind areas with rather low innovation capabilities to participate in more applicatory projects is expected to be greater than in basic research. In other words, we want to research whether the executed shift in "bio"-related innovation policies has actually led to a more balanced spatial distribution of Research and Development (R&D) projects and thus hold out the prospect to equate increasing regional disparities proactively.

This elaboration is organised as follows: In section two, we first introduce and demonstrate how the leitmotif transition in innovation policy from biotechnology to bioeconomy developed before we point out in section three empirical evidence about the impact of innovation policy on technologies and regional (peripheral) development. In section four, the idiosyncrasies of bioeconomy as well as the regional assumptions to be expected are illustrated. Subsequently, we describe the data and method we used for our empirical analysis and display our results based on descriptive analysis and regression models. The paper ends with a brief conclusion.

2 The Transition from Biotechnology to Bioeconomy in the German Innovation Policy

The first official global appearance of a bio-based economy idea was carried out by the Organisation for Economic Co-operation and Development (OECD) in the document "Biotechnology for sustainable growth and development" in 2004.[7] As such, this vision of "a concept that uses renewable bioresources, efficient bioprocesses and eco-industrial clusters to produce sustainable bioproducts, jobs and income"[8] was first mentioned in a report about biotechnology. This illustrates that both subjects are not only closely connected, but that the understanding is that bioeconomy would not be feasible without the development of advanced biotechnologies. Subsequently, many countries altered their innovation policy in the form of a transition from biotechnology to bioeconomy.

[6] Cf. Federal Ministry of Education and Research (2017).

[7] Cf. Patermann and Aguilar (2018).

[8] Organisation for Economic Co-operation and Development (2004, p. 4).

This arising subject in global,[9] supranational[10] and national[11] funding schemes aims at the promotion of R&D related to biological resources. In the holistic bioeconomy concept, next to the trans-sectoral technologies of the biotechnology sector, upstream industries such as agriculture and forestry as well as downstream sectors, for instance energy production from biomass or the food industry, should also be engaged in and benefit from the bioeconomy concept.[12] This policy shift raises the question whether and in what ways the spatial patterns of bioeconomy funding changed over time.

Before we give an overview about past empirical studies on the impact of innovation policy, the past "bio-related" funding measures in Germany are briefly presented below in order to illustrate the underlying transition from biotechnology to bioeconomy (see Table 8.1).

Pioneering endeavours of the German biotechnology promotion date back to the period around 1970.[13] Most funding in biotechnology largely favoured basic research by public institutions and aimed at the generation of scientific knowledge.[14] However, in respect of bio-related funding in Germany, the initial systemic approach to commercialise biotechnological procedures was proposed in 1995 in the form of the BioRegio contest. Subsequent programmes implemented by the German government (i.e. BioProfile, BioChance, BioFuture) somewhat softened the strict focal point on biotechnology. In 2001, the German Federal Ministry of Education and Research introduced another funding measure ("Framework Programme Biotechnology" (FPB)) to not only foster biotechnology independently of their location (contrasting to BioRegio and BioProfile), but also biosciences in general. Ultimately, this programme merged in 2010 into the current "National Research Strategy Bioeconomy 2030" (NFSB), which involves six resorts[15] and hence, emphasises the concept's holistic character.[16] The introduced blueprint was subsequently complemented by the "National Policy Strategy Bioeconomy" of the Federal Ministry of Food and Agriculture,[17] which also interfaces with other political schemes such as "Action plans for the material and energetic use of renewable raw materials" (2009/2010) or the "Forest strategy 2020" (2011). Eventually in 2020, the Research

[9] Cf. Organisation for Economic Co-operation and Development (2009).

[10] Cf. European Commission (2012).

[11] Cf. Federal Ministry of Education and Research (2010), The White House (2012), Ministry of Economic Affairs and Employment (2014) and HM Government (2018).

[12] Cf. Federal Ministry of Education and Research (2010) and European Commission (2012).

[13] Cf. Warmuth (1991), Federal Ministry of Education and Research (2011) and Schüler (2016).

[14] Cf. Warmuth (1991).

[15] Under the leadership of the Federal Ministry of Education and Research, following ministries are responsible for the implementation: Federal Ministry for Economic Affairs and Energy, Federal Ministry of Food and Agriculture, Feral Ministry for the Environment, Nature Conservation and Nuclear Safety, Federal Ministry for Economic Cooperation and Development and the Federal Foreign Office.

[16] Cf. Hüsing et al. (2017).

[17] Cf. Federal Ministry of Food and Agriculture (2014).

Table 8.1 Dedicated bio-related funding programmes since 1997

Leit-motif	Time-period	Funding Programme	Purpose & Implementation
Biotechnology	1979-1983	"Benefit plan Biotechnology 1979-1983"	1st programme to foster narrowly defined biotech.
	1985-1988	"Applied Biology and Biotechnology"	Mainly basic research
	1997-2005	"BioRegio"	Fostering biotechnology in 4 regions
	1999-2007	"BioProfile"	Fostering biotechnology "profiles" in 3 regions
	2001-2010 (into NFSB)	"Framework Programme Biotechnology"	Fostering Biotechnology and related biosciences
⇩	since 2006 (continued 2014)	"Hightech-Strategy"	Priorities on innovations; since 2014 including non-technical solutions
Bioeconomy	since 2010	"National Research Strategy Bioeconomy" (NFSB) connected to the Hightech-Strategy	Conceptualisation of a bioeconomy in Germany
	since 2014	"National Policy Strategy Bioeconomy"	Connecting multiple policy areas in the bioeconomy framework
	expected in 2019	Bundling of Research and Policy Strategy Bioeconomy	Pursuit of a mutual "Bio-Agenda"

Source: own draft according to German Bundestag (1990), Staehler et al. (2006), Federal Ministry of Education and Research (2011) and Schüler (2016)

and Policy Strategies were bundled to form an overall strategy.[18] It can therefore be stated that there is clear evidence in Germany of a theme shift "from a biotechnology-centric vision to an economic activity that spreads across several key sectors and policy families".[19]

3 The Impact of Innovation Policy on Regional Development

While there is little literature about empirical studies on the impact of regionalised innovation policies, we find several analyses on the effects of innovation policy in general. Prominently, Mariana Mazzucato showed that the long-term vision of governmental support is inherently underestimated, since they invest in risky research and technologies that would have been avoided by private actors.[20] Eventually, this predominantly leads to the advancement of technologies and thus, to the

[18] Cf. Federal Ministry of Education and Research and Federal Ministry of Food and Agriculture (2020).

[19] Organisation for Economic Co-operation and Development (2018, p. 11).

[20] Cf. Mazzucato (2014).

welfare of companies. Although there are cases, which show failures or wrong chosen paths and thus illustrate the difficile nature of innovation policy, we find many diverse scholars that vindicate the policy-driven economy approach.

Some studies on specific innovations or branches, furthermore, illustrate the extent of policy-driven approaches. E.g. William Lazonick and Öner Tulum showed that the ground-breaking innovations founding biotechnology were initiated and supported by government funds.[21] Additionally, using examples of basic innovations of "Apple" products Mazzucato demonstrated that most of the innovative components originated from publicly funded technologies.[22] In another example, she traced the development of certain pharmaceutical drugs which oftentimes came from R&D subsidies. Erik Arnold concluded in his study the measurable long-term effects of multiple funding studies or programmes stimulating several individual fields (e.g. brain research, O_3 research) or whole branches (e.g. automotive industry, Information and communications technology (ICT)).[23] Altogether, Philippe Aghion, Julian Boulanger and Elie Cohen justified state intervention by providing evidence that, on the one hand, incessant laissez-faire would lead to environmental failures and, on the other hand, sectoral policy has positive effects when appropriately applied.[24] For instance, policy measures have greater impact if they are practised decentralised and complemented with further instruments, such as taxes for industries that produce inadvertent negative externalities. Wolf-Hendrick Uhlbach, Pierre-Alexandre Balland and Thomas Scherngell inferred similar results and stressed the necessity of appropriate allocation of public R&D subsidies in terms of the relatedness level. They found the EU-support was most effective in regions with neither too low nor too highly related knowledge existent.[25]

Moreover, there are indications that co-operative public R&D is especially beneficiary at both, the microeconomic and the macroeconomic level. Dirk Czarnitzki and Andreas Fier as well as Czarnitzki, Bernd Ebersberger and Fier focussed on the firm level and found statistical affirmation that R&D networks are more likely to generate patents than sole R&D.[26] In addition, networks with financial support perform better than collaborations without governmental aid. A study on Japanese research consortia leads to comparable results, which concentrated on firm's outcomes such as research productivity. Particularly the aspect of increased knowledge spillovers within the networks illustrates the importance of collective

[21] Cf. Lazonick and Tulum (2011).

[22] Cf. Mazzucato (2014).

[23] Cf. Arnold (2012).

[24] Cf. Aghion et al. (2011).

[25] Cf. Uhlbach et al. (2017).

[26] Cf. Czarnitzki and Fier (2003) and Czarnitzki et al. (2007).

learning.[27] On the macroeconomic tier, regions with low innovation capacity are more probable to take advantage of subsidised R&D projects, notably in partnership with external research institutions.[28]

Altogether, most empirical analyses indicate a positive impact of governmental-led initiatives to foster specific technologies, branches or regions. However, the policy design needs to fit the targets set. That means, the one-size-fits-all perspective, in which a top-down approach is applied independently of the region's characteristics, is obsolete and shows that policies should be more differentiated and adapted to regional conditions.[29]

In order to foster peripheral regions and trigger their innovation potential, empirical studies indicate that the policy interventions applied have to build up upon existing endowments.[30] Some studies go one step further and point out that peripheral regions offer advantages over agglomerations in many areas. For instance, lower negative externalities (e.g. negative spillovers) and strong in-house capabilities might be important determinants that lead to a faster growth outside of agglomerations.[31] Other factors such as the employee's loyalty in addition to the lower salaries, close relationships to local institutions as well as the affinity to the location can play a vital role for the company's progress.[32]

With that said, subsequently the starting points are elaborated that have been identified by theoretical and empirical works, which play or could play a vital role for innovation policy designs and the improvement of more rural region's economic performance.

4 Bioeconomy's Capabilities to Reduce Regional Disparities

The arising bioeconomy concept claims to possess the potential of involving rural and peripheral regions by its extensive nature with a focus on new production mechanisms regarding biological resources or novel procedures in traditional branches such as food and feed industries or paper and pulp industries. Next to Germany's and the EU's more general endeavours to balance interregional inequalities mentioned above, both respective bioeconomy strategies integrate the development of peripheral regions. While the EU's strategy particularly refers "to support local bioeconomy development (rural, coastal, urban) via Commission instruments and programmes",[33] the Federal Ministry of Food and Agriculture

[27] Cf. Branstetter and Sakakibara (1998).

[28] Cf. Broekel (2015).

[29] Cf. Tödtling andTrippl (2005).

[30] Cf. Aghion et al. (2011), Marrocu et al. (2013) and Carvalho and Vale (2018).

[31] Cf. Grillitsch and Nilsson (2017).

[32] Cf. Eder and Trippl (2019).

[33] European Commission (2018, p. 18).

highlights the demand for "securing and creating employment and added value, especially in rural areas".[34]

By implication, the issue of a concrete and comprehensive implementation of this political intention comes to the fore. At first, the most apparent fact is that the majority of the upstream industries involved in the production of biological raw materials, such as agriculture and forestry, are located in rural areas. The same applies to some downstream industries, notably food and feed, the chemical industry, textiles as well as the production of energy by using biomass. Those sectors are often characterised by their rather low technological requirements, although they already underwent modernisation by applying modern technologies. Modern policy recognised the unfeasibility and inefficiency of trying to create strong high-tech sectors in any location whatsoever and hence, underscores the need to transmit knowledge from the core into structurally weak regions, according to the precept of "adoption, adaption and diffusion"[35] of (external) knowledge. Thus, the aspiration is not "to leverage existing strengths, [but instead] to identify hidden opportunities and to generate novel platforms upon which regions can build competitive advantage in high value-added activities"[36] as Balland, Ron Boschma, Joan Crespo and David L. Rigby put it. In other words, in place of specialising in already existing dominant industries, the endeavours rather ought to be directed towards diversifying the prevailing structural conditions.

From a theoretical perspective, Robert Hassink conceptualises the risk of a strong specialisation by a perilous dependence on one or few industries caused by an absence of renewal tendencies.[37] Subsequently, a decreasing innovation potential along one or a small number of technological paths increases the susceptibility of external shocks and might result in a lock-in, which ultimately leads to path exhaustion.[38] To avoid these scenarios, current policies pursuing the bioeconomy concept aim at a diversification of incumbent trajectories and path renewal, respectively. Philip McCann and Raquel Ortega-Argilés clarify "that very few regions make fundamental structural or sectoral shifts in the short- to medium-term"[39] and thereby, illustrate the relevance of regional branching. In the context of bioeconomy, it means that in rural and peripheral areas existing endowments in low technology sectors possess the potential to enrich their local capabilities with exogenous developed general- purpose technologies (GPT), particularly the biotechnology. Moreover, these very GPT are due to their knowledge bases, connoted as "analytical knowledge",[40] predestined for the implementation in geographically distant

[34] Federal Ministry of Food and Agriculture (2014, p. 20).

[35] McCann and Ortega-Argilés (2015, p. 1299).

[36] Balland et al. (2019, p. 1).

[37] Cf. Hassink (2010).

[38] Cf. Isaksen (2015).

[39] McCann and Ortega-Argilés (2015, 1296).

[40] Asheim (2007, p. 224 f).

industries.[41] The possibility to codify and formalise the biotechnological knowledge enables the opportunity for traditional branches to transfer extant expertise over long distances, to employ them in new ways and through this, renew or yet create regional trajectories. Especially for structurally weak regions, which are often characterised by small and medium size enterprises (SMEs) without or with few own R&D establishments, are external and public knowledge remarkably viable.[42] In the context of forest-related strategies, Sylvie Albert stresses the attention for rural areas "to perfect their 'outside-in' thinking skills, relating information about development in the external world to what is going on internally".[43] This does not only apply to forest management, but it also includes several other facets of bioeconomy and underlines the beneficial nature of complementarity between exogenous and endogenous knowledge.[44]

Luís Carvalho and Mário Vale propose peripheral regions to diversify actual structures by "bricolage", which focusses on agency and available local resources.[45] In general, bricolage "connote[s] resourcefulness and improvisation on the part of involved actors"[46] and was founded on the development of the wind turbine path in Denmark. In fact, the mainstays of the bricolage conception are similar to some basic notions of bioeconomy strategies. Namely, local actors who maintain knowledge about their site, resources, institutions, or markets as well as linkages to other relevant actors play a crucial role to allocate available resources efficiently. Natural resources themselves also have a great significance. As decisive elements within bioeconomy, rural regions produce most of the biomass to be processed. Furthermore, the Danish wind turbine sector was financially supported by the government in order to foster the new path development.[47] The same applies to Arne Isaksen's case and, of course, the heavily funded bioeconomy.[48] Eventually, the fact that the developers of the wind turbines were not necessarily the same individuals, which utilised the technology for commercial reasons, emphasises the potential to apply external generated knowledge adapted to the regional environment and capabilities, for example, to steadily improve low-tech industries. This type of diversification, based on Joseph Schumpeter's considerations on recombining existing resources, also reflects the deliberation on "self-discovery".[49,50] This notion stresses the necessity "to allow for experimentalism in order to discover what works in what

[41] Cf. Asheim et al. (2011).
[42] Cf. Soete and ter Weel (1999) and Isaksen (2015).
[43] Albert (2007, p. 65 f).
[44] Cf. Bugge et al. (2016).
[45] Cf. Carvalho and Vale (2018).
[46] Garud and Karnøe (2003, p. 278).
[47] Cf. Garud and Karnøe (2003).
[48] Cf. Isaksen (2015).
[49] Hausmann and Rodrik (2003, p. 605).
[50] Cf. Schumpeter (2005 [1942]).

context"[51] and to move "beyond [. . .] natural-resource based products",[52] which is particularly essential for rural and structurally weak regions. In general, innovation opportunities for peripheral regions successively improve due to the availability of external knowledge (via modern ICT), growing negative externalities in agglomerations as well as local agency and internal knowledge about the respective site.[53] In other words, regions that often are perceived and labelled as provider for natural resources and location for space-intensive industries might have a better chance of going beyond this stigma and becoming stronger economic actors themselves under the bioeconomy leitmotif.

Altogether, the depicted development in innovation policy (Sect. 2) along with the assigned attributes of the bioeconomy (Sect. 4) leads to the following hypothesis:

H1: Rural peripheries and structurally weak regions with a traditional and less knowledge-intensive sectoral basis are favoured by the funding shift from biotechnology to the broader bioeconomy concept.

5 Bioeconomy's Structure

Since one objective of this study is to trace the transformation from biotechnology towards bioeconomy, it is necessary to distinguish between the initial focussed funding by sector and the additional dimensions of bio-related support, which were conceptualised by the bioeconomy strategy, i.e. input and output dimensions as well as the socio-economic framework. This is why the individual components of the bioeconomy concept are henceforth categorised and designated as follows (see Fig. 8.1):

- **biotechnology nucleus**: green, red and white biotechnology,
- **bioeconomy shell**: input and output dimensions as well as socio-economic framework.

The biotechnology nucleus and the bioeconomy shell jointly represent the bioeconomy concept.

Within the study, data and conceptions from various actors involved with the bioeconomy concept have been collected and systemised according to their opinions within a disaggregation of bioeconomy along the value chain of biological resources. Since bioeconomy is "largely driven by policy action and the contents of bioeconomy strategies worldwide",[54] the political vision has mainly determined

[51] McCann and Ortega-Argilés (2013, p. 208).

[52] Hausmann and Rodrik (2003, p. 605).

[53] Cf. Grillitsch and Nilsson (2017).

[54] Viaggi (2016, p. 105).

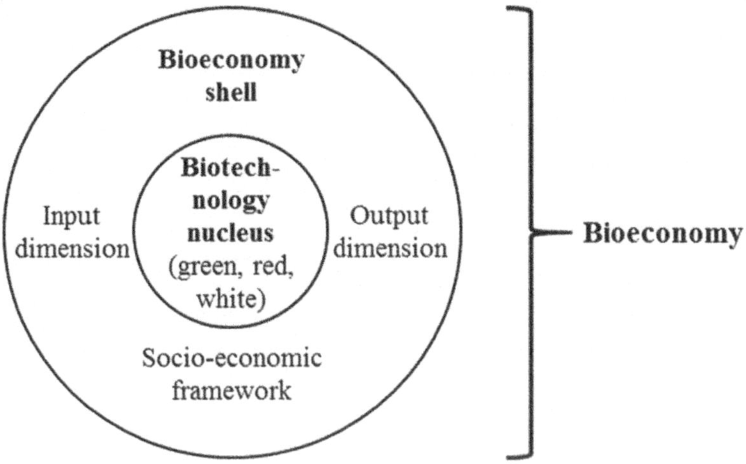

Fig. 8.1 Illustration of the bioeconomy components and structure (Source: own draft)

the definition at hand. Thus, the delineation developed represents a wide-ranging definition, similar to the German bioeconomy strategy (NFSB). The four mentioned mainstays of the bioeconomy concept thus cover different stages of the value chain (see Table 8.2).

6 Data and Method

In order to examine the posed hypothesis, a distinct database has been built containing all funded projects by the federal ministries, which are captured in the German subsidy database called "Förderkatalog".[55,56] By April 2017, the dataset comprised 191,347 projects with valid information, the earliest entry from 1968. The content of this data is structured based on the applied funding measures and has superordinate topics such as "A—Health research and health economy", "D—Food, Agriculture and Consumer Protection", or "E—Energy research and technologies". This systematic is refined by two further tiers. Eventually, the issue "B—

[55] According to a statement of the Federal Ministry of Education and Research, the database contains approximately 95% of all R&D projects funded by their ministry (with increasing tendency). However, it is the responsibility of the other departments (e.g. Federal Ministry of Food and Agriculture, Federal Ministry for Economic Affairs and Energy) to record their projects and the data suggests that only a fraction of the ministries' projects have been entered into the database. Yet, the Federal Ministry of Education and Research is in charge of not only implementing the biotechnology and bioeconomy strategies, but also accounts for approximately 58% of total R&D expenditure in Germany (Federal Ministry of Education and Research, 2017) and therefore takes over the lion's share of all funding. Thus, this database is sufficient in order to make empirical statements about the knowledge-driven bioeconomy funding landscape.

[56] Cf. Federal Ministry of Education and Research.

Table 8.2 Structure of the bioeconomy

	Dimension	Components
Value Chain ↓	**Input dimension** (production)	agriculture & forestry climate & environmental protection
Value Chain ↓	**Biotechnology** (processing)	green biotechnology (agriculture and aquaculture, e.g. plant and animal breeding on land and in water) red biotechnology (medicine, e.g. biopharmaceuticals, human genes) white biotechnology (industry and environment, e.g. sewage treatment, new materials)
Value Chain ↓	**Output dimension** (material utilisation of biomass)	products and materials energy & fuels food & feed
	Socio-economic framework including immaterial matters such as	coherent policy viable financing for bioeconomy practicing companies platforms and arrangements for networking education and qualification sensitisation of the society

Source: own conception

Bioeconomy" is aggregated in its own category. However, there are two issues, which need to be taken into account during analysis. On the one hand, the segment "B" includes projects that date from far before the official bioeconomy policy concept was formulated—the first record in this category is dated 1968 while the start of the bioeconomy strategy was in 2010. On the other hand, it is apparent that numerous topics or projects within several other classes such as "EB1920—Energetic use of biomass", "GC2060—Organic electronics", or "KA1210—Nanobiotechnology" can clearly be assigned to bioeconomy, but are not covered by this class. On that account, we considered it necessary to integrate all projects de facto related to the bioeconomy approach.

The substance of the database used is openly accessible and offers valid information about the temporal horizon, the monetary investment, names of grantees and executing organisations along with their respective location and the collaboration partners in the case of joint projects. Most relevant for the identification process, two types of information about each undertaking's topic have been diagnosed. In addition to the Federal Ministry of Education and Research's internal classification that indicate the subject area, the title of the project provides genuine indications about its content.

Given these circumstances and based on the Federal Ministry of Education and Research classification,[57] we first categorised the dataset into three divisions (see Fig. 8.2):

[i]. classes that ascertained belong to bioeconomy,
[ii]. classes that only partially belong to bioeconomy and,
[iii]. further categories that are unlikely to contain bioeconomy projects.[58]

Consequently, with common text-mining techniques (e.g. removing whitespaces and stop words) the project titles were simplified. Considering the main principle of bioeconomy, namely the involvement of biological material, it is in our opinion an appropriate measure to draw on this basic idea and hence, to create a collection of biomass-connected terms and expressions. For that reason, we detected all bio-related phrases that appeared in at least five project titles and had a unique stem within the division [i]. The result was a set of 355 terms. With this array of phrases, we calculated the amount of bioeconomy-relevant terms in each project title and applied a straightforward keyword search to identify relevant projects in [ii] and [iii]. Using these calculations, after gradual refinement of the adjusting parameters, we applied a stepwise procedure to select projects relevant to bioeconomy:

1. Classification of subdivisions from the subsidy database with regard to their relatedness to bioeconomy—as part of which all projects of the subdivision [i] were selected and added to the database.
2. Application of distinct thresholds.
   (a) Within the division [ii], we chose all projects that featured at least two keywords.
   (b) Due to particular bio-related projects in unanticipated classes [iii], the counting threshold value was set to four.

In order to categorise the bioeconomy projects into the subsections mentioned above (see Table 8.2), we determined groups of the Federal Ministry of Education and Research's internal classification, which are clearly assignable to one of the previously worked out bioeconomy sections along the value chain. A significant proportion, however, had to be attributed by hand, which also served as a result review and occasionally led to the identification of unfitting projects that were subsequently eliminated from the database. A schematic visualisation of the process is depicted in Fig. 8.2. By employing this method, 16,500 projects could be identified as relevant to bioeconomy within the time span of 1995–2015 and thus,

[57] It should be noted that this internal system is in constant revision and thus retroactively changes over time.

[58] We excluded the funding area "C—Civil security research" entirely considering that biological warfare agents are not part of the bioeconomy notion.

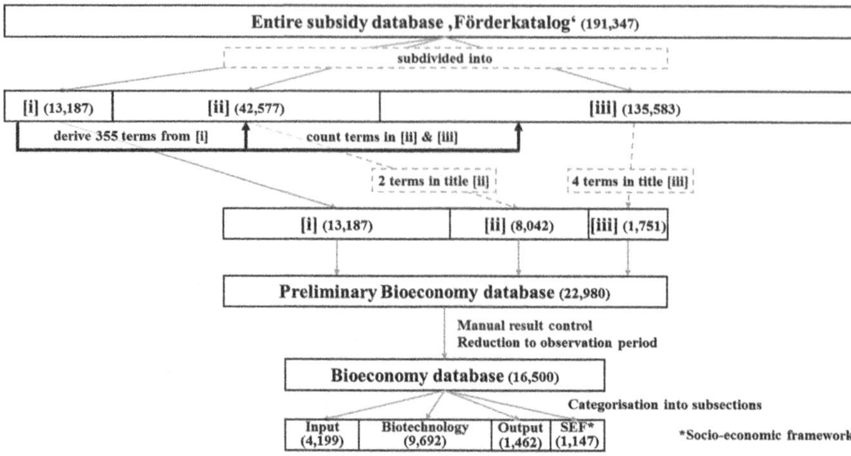

Fig. 8.2 Schematic process of the database derivation (not proportional) (Source: own draft)

build the final dataset for the analysis. Table 8.3 lists the basic descriptive statistics of the decisive data.

With these data, we are able to calculate project count and project funding of each bioeconomy dimension to any given time on any regional level. For this study, we work on the "Kreis"-level (district), equivalent to EU's NUTS-3-tier. Additionally, we employ various regional data from the Regionaldatenbank Deutschland (Regional database Germany) provided by the Federal and State Statistical Offices Germany (2019) to capture characteristics of the observation units.[59] We include the following regional determinants as independent variables for the econometrical model:

- Number of people employed (EMP).
- Gross domestic product per employee (GDP).
- Unemployed people per inhabitant (UNEMP).
- Population density (POPDENS).
- Agriculture production land & forest area (AGRI).
- Employees in knowledge-intensive industries (KNOW).[60]

Furthermore, we add a dichotomous variable called EAST to control the bias in funding which favours Eastern German territories due to the intention to accelerate the catching-up process.

[59] Cf. Federal and State Statistical Offices Germany (2019).

[60] This variable is constructed based on the definition of the INKAR database (BBSR), suggesting that the industrial sectors 20, 21 & 26–30 in the WZ 2008 classification of the Federal Statistical Office of Germany are characterised as high-tech industries.

Table 8.3 Comparative figures of the funding-datasets, 1995–2015

	Full database	Bioeconomy concept	Biotechnology Nucleus	Bioeconomy Shell
Number of projects	114,448	16,500	9692	6808
Av. funding per proj. in €	546,911	426,936	497,838	325,999
Median funding per proj. in €	208,685	235,916	277,536	176,150
Av. proj. duration (days)	995	1118	1144	1081
Median proj. duration (days)	1095	1095	1095	1095
Share of joint projects	0.59	0.72	0.77	0.64
Share of public organisations	NA	0.66	0.65	0.68

Source: own calculations based on Federal Ministry of Education and Research (n.d.)

To test the elaborated hypothesis, we conduct four comparing regressions with varying dependent variables, specifically (i) overall project count, (ii) number of bioeconomy projects, (iii) biotechnology projects and eventually, (iv) count of projects within the bioeconomy shell. For the purpose of tracking the transformation of applied regional innovation policies, analyses are performed taking into account different time intervals. I.e., first we estimate which regional and structural parameter were vital to acquire projects in 1995–2001, corresponding to the era of particular biotechnology funding. Second, based on the same variables, comparative examinations are executed with data from the most recent time interval falling within the period mainly after the introduction of the bioeconomy strategy, from 2009–2015. Since the significance of previous biotechnology knowledge and the specialisation of the region is one integral component to evaluate path dependency, the share of obtained biotechnology projects on the total number of projects in the preceding period of seven years is included into the model (BT_{t-1}). Further regressors contain the regional data from the last year of each period that is 2001 and 2015, respectively.[61]

As apparent, any dependent variable whatsoever represents count data. This implies that the error term of the regression will not be normally distributed. For that reason, we employ generalised linear models. In order to avoid overdispersion, we neglect the Poisson Regression Model and engage Negative Binomial Models for all estimations.[62]

[61] Due to the lack of data regarding the working population in knowledge-intensive industries, we used the earliest available data from 2009.

[62] Cf. Zeileis et al. (2008).

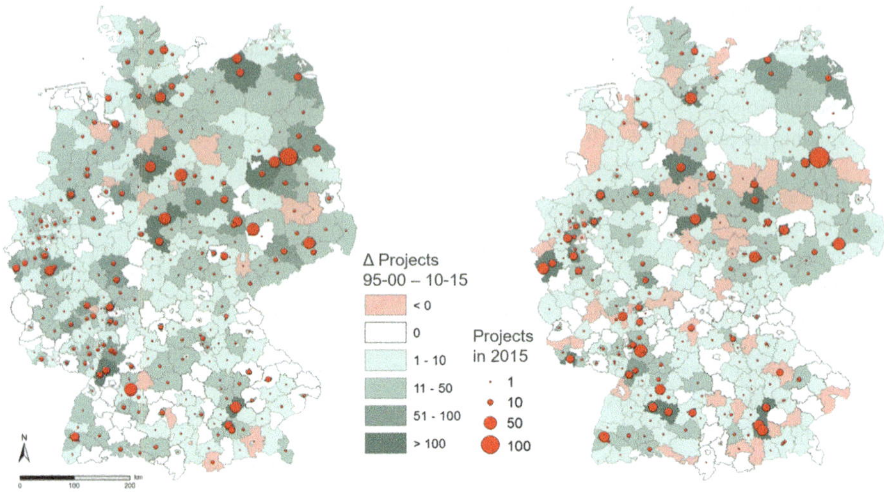

Fig. 8.3 Development of the funding within the bioeconomy shell (left) and the biotechnology nucleus (right) (Source: own calculations based on Federal Ministry of Education and Research (n. d.))

7 Results

A number of depicted statistics about the various datasets in Table 8.3 reveal indications about some underlying differences between both mere biotechnology funding and the added dimensions of bioeconomy. Within the observation period, we find a considerable disparity regarding the average or median subsidisation. While the average grant in the biotechnology is larger than 150,000 € compared to the project volume in the bioeconomy shell, the gap between the median projects is still quite large at 100,000 €. Furthermore, the number of joint projects is noticeably higher in the biotechnology sector and in either of bioeconomy components higher than within all projects in the database. One basic notion of bioeconomy is the implementation of biotechnological procedures into traditional branches as well as knowledge diffusion in general, which most likely ought to occur in a surrounding of collaborations between various actors. Hence, a greater number of concerted projects might indicate that this approach is applied in reality. If one looks at the spatial distribution of the projects, it is recognisable that, for one thing, regions in the north of Germany and, otherwise, the outskirts of agglomerations (e.g. around Berlin, Hamburg and Munich) have received more attention since the introduction of the NFSB (Fig. 8.3).

The simple (visual) comparison of the different datasets shows some structural dissimilarities, which require econometrical analyses to identify and verify the underlying regional implications induced by the policy transition from sectoral to systemic innovation policies. Table 8.4 summarises descriptive statistics about the variables used in the models.

Table 8.4 Descriptive statistics of the variables

	Mean		Standard deviation		Minimum		Maximum		No. of observations
	1995–2001	2009–2015	1995–2001	2009–2015	1995–2001	2009–2015	1995–2001	2009–2015	
Dependent variables (no. of projects)									
All projects	56.83	136.13	126.36	241.08	0	4	1034	1745	400
Bioeconomy concept	6.82	19.39	16.89	40.35	0	0	124	238	400
Biotechnology nucleus	4.62	9.61	12	22.76	0	0	90	173	400
Bioeconomy shell	2.20	9.78	5.59	20.47	0	0	42	149	400
Explanatory variables									
EMP in 1000 ppl.	95.53	103.06	100.32	112.59	19.90	19.80	1055.70	1203.90	400
UNEMP	9.02	6.01	4.87	2.78	2.60	1.30	23.30	15.10	400
GDP p. emp. in 1000 €	50.20	65.71	9.90	11.03	33.24	49.04	117.11	143.02	400
POPDENS	5.14	5.20	6.49	6.73	0.42	0.36	39.52	46.68	400
AGRI in 1000 ha	74.03	73.39	63.21	63	1.06	1.03	455.76	451.15	400
KNOW	10.79	10.57	7.21	7.76	0	0.40	42.80	57	400
BT_{t-1}	0.08	0.09	0.14	0.12	0	0	1	0.80	400

Source: own calculations

Table 8.5 Results of the Negative Binomial Regression, first period from 1995–2001

	Overall Projects		Bioeconomy concept		Biotechnology nucleus		Bioeconomy shell	
Intercept	1.039	*	−0.958		−1.552	*	−1.446	
EMP	0.009	***	0.058	***	0.086	***	0.149	***
UNEMP	−0.001		−0.008		−0.033	*	0.021	
GDP	0.001	**	0.013	*	0.020	*	0.027	
POP	0.003	***	0.019	*	0.034	**	0.044	
AGRI	−0.001	*	−0.008		−0.007		−0.035	
KNOW	0.000		−0.014	*	−0.012		−0.064	**
BT_{t-1}	0.001	**	0.044	***	0.074	***	0.091	***
EAST	0.004	***	0.023	*	0.043	**	0.044	
N	400		400		400		400	
AIC	3505.5		1857.4		1557.2		1217.2	

Source: own calculations
Significance levels: >0.05 ( ); <=0.05 (*); <=0.01 (**); <=0.001 (***)

In order to estimate the model, we excluded the district Berlin as a statistical outlier, which distorted the results to an unreasonable amount so that our calculations are based on the 400 remaining districts in Germany. Table 8.5 shows the estimations for the first period, which are calculated on the basis of the data from 1995–2001, the years from the beginning of the BioRegio contest and the following years. The coefficients were standardised due to their differing scales of measurement.

Hardly surprising, there is evidence that the number of people employed in a region is positively related to the quantity of projects of any kind. Additionally, existing or previous knowledge from earlier biotechnology projects has a positive linkage in every model. In general, the estimation of bioeconomy as a whole and the mere biotechnology does not vary greatly, which can be explained by the fact that the share of biotechnology measures up to 68% of the entire bioeconomy. Yet, one identifies some structural dissimilarities between both models. While a higher unemployment rate is negatively associated with the acquisition of biotechnology projects, this parameter has no significant impact on bioeconomy or the bioeconomy shell. The second difference is that projects in bioeconomy and the bioeconomy shell are more often carried out in regions where the proportion of people employed in knowledge-intensive branches is lower. Moreover, the estimation dealing with the projects within the bioeconomy shell reveals that, unlike the other models, the GDP per employee as well as the population density has no significant influence. This indicates that less densely populated districts tend to participate more often in R&D projects related to the bioeconomy shell than in biotechnology or overall projects. This demonstrates that also low-tech connoted industries are concerned in these projects. In other words, biotechnology and the overall public R&D are more likely to be executed in agglomerations, whereas R&D modelled on the bioeconomy approach are also performed in structurally weaker districts. Thus, the theoretical considerations, which led to the assumption that also rural and structurally weak

Table 8.6 Results of Negative Binomial Regression, second period from 2009–2015

	Overall projects		Bioeconomy concept		Biotechnology nucleus		Bioeconomy SHELL	
Intercept	3.019	***	0.548		−0.853		0.375	
EMP	0.004	***	0.021	***	0.046	***	0.035	***
UNEMP	0.000		−0.000		−0.003		0.001	
GDP	0.000		0.002		0.002		0.004	
POP	0.001	***	0.008	**	0.021	***	0.011	
AGRI	−0.000		−0.001		−0.002		0.000	
KNOW	−0.000		−0.005	**	−0.002		−0.014	***
BT_{t-1}	0.001	***	0.017	***	0.041	***	0.025	***
EAST	0.001	**	0.007	***	0.016	***	0.012	**
N	400		400		400		400	
AIC	4324.6		2725.9		2030		2329.9	

Source: own calculations
Significance levels: >0.05 ( ); <=0.05 (*); <=0.01 (**); <= 0.001 (***)

areas benefit from the transition of leitmotifs and applied policies, are supported by the results of the regression analysis.

The results from the first stretch of time are solidified by the estimations in the most recent period, which acknowledges the introduction of the dedicated bioeconomy strategy (NFSB) and show similar patterns as in the first period (see Table 8.6). However, since the share of the bioeconomy shell rose up to an equilibrate level of 50% of the entire bioeconomy, the extent of some determinates is diluted. Interestingly, compared to the former analysis, the determinant GDP lost its impact in any estimation. This might indicate the policy shift from a sectoral view with a strong focus to "strengthen the strong" towards a regionalised innovation policy that aims at balancing the regional economic structure. R&D within the bioeconomy shell is still located independently of population density. Although the coefficient in the bioeconomy decreased noticeably and thus points to a losing significance, it must be stated that nearly all coefficients fairly declined—most likely due to the massive gain in importance in the matter of project count from 2727 R&D projects between 1995 and 2001 to 7755 undertakings from 2009 to 2015 (the same applies to the general amount of projects independently of the field—from 22,732 to 54,452). The structural change within bioeconomy with a decreasing share of the biotechnology nucleus as well as the growing attention in terms of increasing bioeconomy projects reflects consequently the relevance of the regional implications.

8 Conclusion

This article explored the relationship between the leitmotif change from biotechnology to bioeconomy and the spatial distribution of bio-related R&D projects. The question whether and to what extent the emerging scheme bioeconomy is able to contribute to its intended objectives is highly controversial[63] and requires research on micro and macro levels. One goal according to several governmental strategy papers is the improvement of regional development in more peripheral regions by revitalising traditional branches.

Therefore, in the elaboration we proposed a disaggregation of bioeconomy into four main pillars, which are oriented at the value chain of the integral component of the bio-based economy, by name the biological resources. By means of text-mining methods, we were able to detect a consistent database that is more in line with the reality of the knowledge-based bioeconomy than previous studies that aim at the measurement of its economic scope and thereby, neglect certain segments, e.g. the socio-economic framework. In contrast, this approach disregards the comprehensive character of bioeconomy and focuses on the element that is crucial for the development of the concept. In order to achieve far-reaching changes both in an economic sense and in a matter of ecological sustainability to prevent the further exploitation of the ecosystem and nature, respectively, knowledge-driven solutions are inevitable.

The notion of the implementation of exogenous knowledge sources into bio-related industries, both in upstream and downstream industries, promises innovative solutions in order to diversify the existent economy either into related or unrelated branches, for instance along an existing trajectory or via bricolage and self-discovery adjusted to local capabilities and resources. Thus, this study showed that the R&D in the new dimensions of the bioeconomy shell, namely the input and output dimension as well as the socio-economic framework, were in contrast to biotechnology activities already located independently of its population density, even before the explicit bioeconomy strategy was introduced. Moreover, addressed dimensions are favoured in regions with a smaller share of knowledge-intensive industries. With respect to the structural change within bioeconomy at the expense of the biotechnology funding, the increasing attention of the bioeconomy shell is having a substantial impact on regional development. The opportunity to innovate off the beaten track, i.e. outside core regions as well as apart from a focus on high-tech sectors, corresponds with the theoretical considerations on modern innovation policies. These very policies stress the importance of diversifying the economic status quo, which aim at viable and long-lasting solutions by fostering linkages to create synergies between actors and sectors to prevent lock-ins and to counter regional disparities.

The role of the cross-cutting biotechnology, however, should not be underestimated, since it is supposed to be the initiator of innovations in all concerned

[63] Cf. Birch (2017) and Frenken (2017).

industries. Additionally, other general-purpose technologies, such as nanotechnology, digitalisation and artificial intelligence, are successively applied in the bioeconomy and drive its development. Overall, the diffusion and knowledge transfer into both geographically and technological distant sectors is essential to utilise available bio-themed innovation potential in any region. Because of its formalised and codified character, the biotechnology is particularly suitable to meet this requirement. It should also not be forgotten that although the share of biotechnology in the total number of projects has declined, the monetary distribution is still clearly dominated by biotechnology projects for various reasons. The estimations presented validate the crucial function of the biotechnology to attract further public R&D support and emphasise the path- and place-dependency. Whether and to what intensity collaboration between actors in the biotechnology and the bioeconomic shell are existent and fruitful is a starting point for further research.

Beyond that, it is vital and necessary to estimate the impact of undertaken policy interventions. In this paper, only the input dimension in the form of publicly funded R&D projects was considered. Studies that aim at the actual and quantifiable significance of mission-oriented public efforts to create new markets and paths are crucial for the evaluation of policy measures and the rationale to go beyond market-fixing approaches. Therefore, a possible approach to conduct research at this level would be to include an output dimension, such as patent data.

References

Aghion, P., Boulanger, J., & Cohen, E. (2011). Rethinking industrial policy. *Bruegel Policy Brief, 2011*(4), 1–8.

Albert, S. (2007). Transition to a bio-economy – A community development strategy discussion. *Journal of Rural and Community Development, 2*(2), 64–83.

Arnold, E. (2012). Understanding long-term impacts of R&D funding: The EU framework programme. *Research Evaluation, 21*(5), 332–343.

Asheim, B. T. (2007). Differentiated knowledge bases and varieties of regional innovation systems. *Innovation: The European Journal of Social Science Research, 20*(3), 223–241.

Asheim, B. T., Boschma, R., & Cooke, P. (2011). Constructing regional advantage. *Regional Studies, 45*(7), 893–904.

Balland, P.-A., Boschma, R., Crespo, J., & Rigby, D. L. (2019). Smart specialization policy in the European Union—Relatedness, knowledge complexity and regional diversification. *Regional Studies, 53*(9), 1252–1268.

Birch, K. (2017). The problem of bio-concepts. *Cultural Studies of Science Education, 12*(4), 915–927.

Branstetter, L., & Sakakibara, M. (1998). Japanese research consortia: A microeconometric analysis of industrial policy. *The Journal of Industrial Economics, 46*(2), 207–233.

Broekel, T. (2015). Do cooperative research and development (R&D) subsidies stimulate regional innovation efficiency? *Regional Studies, 49*(7), 1087–1110.

Bugge, M. M., Hansen, T., & Klitkou, A. (2016). What is the bioeconomy? *Sustainability, 8*(7), 691–713.

Carvalho, L., & Vale, M. (2018). Biotech by bricolage? *Cambridge Journal of Regions, Economy and Society, 11*(2), 275–295.

Czarnitzki, D., Ebersberger, B., & Fier, A. (2007). The relationship between R&D collaboration, subsidies and R&D performance. *Journal of Applied Econometrics, 22*(7), 1347–1366.

Czarnitzki, D., & Fier, A. (2003). Publicly funded R&D collaborations and patent outcome in Germany. *ZEW Discussion Papers, 3*(24), 1–29.

Eder, J., & Trippl, M. (2019). Innovation in the periphery: Compensation and exploitation strategies. *Growth and Change, 6*(2), 1–21.

European Commission (EC). (2012). *Innovating for sustainable growth: A bioeconomy for Europe.*

European Commission (EC). (2018). *A sustainable bioeconomy for Europe: Strengthening the connection between economy, society and the environment. Updated bioeconomy strategy.*

Federal and State Statistical Offices Germany. (2019). *Regionaldatenbank Deutschland.* Retrieved January 24, 2019, from, https://www.regionalstatistik.de/genesis/online/

Federal Ministry of Education and Research. (n.d.). *Förderkatalog.* Retrieved April 5, 2017, from, https://foerderportal.bund.de/foekat/jsp/StartAction.do?actionMode=list

Federal Ministry of Education and Research. (2010). *Nationale Forschungsstrategie BioÖkonomie 2030. Unser Weg zu einer bio-basierten Wirtschaft.* Retrieved May 15, 2017, from, https://www.bmbf.de/pub/Nationale_Forschungsstrategie_Biooekonomie_2030.pdf

Federal Ministry of Education and Research. (2011). *Ergebnisbericht: Evaluation des Rahmenprogramms Biotechnologie.* Retrieved June 22, 2018, from, https://www.bmbf.de/pub/evaluation_rahmenprogramm_biotechnologie.pdf

Federal Ministry of Education and Research. (2017). *Fortschritt durch Forschung und Innovation. Bericht zur Umsetzung der Hightech-Strategie.* Retrieved July 20, 2020, from, https://www.kmc-chemnitz.de/files/cto_layout/img/Inhalt/download/2017_Fortschritt_durch_Forschung_und_Innovation.pdf

Federal Ministry of Education and Research; Federal Ministry of Food and Agriculture. (2020). *Nationale Bioökonomiestrategie - Kabinettversion.* Retrieved July 27, 2020, from, https://www.bmbf.de/files/bio%c3%b6konomiestrategie%20kabinett.pdf

Federal Ministry of Food and Agriculture. (2014). *Nationale Politikstrategie Bioökonomie. Nachwachsende Ressourcen und biotechnologische Verfahren als Basis für Ernährung, Industrie und Energie.* Retrieved July 20, 2020, from, https://biooekonomie.de/sites/default/files/publications/npsb_0.pdf

Federal Ministry of the Interior, Building and Community. (2019). *Unser Plan für Deutschland. Gleichwertige Lebensverhältnisse überall.* Retrieved September 13, 2019, from, https://www.bmi.bund.de/SharedDocs/downloads/DE/veroeffentlichungen/themen/heimat-integration/gleichwertige-lebensverhaeltnisse/unser-plan-fuer-deutschland-langversion-kom-gl.pdf;jsessionid=7E9351E74C1F078BF3450E700E76311F.2_cid287?__blob=publicationFile&v=4

Frenken, K. (2017). *A complexity-theoretic perspective on innovation policy. In Complexity, governance & networks – special issue: Complexity, innovation and policy* (pp. 35–47). University of Bamberg Press.

Garud, R., & Karnøe, P. (2003). Bricolage versus breakthrough. *Research Policy, 32*(2), 277–300.

German Bundestag. (1990). *Faktenbericht 1990 zum Bundesbericht Forschung 1988.* Printed Matter 11/6886.

Grillitsch, M., & Nilsson, M. (2017). Firm performance in the periphery: On the relation between firm-internal knowledge and local knowledge spillovers. *Regional Studies, 51*(8), 1219–1231.

Halle Institute for Economic Research – Member of the Leibniz Association. (2019). *Vereintes Land - drei Jahrzehnte nach dem Mauerfall.* Retrieved October 25, 2019, from, https://www.iwh-halle.de/publikationen/detail/Publication/vereintes-land-drei-jahrzehnte-nach-dem-mauerfall/

Hassink, R. (2010). Regional Resilience. *Cambridge Journal of Regions, Economy and Society, 3*(1), 45–58.

Hausmann, R., & Rodrik, D. (2003). Economic development as self-discovery. *Journal of Development Economics, 72*(2), 603–633.

HM Government. (2018). *Growing the bioeconomy. Improving lives and strengthening our economy: A national bioeconomy strategy to 2030*. Retrieved January 21, 2019, from, https://assets.publishing.service.gov.uk/government/uploads/system/uploads/attachment_data/file/761856/181205_BEIS_Growing_the_Bioeconomy__Web_SP_.pdf

Hüsing, B., Kulicke, M., Wydra, S., Stahlecker, T., Aichinger, H., & Meyer, N. (2017). *Evaluation der "Nationalen Forschungsstrategie BioÖkonomie 2030"*. Retrieved July 20, 2020, from, https://www.isi.fraunhofer.de/content/dam/isi/dokumente/cct/2017/Evaluation_NFSB_Abschlussbericht.pdf

Isaksen, A. (2015). Industrial development in thin regions. *Journal of Economic Geography, 15*(3), 585–600.

Koschatzky, K. (2005). The regionalization of innovation policy: New options for regional change? In G. Fuchs & P. Shapira (Eds.), *Rethinking regional innovation and change. Path dependency or regional breakthrough?* Economics of Science, Technology and Innovation (Vol. 30, pp. 291–312).

Lazonick, W., & Tulum, Ö. (2011). US biopharmaceutical finance and the sustainability of the biotech business model. *Research Policy, 40*(9), 1170–1187.

Marrocu, E., Paci, R., & Usai, S. (2013). Proximity, networking and knowledge production in Europe. *Technological Forecasting and Social Change, 80*(8), 1484–1498.

Mazzucato, M. (2014). *The entrepreneurial state: Debunking public vs. private sector myths. Penguin Books.*

McCann, P., & Ortega-Argilés, R. (2013). Modern regional innovation policy. *Cambridge Journal of Regions, Economy and Society, 6*(2), 187–216.

McCann, P., & Ortega-Argilés, R. (2015). Smart specialization, regional growth and applications to European Union cohesion policy. *Regional Studies, 49*(8), 1291–1302.

Ministry of Economic Affairs and Employment (MEAE). (2014). *The finnish bioeconomy strategy*. Retrieved January 21, 2019, from, http://biotalous.fi/wp-content/uploads/2014/08/The_Finnish_Bioeconomy_Strategy_110620141.pdf

Organisation for Economic Co-operation and Development (OECD). (2004). *Biotechnology for sustainable growth and development*. Retrieved February 2, 2018, from, https://www.oecd.org/sti/biotech/23536372.pdf

Organisation for Economic Co-operation and Development (OECD). (2009). *The bioeconomy to 2030. Designing a policy agenda*. Retrieved July 22, 2020, from, https://read.oecd-ilibrary.org/economics/the-bioeconomy-to-2030_9789264056886-en#page1

Organisation for Economic Co-operation and Development (OECD). (2011). *Regions and innovation policy*.

Organisation for Economic Co-operation and Development (OECD). (2018): Meeting policy challenges for a sustainable bioeconomy.

Patermann, C., & Aguilar, A. (2018). The origins of the bioeconomy in the European Union. *New Biotechnology, 40*(Pt A), 20–24.

Schüler, J. (2016). *Die Biotechnologie-Industrie*. Springer Spektrum, Berlin, Heidelberg.

Schumpeter, J. A. (2005). *Capitalism, socialism and democracy* [1942].

Soete, L. L. G., & ter Weel, B. J. (1999). Innovation, knowledge creation and technology policy. *De Economist, 147*(3), 293–310.

Staehler, T., Dohse, D., Cooke, P. (2006). *Evaluation der Fördermaßnahmen BioRegio und BioProfile*. Retrieved September 5, 2018, from, https://www.e-fi.de/fileadmin/Evaluationsstudien/Evaluation_der_Foerdermassnahmen_BioRegio_und_BioProfile.pdf

The White House. (2012). *National bioeconomy blueprint*.

Tödtling, F., & Trippl, M. (2005). One size fits all? Towards a differentiated regional innovation policy approach. *Research Policy, 34*(8), 1203–1219.

Uhlbach, W.-H., Balland, P.-A., & Scherngell, T. (2017). R&D policy and technological trajectories of regions: Evidence from the EU framework programmes. In *Working Paper Series: Papers in Evolutionary Economic Geography* #17.22 (pp. 1–21).

Viaggi, D. (2016). Towards an economics of the bioeconomy: Four years later. *Bio-based and Applied Economics, 5*(2), 101–112.

Warmuth, E. (1991). Biotechnology 2000: A new German R&D programme. *Research Evaluation, 1*(2), 79–88.

Zeileis, A., Kleiber, C., & Jackman, S. (2008). Regression models for count data in R. *Journal of Statistical Software*, 27/8.

Problem Structures of Bioenergy Policy in the Power and Heat Sector in Germany

9

Katrin Beer

Abstract

The nature of problems and how they are perceived (problem structures) are decisive factors for the course and outcome of a political process. For this reason, the analysis of the properties of problems is a subject of the scientific sub-discipline policy analysis. Problems can be classified according to different categories, such as complexity (simple, complex, wicked, superwicked problems), long-term or short-term nature or based on their importance for the economy. This article initially gives an overview of different theoretical classification systems from diverse disciplines such as political science, management science, economics and design. Following this, the problem structures of bioenergy policy in the power and heat sector in Germany are analyzed based on these theoretical considerations and based on the results of an empirical case study, which has been conducted in the frame of the research project "Bio-Ecopoli". The aim of the contribution is to uncover causal mechanisms in political processes of bioenergy policy in Germany and to better understand this regulatory area, which is regarded as a sub-area of bioeconomy policy here. The analysis shows that problems of bioenergy policy can be classified as wicked problems due to their high level of complexity, the unclear definition of the problem and the differing views on suitable political solutions. According to

The original version of the chapter was revised. Fig. 9.2, Fig. 9.3 and Fig. 9.4 have been replaced with the correct versions. The correction to this chapter can be found at https://doi.org/10.1007/978-3-030-87402-5_21

K. Beer (✉)
Department of Political Science and Sustainable Development, University of Magdeburg, Magdeburg, Germany
e-mail: katrin.beer@ovgu.de

D. Lanzerath et al. (eds.), *Bioeconomy and Sustainability*,
https://doi.org/10.1007/978-3-030-87402-5_9

the results of the analysis, three main factors will influence political processes of bioenergy policy in the future: Acceptance, distribution of power and framing.

Keywords

Bioenergy policy · Wicked problems · Policy analysis

1 Introduction: Problems in Bioeconomy Policy

This article deals with the role of problem structures in the concrete design of bioeconomy policy in Germany from a political science perspective and aims to highlight specific characteristics of this policy field. Thus, some potential conflicts as well as starting points for a targeted and effective bioeconomy policy will be identified.

The central object of research in political science is the way in which conflicts of interest are (or should be) negotiated in a society and how social cohabitation is (or can be) organised by binding regulations.[1] In this sense, the term "bioeconomy policy" refers to the totality of political measures (*policy*), political processes (*politics*) and political structures (*polity*) that relate to the political regulation of the bioeconomy.[2] In the broadest sense, it thus includes all collectively binding regulations that regulate the economic use (material and energetic) of biomass. There is an overlap with policy fields that address the origin of biomass, i.e. its production (agricultural policy, forestry policy) or the use of residual and waste materials (waste policy, wastewater policy), as well as with policy fields that regulate production processes (industrial policy, economic policy) or the use of concrete end products (energy policy, plastics policy).

A political problem is always present when real social facts do not correspond to the actors' normative ideas of living together in a society.[3] For the description of characteristics of political problems, the term "problem structures" has become established in policy analysis, which is the branch of political science that deals with concrete political processes and political measures in the various policy fields.[4] This article analyses the problem structures of bioeconomy policy using the example of bioenergy policy in the power and heat sector in Germany and aims to answer the following question: Which characteristic problem structures can be identified and to what extent do they influence the political processes (*politics*) and contents (*policy*) in this policy area? The data basis is provided by qualitative case studies that were carried out at the Otto-von-Guericke University Magdeburg within the framework of the joint project Bio-Ökopoli, funded by the Federal Ministry of Education and Research (BMBF).[5]

[1] Cf. Kevenhörster (2008, pp. 13–24) .

[2] Cf. Blum and Schubert (2018, pp. 9–12).

[3] Cf. Reiter and Töller (2014, p. 26).

[4] Cf. Böcher and Töller (2012b, pp. 19–26) and Reiter and Töller (2014, pp. 25, 94–106).

[5] Cf. Beer et al. (2018) and FernUniversität in Hagen (2019).

2 Problem Structures: Analysing the Characteristics of Problems

The focus of policy analysis lies on the examination of concrete policy content. The term "policy" can stand for a single political measure on the one hand, and for a policy field as a whole on the other hand.[6] In this paper, the role of problem structures of bioenergy policy in the political process and their influence on the concrete contents of political measures are considered from a policy analysis perspective.

Policy analysis research can be divided into Y- and X-centred studies. Y-centred studies focus on a specific policy as the result of a political process (*dependent variable*) and aim to explain their content design by considering different possible causes (*independent variables*) and identifying the determining factors. X-centred studies, on the other hand, focus on one explanatory factor and ask what different effects this variable can have. Studies on the effects of individual policies are also typical in policy analysis (Fig. 9.1).[7]

For the analysis of politics and policies, various theoretical approaches can be used in policy analysis, such as the policy cycle, the political process inherent dynamics approach (PIDA),[8] the multiple streams approach (MSA), the garbage-can model, the Institutional Analysis and Development Framework (IAD) or the actor-centred institutionalism (AZI). These approaches are based on different basic assumptions regarding problem-solving in political processes.[9] For the description and analysis of problems, this paper mainly draws on three approaches: The

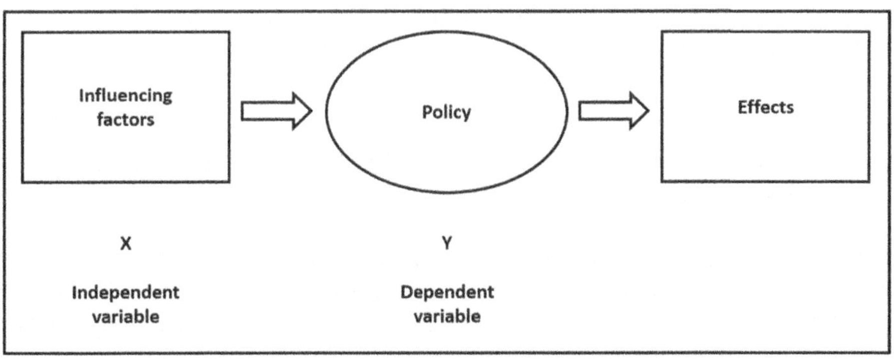

Fig. 9.1 The political process, own illustration

[6]Cf. Blum and Schubert (2018, pp. 10–12).

[7]Cf. Ibid., 39–43 and Reiter and Töller (2014, pp. 15–26).

[8]Cf. Böcher and Töller (2015). German name for PIDA: *Ansatz eigendynamischer politischer Prozesse* (AEP).

[9]Cf. Böcher and Töller (2012b, pp. 178–199) and Reiter and Töller (2014, pp. 26–106).

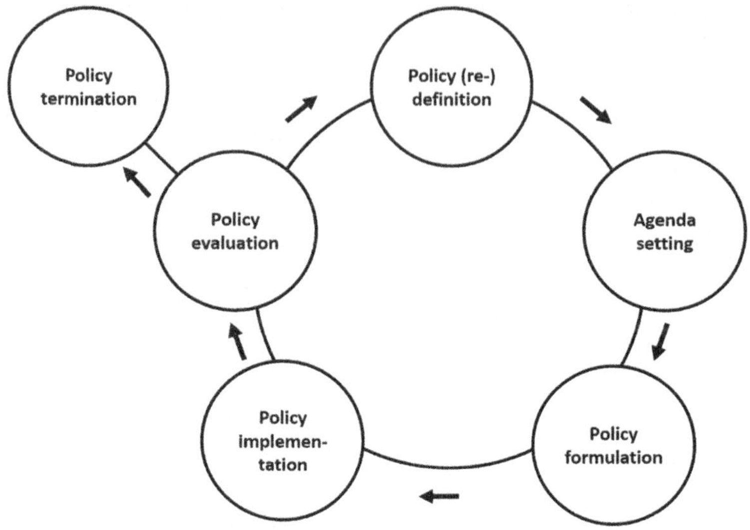

Fig. 9.2 The policy cycle, own illustration based on Jann and Wegrich (2014, p. 106)

theoretical basis for the analysis is formed by the phases of the policy cycle (first approach)[10] and by the categorisation of problems in the explanatory factor problem structures in the PIDA (second approach).[11] These considerations are supplemented by theoretical reflections in which problems are divided into three types according to their complexity (third approach): *Simple (type 1)*, *complex (type 2)* and *wicked problems (type 3)*.[12] This distinction into types is supplemented by further considerations of solution options and possible effects of solutions.

2.1 Political Processes as Problem-Solving Processes: The Policy Cycle

Political processes can be understood as problem-solving processes in which political actors find the best possible solutions to political problems. This basic assumption is the basis of the policy cycle, a popular approach that is often used in policy analysis (Fig. 9.2).[13]

With the policy cycle, political processes can be pre-structured for analysis by dividing them into five conceptual phases, which can lead to a problem solution. The policy cycle assumes that problems must be defined in a first step (*Problem*

[10] Cf. Jann and Wegrich (2014).

[11] Cf. Böcher and Töller (2012a, 2012b) and Reiter and Töller (2014).

[12] Cf. Roberts (2000).

[13] Cf. Jann and Wegrich (2014).

definition) so that they can be processed and solved in the subsequent phases. A further prerequisite for problem solving is that political actors put a problem on the political agenda (*Agenda setting*). However, the approach does not provide any analytical tools for the analysis of the factors that determine whether and why this happens. *Policy formulation* takes place in the third phase of the policy cycle and *Policy implementation* follows in the fourth phase. Whether a policy has the desired effects must be verified by a *Policy evaluation* in the fifth phase. The evaluation can determine whether the problem has been solved. If so, the policy cycle ends (*Policy termination*). If the problem could not be solved, the cycle starts again with a redefinition of the problem in phase 1 (*Problem redefinition*).[14]

There are three main points of criticism of the policy cycle: first, the ideal-typical phases of a political process do not occur in an orderly sequence in reality; second, the approach only offers the possibility of structuring phenomena, but not of explaining them; and third, the approach carries a problem-solving bias in itself, since it places problem-solving at the centre of consideration. However, other factors can also have a decisive influence on political processes.[15] Despite these weaknesses, the policy cycle remains a useful tool for the analysis of political problems, their characteristics and their role in political processes.

2.2 Problem Structures in the Political Process Inherent Dynamics Approach (PIDA)

The PIDA[16] aims to uncover causal mechanisms in political processes. It was developed by Michael Böcher and Annette Elisabeth Töller for the analysis of political processes and measures of environmental policy and it challenges the basic assumption of problem-solving orientation in political processes.[17] As a counter-concept, it assumes an interplay of five explanatory factors that can have a decisive influence on politics and policies. According to PIDA, each individual explanatory factor as well as the political processes as a whole is subject to inherent dynamics that result from the mutual influence of the five explanatory factors (see Fig. 9.3). Decided policies in turn have an impact on the following political processes.

The characteristics of problems (*Problem structures*) and the discussion of possible solution options (*Instrument alternatives*) are two of these five explanatory factors. However, PIDA focuses primarily on the explanatory factor *Actors and their actions*. Actors can be influenced by political framework conditions (*Institutions*) and suddenly occurring events (*Situational aspects*). Overall, PIDA comprises these

[14] Cf. Böcher and Töller (2012b, pp. 181–184) and Jann and Wegrich (2014).

[15] Cf. Böcher and Töller (2012b, pp. 182 f).

[16] Cf. Böcher and Töller (2015).

[17] Cf. Böcher and Töller (2012a, 2012b) and Reiter and Töller (2014).

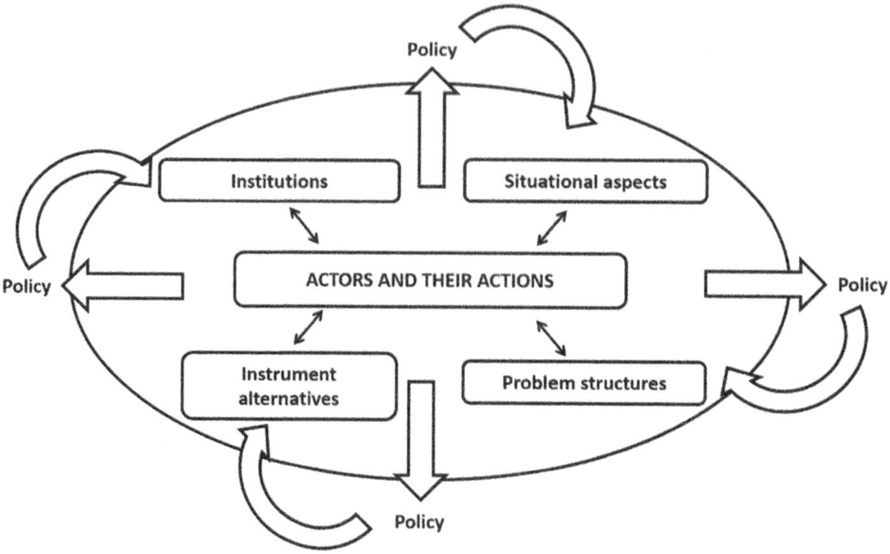

Fig. 9.3 The political process inherent dynamics approach (PIDA), own illustration based on Böcher and Töller (2015, p. 10)

five explanatory factors, which influence each other and determine the course of political processes and the policies that emerge from them.[18]

The analysis of problem structures aims to explain the course of political processes on the one hand. On the other hand, it aims to find answers to the question of the extent to which policy fields differ structurally from one another with regard to the specific characteristics of their problems.[19] As far as the results of a political process are concerned, policy analysis distinguishes between *output* (concrete political measure, policy) and *outcome* (effect of the policy). The analysis of problem structures in PIDA includes the consideration of the characteristics of a problem and possible options for its solution (policies, output) as well as the consequences of possible solutions (outcome).[20]

2.2.1 Analysis of the Characteristics of a Problem

Problems can be differentiated along different dimensions, such as complexity, visibility, characteristics of the problem-causers or even on the basis of possible approaches to solving the problem. For political processes and the design of policies, additional aspects play a decisive role: the short- or long-term nature of a problem, whether there is reliable or uncertain knowledge about the nature of the problem, and whether or not there is a uniform definition of the problem. Furthermore, problems

[18]Cf. Böcher and Töller (2012b, pp. 178–199) and Reiter and Töller (2014, pp. 87–106).

[19]Cf. ibid., 89–98.

[20]Cf. ibid., 25.

Table 9.1 Dimensions of problems, own overview based on Böcher and Töller (2012b, pp. 89–98)

• Visibility • Clarity • Importance for the economy and competition • Number, diversity and social significance of the causes of the problems • Specific actors can be identified as the cause of the problem • Information situation: Scientific knowledge of the problem and solutions • Availability of solutions • Redistribution of costs and benefits (distributive/redistributive) • Time pressure/pressure to act despite uncertainty	• Public goods/private goods • Rivalry for resources • Exclusion/affection/organisation/interests of specific actors or groups • Time dimension: Short-term, long-term problems and actor interests • Uncertainty with regard to the causes of problems and effects of policies • Persistence of (environmental) problems • Cross-sectional character • Responsibility, political level • Acceptance of solution options • Instrumentalisation of existing knowledge in case of uncertain knowledge situation

can be distinguished according to whether a clear identification of the originators is possible or whether there are diffuse connections between supposed cause and supposed effect.[21]

When there is a large amount of diffuse causes and when technical solutions are missing or differently assessed, it is particularly difficult to deal with problems. Such problem structures are typical for environmental policy and they are referred to as persistent environmental problems.[22] These include, for example, problems with a cross-sectional character, such as climate protection, land use or the protection of biodiversity, which require an integrated and intersectoral solution through the coordinated cooperation of different social groups or political departments. These problems with diffuse cause-and-effect relationships can be exacerbated by the division of policy into sectors.[23] With regard to such persistent environmental problems, the coordination of the sectors involved (*policy integration*) therefore plays an increasingly important role in environmental policy, although this is extremely difficult due to the individual approaches and interests of the departments involved (Table 9.1).[24]

2.2.2 Analysis of Possible Solution Options

According to the considerations above, problems can be typified on the one hand by the problem definition and on the other hand by possible solutions. It may be necessary to make political decisions, although little is known about the characteristics of a problem and the possible consequences of individual political

[21] Cf. Böcher and Töller (2012b, pp. 89–94).

[22] Cf. ibid., 89–95 and Jänicke and Volkery (2001).

[23] Cf. Böcher and Töller (2012b, p. 95).

[24] Cf. ibid., 71–74.

Table 9.2 Dimensions of impacts, own overview based on Böcher and Töller (2012b, 89–98)

• Distributive/redistributive • Known/unknown • Desired/unwanted	• Direct/indirect • Second order problems

measures.[25] Which solution options are discussed in a political process and which are ultimately implemented in the form of a policy with certain political instruments depends, among other things, on the understanding of the problem by the actors involved. The higher the complexity of a problem, the more difficult it is to negotiate suitable solutions. If simple, technology-oriented, sectoral solutions that are recognised by all actors are not available for the solution of a problem, the problem is politically difficult to communicate and it will therefore more likely be avoided by political actors. If solution options require a change in personal lifestyles, the associated problems also tend to not be addressed.[26]

If there is a high pressure to act, which makes it impossible to not address a complex problem, *symbolic politics* can be the consequence. A high degree of complexity and uncertainty when dealing with political problems tends to lead to value-based action by actors, since it is hardly possible to deal with problems in a knowledge-based, objective and rational manner. In such situations, political actors select and instrumentalise existing, sometimes contradictory scientific findings according to their own interests. The perception of problems and solutions can change dynamically: Solutions often become a problem themselves and later again develop into solutions to other problems.[27]

2.2.3 Analysis of the Consequences of Possible Problem Solutions

Problem solutions in terms of political measures can be divided into *redistributive* and *distributive policies* with regard to their effects (*outcome*). This distinction refers to a redistribution of costs and benefits among social groups that are reorganised by a political measure. With redistributive policies, costs and benefits are redistributed in such a way that one social group is better off at the expense of another social group. Distributive policies improve the position of social groups without another group having to bear higher costs. Furthermore, solution options can be categorised according to whether and which (un)known (un)desired or (in)direct effects can be expected and whether the solution itself leads to a new problem (*second-order problem*, see Table 9.2).[28]

[25] Cf. ibid., 93.

[26] Cf. ibid., 194–199.

[27] Cf. ibid.

[28] Cf. ibid., 89 f. and Beer et al. (2018, p. 34).

Table 9.3 Types of problems according to Roberts

Type 1: Simple problems	• Consensus on problem definition • Consensus on solution options • Little or no conflict between actors involved • Simple and fast problem solving with known and standardised procedures possible
Type 2: Complex problems	• Consensus on problem definition • Dissent on solution options • Conflict between actors involved with regard to appropriate problem solutions • Complex problem-solving process due to different points of view
Type 3: Wicked problems	• Dissent on problem definition • Dissent on solution options • High level of conflict between the actors involved in terms of understanding the problem and finding appropriate solutions • Problems and solutions are constantly being redefined due to political restrictions and scarce resources • Preferences of society are constantly shifting, interested parties come and go

2.3 Types of Problems: Simple, Complex, Wicked Problems

Problems can be typified by their degree of complexity. Nancy Roberts[29] suggests a division into three types of problems: Type 1 comprises *simple problems* for which there is consensus on the problem definition and appropriate solutions. Type 2 includes *complex problems* for which there is consensus only on the problem definition but there are different opinions on appropriate solutions. Type 3 problems (*wicked problems*) are those for which there is no consensus neither on the problem definition nor on possible solutions, which further increases the degree of complexity and creates a greater potential for conflict (see Table 9.3).[30]

According to Roberts, wicked problems are characterised by the following properties: There is no common understanding of what the problem actually is, the problem thus is not clearly defined. Because of the unclear definition of the problem, there are different framings of the problem and different opinions about what might be appropriate solutions. The problem-solving process is also complex because the framework conditions are constantly changing; and the framework conditions are constantly changing because the actors involved in the process change, because involved actors change their opinions or the "rules of the game", or because communication between the individual actors is not successful.

Jeffrey Conklin and William Weil[31] stress that *social acceptance* plays a decisive role in solving wicked problems and dealing with them. Due to the vagueness of the

[29] Cf. Roberts (2000).

[30] Cf. ibid., 1 f.

[31] Cf. Conklin and Weil (2007).

problem definition, a solution must be found with which all involved actors are as satisfied as possible, even if they have different views on what the problem actually is. Roberts points out that wicked problems cannot be solved with traditional problem-solving strategies. She distinguishes three strategies for dealing with wicked problems along the distribution of power among the stakeholders: Authoritative, collaborative and competitive strategies.[32] According to these considerations, the decisive factors in negotiation processes for wicked problems are not primarily problem definitions and problem solutions, but the *interests of involved actors* and the *distribution of power*.

From the above considerations it becomes clear that problem structures have an influence on whether problems are addressed at all and on whether they are placed on the political agenda. Problems that cannot be solved easily with available solutions are politically difficult to communicate. The same applies to problems that require a change of lifestyle to solve them. Complicated problems without simple solutions that lead to social conflicts of interest are more likely to be avoided or to lead to symbolic politics. Such problems are also known as *wicked problems* and are considered as typical for environmental policy.[33]

3 Problem Structures of Bioenergy Policy in Germany

Within the framework of the Bio-Ökopoli research project,[34] qualitative case studies on bioeconomy policy have been carried out in the subject areas of bioplastics, biofuels (transport sector) and bioenergy (power and heat sector). The analysis was based on PIDA.

Selected policies (*cases*) in these three thematic fields (*case groups*) were the starting point for the analysis of political processes that can be assigned to bioeconomy policy. Cases on European, national and municipal level have first been analysed in a case-related analysis and after that they have been compared in a cross-case analysis. Based on the results, causal mechanisms could be identified and inherent dynamics could be shown. Thus, the course of political processes can now be explained based on empirical results for the individual selected cases, for the case groups and, to some extent, for bioeconomy policy as a whole.[35] The starting point for the following explanations are the case studies in the field of bioenergy (power and heat) at national level. This comprises the results of case studies on the Renewable Energy Sources Act (*Erneuerbare-Energien-Gesetz,* EEG) and the Renewable Energies Heat Act (*Erneuerbare-Energien-Wärmegesetz,* EEWärmeG).[36]

[32] Cf. Roberts (2000, pp. 3–7).

[33] Cf. Roberts (2000), Balint et al. (2011) and Böcher and Töller (2012b, p. 198).

[34] Cf. Beer et al. (2018) and FernUniversität in Hagen (2019).

[35] Perbandt et al. (2021), Vogelpohl et al. (2021), Töller et al. (2021), Otto et al. (2021) and Böcher et al. (2020).

[36] Cf. Beer et al. (2018) and FernUniversität in Hagen (2019).

Table 9.4 Expert interviews for the case group bioenergy at national level (Bio-Ökopoli)

	Affiliation of interviewed expert	Social group
1	Bundesministerium für Ernährung und Landwirtschaft (BMEL) (Federal Ministry of Food and Agriculture)	Politics/ administration
2	Bundesverband Bioenergie e. V. (BBE) (Bioenergy Association)	Economy/ industry
3	BiogasRat+ e. V. (Biogas Association)	Economy/ industry
4	Fachverband BIOGAS (German Biogas Association)	Economy/ industry
5	Fichtner GmbH & Co. KG (Engineering Company)	Economy/ industry
6	Freelance engineer in the field of wastewater treatment	Economy/ industry
7	Naturschutzbund Deutschland e. V. (NABU) (Nature and Biodiversity Conservation Union)	Civil society/ NGOs
8	Kuratorium für Technik und Bauwesen in der Landwirtschaft e. V. (KTBL) (Advisory Board in the Field of Agriculture)	Academia/ consulting
9	Institut für ZukunftsEnergie- und Stoffstromsysteme gGmbH (IZES) (Research Institute)	Academia/ consulting
10	Hochschule Zittau/Görlitz (HSZG) (Zittau/Goerlitz University of Applied Sciences)	Academia/ consulting
11	Forstfachverlag GmbH & co. KG (Specialist Publisher)	Others

For each selected case, document analyses of primary sources and theory-based qualitative content analyses of guideline-based expert interviews have been conducted.[37] The expert interviews were designed to reflect the diverse perspectives and interests of bioenergy stakeholders from different social groups (see Table 9.4).

3.1 Bioenergy Policy as Part of Energy Policy

Bioenergy policy can be conceptually understood as a sub-area of bioeconomy policy. However, as far as political practice and binding political measures are concerned, bioeconomy policy as a policy field has hardly been institutionalised in Germany to date, if at all. Although there are bioeconomy strategies[38] of the German Federal Government, the regulation of those sectors of the economy which today produce, process and use biomass materially or energetically takes place in other, more traditional policy fields, such as agricultural policy, forestry policy, waste management policy or energy policy.[39]

[37] Cf. Beer et al. (2018) for methodological details.

[38] Cf. Federal Ministry of Education and Research (2010), Federal Ministry of Food and Agriculture (2014) and Federal Ministry of Education and Research and Federal Ministry of Food and Agriculture (2020).

[39] Cf. Beer et al. (2018).

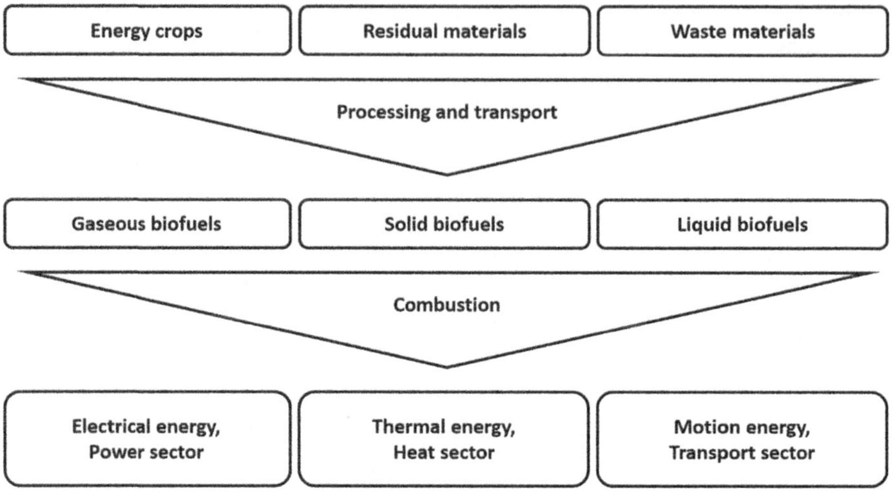

Fig. 9.4 Biomass supply chain in the energy sector, own illustration based on Kaltschmitt et al. (2016, 4)

In the Bio-Ökopoli research project, bioenergy policy is hence understood as a sub-area of German energy policy for the empirical case studies. The focus is thus neither on regulating the origin of biomass nor on regulating the processing of biomass into *solid*, *liquid* or *gaseous biofuels*, but rather on regulating the energetic use of biomass at the end of the supply chain (see Fig. 9.4). The policy field of bioenergy policy can be further subdivided into the three sectors (*power*, *heat* and *transport*) according to the corresponding energy use.[40]

In Germany, energy policy and climate protection policy are closely interlinked.[41] According to its Climate Action Plan 2050,[42] the German government aims to achieve greenhouse gas neutrality to the greatest possible extent by 2050 in order to make its contribution to achieving the climate protection goals set out in the Paris Agreement. By 2030, greenhouse gas emissions in Germany are to be reduced by 55% (base year 1990). In order to achieve this goal, the energy industry has to develop a greenhouse gas-neutral energy system by increasing efficiency and shifting energy production to renewable energies (*Energiewende*). However, the Sixth Energy Transition Monitoring Report[43] shows that it is unlikely that Germany will achieve the interim targets for 2020 and that there is therefore considerable need for action in energy and climate policy.

---

[40]Cf. ibid.

[41]Cf. Federal Ministry for Economic Affairs and Energy (2018).

[42]Cf. Federal Ministry for the Environment, Nature Conservation and Nuclear Safety (2016, pp. 28 f).

[43]Cf. Federal Ministry for Economic Affairs and Energy (2018).

3.2 Bioenergy in the (Renewable) Energy System

Germany's energy system is to be largely converted in the coming decades from a system that is mainly based on non-renewable energies (fossil and nuclear) to a system that is based on renewable energies. Renewable energies include five forms of energy: Solar energy, wind energy, hydro energy, geothermal energy and bioenergy. In order to produce bioenergy from biomass, i.e. organic material originating from *plants*, *animals* or *microorganisms*, the biomass in a first step needs to be converted into *solid*, *liquid* or *gaseous biofuels*. These fuels can then be burned (*combustion*) to generate power (*electrical energy*), heat (*thermal energy*, also including cold) or *motion energy* (see Fig. 9.4).

The process of bioenergy generation releases carbon dioxide (CO_2) into the atmosphere from carbon which has previously been bound in the biomass used. Compared to other renewable energies, bioenergy on the one hand is relatively expensive due to the fuel costs, but on the other hand it offers the advantage that it can be stored and therefore is continuously available. In order for bioenergy to be greenhouse gas-neutral regarding its carbon balance, the same amount of greenhouse gases which are released into the atmosphere by burning biofuels must be bound in a given period, for instance through afforestation or energy crop cultivation.[44]

The importance of bioenergy in the three sectors power, heat and transport differs significantly. The share of renewable energies in gross electricity generation in 2018 was 35.0%. 49.4% of renewables in the power sector came from wind energy, 20.5% from photovoltaics and only 22.7% from bioenergy. The share of bioenergy in total gross electricity generation was 8.1%, whereas the share of bioenergy in heat supply from renewable energies was 86.2% in the same year (geothermal/ environmental heat: 8.6%, solar thermal: 5.2%). In the transport sector, 88% of renewable fuels came from biogenic sources, the other 12% were covered by electrical energy (energy sources not defined in more detail).[45]

3.3 Results of the Case Studies: Problem Structures of Bioenergy Policy

The explanations above illustrate the diversity and complexity of the problem area of bioenergy. The effects of this aspect on the political processes and what other factors characterise the problem structures of bioenergy policy will be explained in more detail below based on the results from the Bio-Ökopoli case studies on the EEG and the EEWärmeG.[46] The key characteristics of bioenergy policy in the power and heat sector in Germany, which were described in the expert interviews, are listed in Table 9.5.

[44] Cf. Kaltschmitt et al. (2016, pp. 1–123).

[45] Cf. Fachargentur Nachwachsende Rohstoffe e. V. (2019).

[46] See Table 9.4: Expert interviews for the case group bioenergy at national level.

Table 9.5 Key characteristics of bioenergy policy in the power and heat sector in Germany

Bioenergy policy in the power sector	Bioenergy policy in the heat sector
Focus of the *Energiewende* in the past	"Sleeping giant"
Mainly gaseous biofuels (biogas, biomethane, wood gas), electricity generation in power plants or combined heat and power units (CHP, more efficient)	Mainly solid biofuels (wood), also biomethane, bioenergy mostly used in the building sector and for industrial process heat
Complex legislation, extremely difficult to comprehend	Legislation less differentiated, more manageable
Advantages: System services, flexible, storable, suitable for baseloads	Advantages: Provides biggest part of renewable energy in this sector, technology available, storable
Critical points: Comparably expensive, decrease of state support, politically not wanted in the power sector, limited availability, biodiversity loss due to monocultures	Critical points: Comparably high particle emissions of solid biofuels, limited availability, conflicts with nature conservation and recreational use of forests
Strong support in the past, then strong criticism of bioenergy due to *Vermaisung*, political adjustment after emotional debate, negative image of bioenergy	Rather neglected in past debates, disadvantages for biofuels compared to fossil fuels in legislation in the building sector
Recent debates and potentials: Energetic use of manure, end of EEG support scheme, lack of support for gasification (cf. concerning wood gas), storage capability and suitability for baseloads of renewables	Recent debates and potentials: Sustainability assessment, climate change adaption in the forestry sector, nature conservation and recreational use of forests versus economic use, biomethane, use of natural gas versus solid biofuels, air pollution

Cf. Summary of expert interview statements in the project Bio-Ökopoli; EEG and EEWärmeG case studies.
Energiewende: German buzzword for the shift of the energy system towards the sustainable use of renewable energies, see chap. 3.1.
Vermaisung: German buzzword for the debate about overcultivation of maize, see chap. 3.3.2.

In the following subchapters, the statements on problem structures of bioenergy policy from the expert interviews are summarised with reference to the theoretical approaches presented above.

3.3.1 Political Support of Bioenergy as Problem Solution

In 2000, Germany adopted the EEG, a political instrument to promote renewable energies in the power sector, which led to a strong upswing of bioenergy (mainly biogas). At that time, the use of biomass for energy production and its cultivation in agriculture and forestry were framed mostly positively. An expansion of the use of bioenergy, especially the generation of power from biogas from agriculturally produced energy crops, was to contribute to rural development, decentralisation and defossilisation of the energy system and thus also to climate protection. Due to its versatility, the promotion of bioenergy was considered a solution strategy that addressed several problems simultaneously at the beginning of the millennium. After its introduction, the EEG has been amended several times and adapted to current developments at relatively short intervals.

In the heating sector, a political measure to promote renewable energies has been introduced several years later with the adoption of the EEWärmeG in 2009. In recent years, numerous other energy policy measures have been introduced at European and national level, which continue to have a direct or indirect impact on the production and use of bioenergy.[47]

In principle, bioenergy can be obtained from a variety of sources, processed into a wide range of biofuels and used in all sectors. In public debate, however, the term bioenergy usually refers to the production of electrical energy from biogas in the power sector or to the use of liquid biofuels[48] in the transport sector. These forms of bioenergy have been promoted by the German government and the European Union, which has led to several direct and indirect negative (environmental) effects in the past 15 years.

3.3.2 Effects of Bioenergy Promotion as a Second-Order Problem

The demand for biofuels in the power and transport sector increased in the 2000s and led to an intensification of energy crop cultivation. This was initially seen as an opportunity for agriculture, but later energy crop cultivation was increasingly perceived and communicated as an ecological and social problem by various social actors, especially environmental associations. Under the buzzwords *Vermaisung*,[49] *Tank-Teller-Debatte*[50] and *iLUC*,[51] the negative impacts of biomass cultivation were increasingly critically discussed.

According to critics, the promotion of bioenergy has led to second-order problems, such as biodiversity loss and soil degradation through monocultures, deforestation for land reclamation, land grabbing and displacement, and land-use competition between food and feed production on the one hand and energy crop cultivation on the other. The criticism in the power sector was primarily related to land use changes in Germany, while in the biofuel sector global interdependencies and indirect land use changes in other parts of the world were criticised in particular, such as rainforest clearing for palm oil plantations in Southeast Asia. It became clear that German and European energy policy had undesirable indirect effects via global

[47] For an overview of the central measures, see Federal Ministry for Economic Affairs and Energy (2018).

[48] Primarily bioethanol, which can be blended with petrol (E10) and biodiesel.

[49] *Vermaisung*: German buzzword for the accusation that energy crop cultivation for electricity generation with biogas leads to overcultivation of maize and hence to biodiversity loss due to monocultures.

[50] *Tank-Teller-Debatte:* German buzzword for the *food versus fuel debate*; referring to the conflicts of interest between food production and energy production from agricultural crops.

[51] *iLUC*: Abbreviation of *indirect land use change*; used as a buzzword in German and English debates in the context of biofuel production for the transport sector; mainly referring to palm oil production.

trade structures. The global interdependencies of biofuel production are another factor contributing to the high complexity of bioenergy policy.

After a veritable bioenergy boom at the beginning of the 2000s, the undesirable effects discussed in the public debate led to a change in the image of bioenergy some 10 years later. At the beginning of the millennium, bioenergy was perceived as a problem solution and as a form of energy production that was desirable from an environmental point of view. Within a few years, however, the image of bioenergy became very negative. This negative image still has an impact on the perception of bioenergy in the public and on political processes and measures of bioenergy policy today.

According to the interviewees, a characteristic feature of bioenergy policy in Germany is that the debate over the years has been conducted increasingly emotionally (value-based) and less on the basis of objective arguments and facts (knowledge-based). Ecologically oriented stakeholders reportedly tend to criticise the energetic use of biomass in a very fundamental way. Experts from the bioenergy sector, on the other hand, accuse environmental associations and environmentally oriented governmental organisations of 'populist propaganda campaigning' against any form of bioenergy use, which, supposedly, cannot be justified from a scientific and fact-oriented point of view. Furthermore, it was pointed out that actors involved in the political process selected and used existing scientific knowledge in such a way that it strengthened their own standpoint and that, in doing so, at least in part, methodologically dubious commissioned studies were being used.

At present, bioenergy as a form of renewable energy is understood in debates as a solution strategy in the context of climate protection, energy system transformation, bioeconomy, rural development and decentralisation on the one hand, but it is also associated with biodiversity loss, soil degradation, rising energy costs, deforestation and displacement on the other hand. German and European Policymakers responded to the critical debates by reducing bioenergy subsidies and by introducing sustainability criteria for the energetic use of biomass.[52] Some of the bioenergy experts interviewed described the introduction of this kind of measures as the politically desired end of bioenergy in Germany.[53] It was stated that bioenergy, especially in Germany, was presented in a very negative light in the public debate, that it was downright demonised and that the criticism to this extent could not be proven by scientific facts. Representatives of the industry warned that this momentum, which is a specific characteristic of the political debate on bioenergy in Germany as compared to other European countries, has led to the development that too little attention was being paid to the positive properties of bioenergy. Because of this, the potential for climate protection, energy system transformation

[52] Cf. Federal Ministry for Economic Affairs and Energy (2018) for an overview of the political measures of the Federal Republic of Germany and the European Union for the regulation of renewable energies.

[53] The interviewees here referred primarily to the EEG (2012) and the EEG (2014) and thus to the power sector.

and sustainable development in general was not being fully exploited, according to the interviewees.

Although the interviewed actors independently described that bioenergy has been perceived somewhat more positively again in Germany in recent years, partly due to changing constellations of actors and new scientific data, the negative image of bioenergy and the relatively uncertain knowledge situation regarding the possible effects of political measures still play a central role in German bioenergy policy.

With regard to bioenergy or the economic use of biomass in general, it can therefore be stated that there are different opinions as to whether the use of biomass should be understood more as a problem or as a solution. In addition, the effects of already introduced solution options (policies promoting bioenergy) also have a strong influence on the current political processes in this policy field. This momentum in turn has an impact on the political debates on bioenergy policy as a whole, since many of the environmental impacts and conflicting goals listed here occur with every form of biomass use.

3.3.3 Bioenergy Policy as a Wicked Problem

The political regulation of bioenergy use is a problem for which there is neither a consensus on the definition of the problem nor a consensus on the solution. It can therefore be classified as a wicked problem (Type 3 according to Roberts).[54] The claim that wicked problems are politically difficult to deal with and tend to be avoided by political actors could be confirmed for the field of bioenergy policy in the case studies.

The analysis of political processes of bioenergy policy with the policy cycle showed that the uncertainties and complexity of bioenergy policy create barriers already in the first phase of the policy cycle, the problem definition. Hence, the issue is not given priority on the political agenda. The interviewed experts confirmed the theoretical claim that political actors rather avoid negatively biased topics. With regard to possible solutions in terms of political measures and their effects, there is still a great deal of uncertainty, even though there are already some practical experiences that can be drawn on.

The empirical study also confirmed the basic assumption of PIDA that political processes are not necessarily pure problem-solving processes, but that they can be strongly influenced by other factors, such as actor interests. The thesis that social acceptance and power issues play a more important role in wicked problems than in simple and complex problems could also be empirically confirmed. According to the results of the case studies, the following three factors in particular are decisive for the further development of bioenergy policy in Germany: *acceptance* (to what extent is the economic use of biomass socially accepted?), *power distribution* (what is the distribution of power among stakeholders?) and *framing* (which framing by which interest groups will prevail?).

[54] Cf. Roberts (2000).

The dissemination of the idea of the bioeconomy as a sustainable bio-based economic system offers actors the opportunity to frame bioenergy again as a solution rather than a problem in the political debate in Germany. It remains to be seen which interest groups and which perceptions of bioenergy will prevail and whether the energetic use of biomass as part of the bioeconomy will experience a new upswing.

4 Summary and Outlook

The aim of this contribution was to outline the characteristic problem structures of bioeconomy policy based on the analysis of political processes in the field of bioenergy policy in the power and heat sector in Germany. The empirical data basis was formed by the results of qualitative case studies, which were carried out within the political science research project Bio-Ökopoli. Using a theory-based qualitative content analysis, primary sources and guideline-based expert interviews were analysed for selected cases of bioeconomy policy. The theoretical framework for the analysis of the problem structures of bioenergy policy was provided by three analytical approaches: (1) the policy cycle,[55] (2) the PIDA with the explanatory factor problem structures[56] according to Böcher and Töller and (3) the problem typification into simple, complex and wicked problems[57] according to Roberts.

With the analysis, characteristic problem structures of bioenergy policy in Germany could be identified. Furthermore, it could be shown that these have a decisive influence on the political processes (*politics*) and the resulting political measures (*policy*) at national level. Furthermore, it was found that political processes of bioenergy policy in Germany can only be explained to a limited extent with the policy cycle, which sees political processes primarily as problem-solving processes. Problems in the field of bioenergy policy, however, are to be regarded as wicked problems due to their high complexity, the unclear definition of the problem and the different views on suitable political solutions. The basic assumption of the PIDA, that political processes are shaped by inherent dynamics and that problem solving is only one of several possible explanatory factors, could be confirmed with the empirical results.

Against the background of these findings, and in view of the cross-sectional character and intersectorality of bioeconomy policy, it should be noted for a future concrete design of political processes and measures that an understanding of political processes as pure problem-solving processes is not appropriate in this policy area. Since both the political regulation of bioenergy policy and bioeconomy policy as a whole are to be classified as wicked problems, traditional solution strategies geared to simple or complex problems with a clear definition of the problem are not effective. In contrast, the social perception and social acceptance of possible problem

[55] Cf. Jann and Wegrich (2014).

[56] Cf. Böcher and Töller (2012b).

[57] Cf. Roberts (2000).

solutions (*acceptance*), the self-interests, power resources and values of the actors involved in negotiating conflicting goals (*power distribution*) and the question of which *framing* of the (energetic) use of biomass will prevail in the political debate are decisive.

References

Balint, P. J., Stewart, R. E., Desai, A., & Walters, L. C. (2011). *Wicked environmental problems. Managing uncertainty and conflict.*

Beer, K., Böcher, M., Bollmann, A., Töller, A. E., & Vogelpohl, T. (2018). *Politische Prozesse der Bioökonomie zwischen Ökonomie und Ökologie. Bio-Ökopoli-Arbeitsbericht 1: Fallauswahl und Übersichtsanalysen.*

Blum, S., & Schubert, K. (2018). *Politikfeldanalyse. Eine Einführung.*

Böcher, M., & Töller, A. E. (2012a). Eigendynamik und Zufall als Triebkräfte der Umweltpolitik? Ein Ansatz zum Erklären umweltpolitischer Entscheidungen. *Zeitschrift für Umweltpolitik und Umweltrecht, 4*, 450–479.

Böcher, M., & Töller, A. E. (2012b). *Umweltpolitik in Deutschland. Eine politikfeldanalytische Einführung.*

Böcher, M., & Töller, A. E. (2015). *Inherent dynamics and chance as drivers in environmental policy? An approach to explaining environmental policy decisions.* Retrieved November 2, 2021, from, https://www.researchgate.net/publication/293145722_Inherent_Dynamics_and_Chance_as_Drivers_in_Environmental_Policy_An_Approach_to_Explaining_Environmental_Policy_Decisions

Böcher, M., Töller, A. E., Perbandt, D., Beer, K., & Vogelpohl, T. (2020). Research trends: Bioeconomy politics and governance. *Forest Policy and Economics, 118*, S.102219. https://doi.org/10.1016/j.forpol.2020.102219

Conklin, J., & Weil, W. (2007). *Wicked problems: Naming the pain in organizations.* Retrieved November 2, 2021, from, http://www.accelinnova.com/docs/wickedproblems.pdf

Fachargentur Nachwachsende Rohstoffe e. V (2019). *Basisdaten Bioenergie Deutschland 2019. Festbrennstoffe, Biokraftstoffe, Biogas.* Retrieved November 2, 2021, from, https://www.fnr.de/fileadmin/allgemein/pdf/broschueren/basisdaten_bioenergie_2019_web.pdf

Federal Ministry for Economic Affairs and Energy. (2018). *Sechster Monitoring-Bericht zur Energiewende. Die Energie der Zukunft. Berichtsjahr 2016.* Retrieved November 2, 2021, from, https://www.bmwi.de/Redaktion/DE/Publikationen/Energie/sechster-monitoring-bericht-zur-energiewende.pdf?__blob=publicationFile&v=18

Federal Ministry of Education and Research. (2010). *Nationale Forschungsstrategie BioÖkonomie 2030. Unser Weg zu einer bio-basierten Wirtschaft.* Retrieved November 2, 2021, from, https://www.bmbf.de/SharedDocs/Publikationen/de/bmbf/pdf/nationale-forschungsstrategie-biooekonomie-2030.pdf

Federal Ministry of Education and Research; Federal Ministry of Food and Agriculture. (2020). *Nationale Bioökonomiestrategie.* Retrieved November 2, 2021, from, https://www.bmbf.de/SharedDocs/Publikationen/de/bmbf/pdf/nationale-biooekonomiestrategie_3341.pdf

Federal Ministry for the Environment, Nature Conservation and Nuclear Safety. (2016): *Klimaschutzplan 2050 – Klimaschutzpolitische Grundsätze und Ziele der Bundesregierung.* Retrieved November 2, 2021, from, https://www.bmu.de/fileadmin/Daten_BMU/Download_PDF/Klimaschutz/klimaschutzplan_2050_bf.pdf

Federal Ministry of Food and Agriculture. (2014). *Nationale Politikstrategie Bioökonomie. Nachwachsende Ressourcen und biotechnologische Verfahren als Basis für Ernährung, Industrie und Energie.* Retrieved November 2, 2021, from, https://biooekonomie.de/sites/default/files/files/2016-09/npsb.pdf

FernUniversität in Hagen. (2019). *Bio-Ökopli. Political processes of bioeconomy between economy and ecology.* Retrieved November 2, 2021, from, https://www.fernuni-hagen.de/bio-oekopoli/en/

Gesetz zur grundlegenden Reform des Erneuerbaren-Energien-Gesetzes und zur Änderung weiterer Bestimmungen des Energiewirtschaftsrechts (EEG 2014). (Act concerning renewable energy sources). In *Bundesgesetzblatt* I/33, 24.07.2014 (pp. 1066–1132).

Gesetz zur Neuregelung des Rechtsrahmens für die Förderung der Stromerzeugung aus erneuerbaren Energien (EEG 2012). (Act concerning renewable energy sources). In *Bundesgesetzblatt* I/42, 04.08.2011 (pp. 1634–1678).

Jänicke, M., & Volkery, A. (2001). Persistente Probleme des Umweltschutzes. In: *Natur und Kultur. Transdisziplinäre Zeitschrift für ökologische Nachhaltigkeit, 2*(2), 45–59.

Jann, W., & Wegrich, K. (2014). Phasenmodelle und Politikprozesse: Der Policy Cycle. In K. Schubert & N. C. Bandelow (Eds.), *Lehrbuch der Politikfeldanalyse* (3rd updated and revised ed., pp. 75–113).

Kaltschmitt, M., Hartmann, H., & Hofbauer, H. (2016). *Energie aus Biomasse. Grundlagen, Techniken und Verfahren* (3rd ed.).

Kevenhörster, P. (2008). *Politikwissenschaft. Band 1: Entscheidungen und Strukturen der Politik* (3rd ed.).

Otto, S., Hildebrandt, J., Will, M., Henn, L., & Beer, K. (2021). Tying up loose ends. Integrating consumers' psychology into a broad interdisciplinary perspective on a circular sustainable bioeconomy. *Journal of Agricultural and Environmental Ethics, 34*(2). https://doi.org/10.1007/s10806-021-09851-6

Perbandt, D., Vogelpohl, T., Beer, K., Töller, A. E., & Böcher, M. (2021). *Zielkonflikte der Bioökonomie. Biobasiertes Wirtschaften im Spannungsfeld von Ökonomie und Ökologie. Lehrbuch. Wiesbaden: Springer nature* (Energie in Naturwissenschaft, Technik, Wirtschaft und Gesellschaft). https://doi.org/10.1007/978-3-658-35093-2

Reiter, R., & Töller, A. E. (2014). *Politikfeldanalyse im Studium. Fragestellungen, Theorien, Methoden.*

Roberts, N. (2000). Wicked problems and network approaches to resolution. *International Public Management Review, 1*(1). Retrieved November 2, 2021, from, https://journals.sfu.ca/ipmr/index.php/ipmr/article/view/175/175

Töller, A. E., Vogelpohl, T., Beer, K., & Böcher, M. (2021). Is bioeconomy policy a policy field? A conceptual framework and empirical findings on the European Union and Germany. *Journal of Environmental Policy & Planning.* https://doi.org/10.1080/1523908X.2021.1893163

Vogelpohl, T., Beer, K., Ewert, B., Perbandt, D., Töller, A. E., & Böcher, M. (2021). Patterns of European bioeconomy policy. Insights from a cross-case study of three policy areas. In *Environmental politics* (pp. S1–S21). https://doi.org/10.1080/09644016.2021.1917827

The Bioeconomy Transformation in the German Rheinische Revier: Stakeholders and Discourses in Media Coverage

10

Sandra Venghaus, Sophia Dieken, and Maria Belka

Abstract

As part of Germany's strategy to phase out coal power, the so-called Coal Commission enlisted by the government proposed the bioeconomy as a more sustainable concept for managing natural resources and directing structural change in the Rheinische Revier in their final report from January 2019. The Rheinische Revier is well equipped for a transformation towards a sustainable bioeconomy due to already existing structures in directly related fields such as agriculture, chemical and energy industry, as well as education and research. However, realizing a sustainable bioeconomy requires major changes to the socio-political structures, especially in terms of a shared and supportive bioeconomy vision among the numerous and diverse stakeholders affected. The initiated transformation in the Rheinische Revier offers a unique opportunity to analyze such stakeholder dynamics and how they shape possible transformation pathways towards a sustainable bioeconomy. In order to generate an understanding of which stakeholders and visions are publically considered in the debate on the bioeconomy transformation, a mixed-method, computer-assisted discourse analysis of German-language media texts between 2010 and 2019 was conducted. The analysis revealed that research and government stakeholders strongly dominated, who promoted visions of the bioeconomy as a growth concept for

S. Venghaus (✉)
Forschungszentrum Jülich, Systems Analysis and Technology Evaluation (IEK-STE), Institute of Energy and Climate Research, Jülich, Germany

School of Business and Economics, RWTH Aachen University, Aachen, Germany
e-mail: s.venghaus@fz-juelich.de

S. Dieken · M. Belka
Forschungszentrum Jülich, Systems Analysis and Technology Evaluation (IEK-STE), Institute of Energy and Climate Research, Jülich, Germany
e-mail: s.dieken@fz-juelich.de; m.belka@fz-juelich.de

D. Lanzerath et al. (eds.), *Bioeconomy and Sustainability*,
https://doi.org/10.1007/978-3-030-87402-5_10

supporting the use of biomass and biotechnologies against the background of global challenges of resource depletion. Stakeholders from the general public were often called upon, but seldom expressed their own visions, which were characterized by more diverse themes and arguments.

Keywords

Sustainable bioeconomy · Discourse analysis · Stakeholder visions

1 Introduction

Germany is faced with considerable challenges of global scale regarding its way of managing natural resources, from climate change and environmental degradation over the technical integration of new technologies to trade conflicts and demands of social justice. However, the German governance system is slow to adapt. Its institutions and structures have developed over decades and even centuries, including the energy-intensive industries which evolved in tandem with and close proximity to areas of lignite mining. However, unsustainable practices and structures of natural resources management are increasingly challenged and abolished, sometimes with perceptible adverse social and economic consequences for the respective regions. Currently, two profound processes of change permute. On the one hand, a rapid structural change process of German lignite mining areas—including the Rheinische Revier as Europe's largest lignite mining area—has been induced and propelled by the German government's decision to phase out coal power. At the same time, the national bioeconomy strategy promotes a transformation towards an economy based on the sustainable use of renewable biological resources and biotechnologies in place of the fossil economy. As part of its climate action strategy, the German government appointed the so-called Coal Commission ("Commission on Growth, Structural Change and Employment") in 2018 in order to develop plans on how to conduct the coal phase-out and mitigate its consequences especially in the lignite mining regions.[1] In its 2019 final report, the commission suggests the further establishment of a bioeconomy as a central element of steering structural change towards a positive future in the affected regions, highlighting its economic, social, and ecological potential.[2] This proposal ties in with federal state-level plans to support the development of the bioeconomy, such as the 2012 strategy by North Rhine-Westphalia (NRW), the state in which the Rheinische Revier is situated.[3] Both processes, structural change and implementation of a bioeconomy, are complex and intertwined, and will be difficult to steer successfully.

[1]Cf. Federal Ministry for the Environment, Nature Conservation and Nuclear Safety 2019.

[2]Cf. Commission "Wachstum, Strukturwandel und Beschäftigung" 2019.

[3]Cf. Ministry of Culture and Science of the state North Rhine-Westphalia (2012).

Specifically, these change processes affect and depend on the numerous and diverse stakeholders involved. However, stakeholders are often not sufficiently considered in mostly techno-economic assessments of structural change. The initiated transformation in the Rheinische Revier offers a unique opportunity to analyze such stakeholder dynamics and how they shape possible pathways from a fossil-based economy towards a sustainable bioeconomy.

With good cause, issues of stakeholder participation and acceptance are identified as critical in transformation strategies[4] as well as scientific research on the topic.[5] The *German Energiewende* has been an educational experience in that local activist groups and alliances of incumbent interest groups have considerably affected the development of transformative policies or individual projects.[6] In this context, transformations are understood as processes of systemic change resulting from stakeholder-initiated institutional changes.[7] Stakeholders as "agents of change"[8] shape the outcome through their different roles, depending on individual interests, values, and visions.[9] However, a multitude of interrelated actors is involved on various decision-making levels and across different sectors and policy fields. Implementing such a complex transformation requires a shared, or at least complementary, vision of the bioeconomy among the stakeholders.[10] Identifying stakeholder visions and the relations among them is thus essential for understanding possible transformation trajectories.

A common approach to investigating stakeholder roles in transformations is a discourse analysis.[11] The "analysis of language that looks at patterns of language across texts as well as the social and cultural contexts in which the texts occur" serves to characterize different world views, the interrelations between social relations and the use of language.[12] With regard to the bioeconomy, there are a number of discourse analyses on different understandings and perceptions of the concept.[13] Research on stakeholder discourses, so far, has been primarily limited to political strategies and statements of few select experts with the exception of a recent German study on citizens' perspectives.[14] In contrast, analyzing media discourses allows for a better understanding of the broader societal debate, revealing which

---

[4]Cf. Federal Ministry of Education and Research (2010), Federal Ministry of Food and Agriculture (2013) and Federal Ministry of Education and Research (2014).

[5]Cf. Hake et al. (2015), Fischer et al. (2016), Editorial Nature Climate Change (2016) and Editorial Nature Energy (2016).

[6]Cf. Hake et al. (2015) and Moe (2016).

[7]Cf. Merkel et al. (2019).

[8]Editorial Nature Climate Change (2016).

[9]Cf. Editorial Nature Energy (2016).

[10]Cf. Wohlfahrt et al. (2019).

[11]Cf. e.g. Leipprand and Flachsland (2018).

[12]Paltridge (2006).

[13]Cf. e.g. Stern et al. (2018) and Leipold and Petit-Boix (2018).

[14]Cf. Hempel et al. (2019).

stakeholders decisively influence the public perception of the bioeconomy. So far, however, little research is available in this regard.[15]

Addressing this research gap, we conducted a computer-assisted, comprehensive media-based discourse analysis to generate an understanding of which stakeholders and discourses are publically portrayed in the debate on the bioeconomy transformation in Germany. Since the bioeconomy transformation is closely intertwined with local structural change, a specific focus was set on the Rheinische Revier. The remainder of this chapter is thus structured as follows. In Sect. 2, the research approach of a mixed-method, computer-assisted discourse analysis is introduced. The results with regard to considerations of stakeholders and discourses are presented in Sect. 3, and discussed in Sect. 4. In Sect. 5, conclusions and suggestions for further research are discussed.

2 Research Approach

2.1 Data Selection and Preparation

In order to identify and characterize stakeholders and their visions of the regional bioeconomy in the Rheinische Revier, a discourse analysis of media texts was conducted. In a first step, relevant texts were identified. Using the lexisnexis research database, German-language media texts were searched for the period from January 1st 2010 to July 31st 2019. The year 2010 lends itself as starting point for the search query, since it marks the publication of the first formal German national bioeconomy research strategy, the "National Research Strategy BioEconomy 2030" by the Federal Ministry of Education and Research. Documents until July 2019 were considered in order to include the debates following the before-mentioned coal commission's final report from January 2019. The relevant search keywords were the terms structural change ("Strukturwandel"), as well as variations of the term bioeconomy ("Bioökonomi*"). The database search for structural change provided more than 10,000 results and the search for bioeconomy produced slightly over 1000 results for the given time range. Duplicates were removed. The resulting media texts for bioeconomy served as the basis for analysis, whereas the results of the structural change query were used to provide indicative results for comparison.

2.2 Analysis of Stakeholders and Their Visions

In order to identify media considerations of stakeholders and their visions in the context of the bioeconomy, a mixed-method, computer-assisted discourse analysis was conducted. To this aim, a bioeconomy-specific coding scheme was developed. The analysis was conducted with support of the software MAXQDA (Version 18),

[15] Cf. as a noteable exception Peltomaa (2018).

which allows the coding and text mining using a framework of categories that can be iteratively developed during the research process. This approach is especially relevant when pursuing an inductive research approach. Moreover, this approach allows combining qualitative and quantitative text analysis.

The coding scheme was developed in two steps. Firstly, based on scientific literature a basic coding frame of known categories was derived. This a priori set of categories included visions of the bioeconomy and stakeholder groups. The bioeconomy concept is considered to draw on three distinct visions, which can be abbreviated to "bio-resource," "bio-technology," and "bio-ecology."[16] The bio-resource vision focuses on upgrading and converting biological raw materials, primarily in agriculture, marine, forestry, and bioenergy. The objectives are both economic growth and sustainability, and further issues include the cascading use of biomass and land use, emphasizing research and innovation more broadly as well as development in rural areas.[17] The bio-technology vision focuses on bio-technology research and commercialization across sectors, the objective of economic growth, and a global scale of competition.[18] The bio-ecology vision focuses on the optimization of energy and nutrient use, the promotion of biodiversity and avoiding soil degradation, preferably on a regional level. Primary objective is sustainability, emphasizing circular and self-sustained production modes as well as organic bio-ecological practices, also preferably at a local level.[19] These visions are not mutually exclusive, but condense certain assumptions and focal points into three specific lines of thought. The above categories on general bioeconomy visions were supplemented by the stakeholder categories. Based on the characterization of bioeconomy supply chains,[20] stakeholders can be grouped according to their position within the bioeconomy, such as biomass production or governance and regulation. These categories served as the basis for the coding scheme development.

In a second step, based on the data the coding scheme was inductively complemented and refined.[21] Thus, code categories were flexible enough to account for new considerations occurring in the texts. The aim of this approach was to develop a coding scheme capable of reflecting differences between stakeholder discourses with regard to the German bioeconomy. Based on this scheme, the media texts were assessed quantitatively, and relevant text segments were analyzed in depth qualitatively. The results are presented in the following section.

[16] Cf. Bugge et al. (2016).

[17] Cf. ibid., 11 f.

[18] Cf. ibid., 10 f.

[19] Cf. ibid., 12 f.

[20] Cf. e.g. Lewandowski (2018).

[21] Cf. Gibbs (2014) and Maxwell and Chmiel (2014).

3 Results

The analysis of stakeholder discourses in German media texts combines quantitative and qualitative methods to provide an in-depth understanding of which stakeholders and which visions prevail in the public perception of the bioeconomy transformation. In the following, the results are presented for the overall media coverage of structural change and bioeconomy, for considerations of bioeconomy stakeholders, and for stakeholder visions presented in media texts, respectively.

3.1 Media Coverage of Structural Change and the Bioeconomy

Overall, the mere amount of media texts about a given topic indicates how prominent it is in the public perception. The search for media texts in the given time range produced more than 10 times as many results for structural change than for bioeconomy. Structural change as a broad economic phenomenon has long been part of public debate and, as such, is significantly more prominent than the relatively new and specific concept of the bioeconomy. So far, the bioeconomy appears to remain a niche topic.

However, the temporal development shows that the number of media texts referring to bioeconomy has increased from 24 in 2010 to 93 in 2018 and already 67 in the first seven months of 2019. Thus, bioeconomy generally receives increasing attention, though starting from a relatively low level.

Interestingly, the analysis revealed little to no overlap between the debates on structural change and bioeconomy. On a strategic level, the bioeconomy has been identified as a potential transformation path for managing structural change in lignite mining regions. Nevertheless, in media texts this connection has rarely been drawn: only 72 texts explicitly mention both terms in the investigated time range. Most of these articles are directly concerned with the local and regional plans for a bioeconomy in the Rheinische Revier, e.g. articles on the work of the regional development agency Innovationsregion Rheinisches Revier GmbH.[22] Thus, discourses of structural change and bioeconomy are barely linked, almost exclusively in the context of North Rhine-Westphalia (NRW). Nonetheless, both discourses are thematically close with texts on the bioeconomy regularly highlighting its contribution to competitiveness and growth especially in rural areas, a point we will return to in Sect. 3.3.

This finding hints at regional differences in the bioeconomy discourse, which is why we also compare media texts with regard to considerations of the Rheinische Revier and the second-largest German lignite mining region, the Lausitzer Revier. Of the texts on structural change, 42% directly refer to the Lausitzer Revier, but only 17% directly refer to the Rheinische Revier or the federal state it is situated in, NRW. In turn, with regard to texts on the bioeconomy, 27% directly refer to the Rheinische

[22] Cf. e.g. Aachener Zeitung (2013). or Speen (2014).

Revier and 5% to the Lausitzer Revier. Often, the bioeconomy is discussed in a national or global context, such as with regard to decisions made by the federal government in Berlin or against the backdrop of the "Global South" or developing countries. Thus, while structural change is discussed primarily in the context of the Eastern German mining region the Lausitzer Revier, the bioeconomy discourse is not as regionally specific and more often related to the West German mining region the Rheinische Revier. The media texts also occasionally refer to the federal states of Bavaria, Baden-Wuerttemberg and Hesse, mostly with regard to research institutions and biotechnology companies located there. Though not involved in the structural change of the German lignite mining regions, these states seek an active role for themselves in the development of a bioeconomy. This supports the finding that discourses on bioeconomy and structural change are largely distinct from each other. In general, the bioeconomy is primarily discussed on national and NRW levels.

Hence, compared to the broader consideration of structural change, media coverage on the bioeconomy is limited in scope and, despite strategic intentions, discourses are largely disconnected. Given the concept's limited space in public perception, the question of which specific stakeholders and visions are considered is essential for understanding who drives the broader societal debate on the bioeconomy and in which direction.

3.2 Stakeholders: Subjects and Objects of Speech

As outlined in the introductory section, stakeholders are influenced by and shape the transformation towards a bioeconomy. However, not all stakeholders are equally prominent in the media debate, and thus certain stakeholders are more active in directing the broader social debate. This is relevant to how the bioeconomy is publically perceived and consequently implemented. Before analyzing which specific discourses the stakeholders support, the results on which stakeholders are considered are presented.

The media texts account for a wide range of stakeholders, both in terms of stakeholder groups identified as relevant, such as governments or biotech start-ups, and individual stakeholders personally named, such as Prof. Joachim von Braun, former chair of the German Bioeconomy Council, an independent advisory body to the German Federal Government. Even beyond the expected bioeconomy stakeholders, e.g. farmers, the chemical and pharmaceutical industry, or political decision-makers, numerous other stakeholders are considered, such as associations of the textile industry or logistics and transport service providers. In principle, representatives of the entire bioeconomy supply chain are portrayed in the media texts—however, to very differing degrees.

By far, the two stakeholder groups most often considered are research organizations and scientists, as well as governmental institutions and politicians. These two groups are specifically accentuated and cited often. For research organizations and scientists, almost two thirds of media texts portray stakeholders from this sphere, which suggests that they are the main drivers of the public perception of the bioeconomy. Furthermore, in most cases these stakeholders are named explicitly and cited with regard to their vision of the bioeconomy or related research projects. A prominent example is the University of Hohenheim, a leading university in agricultural research specializing in the bioeconomy, which is referred to in 10% of all texts (more than, e.g., federal, state and foreign ministries of agriculture combined), mostly reporting on recent scientific contributions or administrative changes. For example, university president Prof. Dr. Stephan Dabbert was quoted on the financing of bioeconomy research and teaching in 2014[23] or on the occasion of the publication of a handbook on the bioeconomy in 2017.[24]

A similarly active scientific stakeholder is the Forschungszentrum Jülich, located in the Rheinische Revier and tasked to provide scientific support for structural change and the implementation of a bioeconomy in the region. It is also referred to in 10% of all texts, with regard to recent scientific contributions and administrative changes, but further with a strong focus on the transformation strategy for the Rheinische Revier. For example, in a 2011 regional newspaper article, the Forschungszentrum Jülich's contribution to creating new jobs in the bioeconomy is discussed against the background of the newly initiated regional development agency Innovationsregion Rheinisches Revier GmbH.[25] Stakeholders from the research sphere, and specifically single key actors such as specialized research institutions, are not only very present in media coverage of the bioeconomy, but also position themselves through numerous quoted statements or detailed reporting on their activities. Thus, research organizations and scientists dominate public perception as reflected in media discourse.

Governmental institutions and politicians are considered in more than half of the texts with many media texts reporting on political developments, such as recent policies and related comments by the politicians responsible. For example, there is extensive coverage of the first Global Bioeconomy Summit in Berlin in 2015, and of the developing NRW strategy for the transformation towards a bioeconomy. Often, the governments of Germany and other European and international countries are referred to (approx. 39% of all texts). In turn, local political institutions and actors are only rarely mentioned and if so mostly in local news reporting their reaction to the implementation of national and regional policies. Hence, especially national political institutions are publically perceived as central bioeconomy actors.

Two ministries are especially prominent: the Federal Ministry of Education and Research and the Federal Ministry of Food and Agriculture. Also, the aforementioned German Bioeconomy Council is regularly covered, as its members are frequently cited as experts on bioeconomy policy. Thus, the government appears to be understood as a relevant bioeconomy stakeholder primarily with regard to

[23] Cf. Jacobs (2014).

[24] Cf. Labo (2017).

[25] Cf. Aachener Zeitung (2011).

issues of research and biomass production. The dominance of stakeholders from the political sphere indicates that the bioeconomy is primarily discussed as a national policy strategy.

In addition to these two groups of stakeholders, numerous further stakeholders are mentioned from the various (biomass production, processing, trading and recycling) stages of bioeconomy supply chains. However, they are almost exclusively referred to in a non-specific way by simply listing industries that are relevant to the bioeconomy. For example, in a 2012 article in the specialist journal CITplus, a list of industries is compiled by stating: "Dazu gehören die Betriebe der Agrarwirtschaft und Forstwirtschaft, der Lebensmittel- und Holzverarbeitenden Industrie einschließlich der Zellstoffproduzenten oder jene, die sich mit der Produktion biobasierter Dämmstoffe, Fasern oder Kunststoffe befassen."[26] Especially bio-technology companies or start-ups[27] as well as the chemical industry[28] are presented as relevant stakeholders. These include the expected large companies, e.g., BASF, Bayer or Monsanto, who are considered as stakeholders affected by or engaging in the bioeconomy.[29] This stakeholder group is mainly considered passively.

In this context, only few specific companies or industry associations are identified and quoted. These references are primarily made in the context of business conferences, e.g. the ACHEMA, a trade fair by the process industry on, among other fields, biotechnology,[30] or research cooperations between academia and business, e.g. again by the University of Hohenheim.[31] However, such reports are almost exclusively found in specialist magazines, so their relevance for the overall public debate can be expected to be limited.

The corporate stakeholder most prominently featured is the German biotechnology company BRAIN AG, which is single-handedly referred to in a quarter of all articles (outpassing both University of Hohenheim and Forschungszentrum Jülich). Most often, BRAIN AG CEO Holger Zinke is quoted on the potential of biotechnology and specific products. In 2015, *DIE ZEIT* even featured a portrait of him and his company.[32] The company is considered often in specialist journals, but it is also regularly reported on in national and local newspapers. Despite its much publicized financial setbacks, BRAIN AG is presented as a trailblazer of the German

[26] CITplus (2012), "This includes companies of agriculture and forestry, food and wood processing industries as well as pulp producers and producers of bio-based insulation, fibers or synthetics" (own translation).

[27] Cf. e.g. Ronzheimer (2018).

[28] Cf. e.g. Kircher and Schwarz (2015).

[29] Cf. e.g. Holdinghausen (2019c).

[30] Cf. Labo (2012).

[31] Cf. Allgemeine Fleischer Zeitung (2016).

[32] Cf. Grefe (2015).

bioeconomy and, given its frequent appearance in media coverage, it can be considered to play a crucial role in shaping the public perception on the bioeconomy.

The only other stakeholder group from the production and processing sphere with a notable presence are farmers, mentioned in approximately 12% of all media texts. However, again in this context stakeholders are mostly referred to in a non-specific way with descriptions of the key role of agriculture. For example, in a 2015 article on a meeting of agricultural ministers at the Global Forum for Food and Agriculture in Berlin, farmers—especially small farmers in developing countries—are characterized as providing the basis for the bioeconomy and food security, while simultaneously profiting economically from an optimization of distribution channels for biomass.[33] In line with other stakeholders from the biomass production and processing stages of bioeconomy supply chains, farming stakeholders are primarily considered passively.

In cases where farming stakeholders are explicitly identified and quoted, it is mostly representatives from international, national, and local associations, e.g. the World Farmers' Organisation,[34] the Deutsche Bauernverband,[35] or the Rheinische Landwirtschaftsverband.[36] These are considered, e.g., in the context of trade fairs in specialist magazines or local stakeholder meetings for the bioeconomy transformation in the Rheinische Revier. Overall, the stakeholder groups who are involved in the production and processing of biomass are often referred to, but only few specific actors are actively involved in the bioeconomy discourse.

Another stakeholder group often considered with references in half of all texts, but rarely contributing to media coverage of the bioeconomy, is the broader society, understood to include individual citizens as well as civil and non-governmental organizations. In the media texts, societal actors are generally considered either as citizens and workers who participate in the developing bioeconomy or as consumers of bioeconomy products. For example, Bavarian agricultural minister Helmut Brunner argues that convincing consumers of biomass-based products is especially important for successfully implementing a bioeconomy.[37] Although this stakeholder group is presented as central to the bioeconomy, it is usually considered only when other stakeholders (e.g., politicians) are referring to it.

Nevertheless, specific environment, development, or consumer associations take part in the media discourse, e.g. a national coalition of several German associations dedicated to the bioeconomy, "Zivilgesellschaftliches Aktionsforum Bioökonomie."[38] They position themselves on the occasion of specific policies or stakeholder meetings. For example, Greenpeace and the anti-genetic engineering initiative "Gendreck weg" are quoted in a 2011 Der Spiegel review of the German

[33] Cf. Werner (2015).

[34] Cf. ibid.

[35] Cf. Georgis (2016).

[36] Cf. Drogowski (2019).

[37] Cf. Bäumel-Schachtner (2016).

[38] Cf. Ronzheimer (2019).

policy "Nationale Forschungsstrategie BioÖkonomie 2030," published the year before.[39] Also, occasionally citizens' initiatives are considered in local newspaper articles on the transformation in the mining regions, such as the coordination association between different local groups in the Rheinische Revier, Zivilgesellschaftlicher Koordinierungskreis Strukturwandel (ZKS), which has organized events with the Forschungszentrum Jülich.[40] However, overall, societal stakeholders' participation in media discourse on the bioeconomy is limited.

The analysis of the stakeholder groups which are referred to in media texts on the bioeconomy revealed a generally broad consideration of a multitude of stakeholders, but only a limited number of them is specifically identified and quoted in terms of their position on the topic. Also, the discourse is strongly dominated by stakeholders from research and (national) politics, who are most often considered and quoted. In terms of stakeholders who produce and process biomass, they are often considered as relevant, but only specific actors such as biotechnology and large chemical/pharmaceutical companies as well as farming associations actively contribute to the discourse. Similarly, while other stakeholders often refer to the central role of civil society, only some associations take part in the discourse on specific occasions. So, the presence of stakeholders in the public perception of the bioeconomy is highly uneven.

3.3 Stakeholder Discourses

The in-depth qualitative analysis of relevant text segments not only produced the bioeconomy discourses considered in the media, but also allowed us to connect stakeholders to their respective visions. In a general context the analysis shows that the bioeconomy is largely portrayed superficially in the media texts, since the term is often only employed as a keyword for policies, research or industry developments that are discussed. Also, many texts provide only selected illustrative examples for what bioeconomy entails, e.g. listing specific products such as "Waschmittel, deren reinigende Enzyme von gentechnisch veränderten Bakterien hergestellt werden; Kunststofftüten aus Mais; Lebensmittelzusätze aus Algen—das alles sind Beispiele für die Bioökonomie."[41] This supports our earlier finding that the bioeconomy so far seems to remain a niche topic in the public perception currently lacking an in-depth debate.

Interestingly, specific incidences can be found which document an awareness of how limited the bioeconomy discourse in Germany is. For example, a 2015 die Tageszeitung article criticizes the way of defining the bioeconomy by listing

[39] Cf. Schwägerl (2011).

[40] Cf. e.g. Drogowski (2019).

[41] Holdinghausen (2019b), "Detergents based on cleaning enzymes produced by genetically modified bacteria; plastic bags made of maize; food additives made of algae – these are all examples of the bioeconomy" (own translation).

products, as described above, and refers to this phenomenon as "Was-es-nicht-alles-gibt-Modus".[42] So even though the discourse is often superficial, critical voices are highlighting exactly this shortcoming.

Furthermore, texts which provide a broader examination of the bioeconomy are often oriented towards official definitions which rest on the use of biological resources and the sustainability paradigm, e.g. "Die Bioökonomie basiert auf der Nutzung biologischer Ressourcen wie beispielsweise nachwachsender Rohstoffe für die Erzeugung verschiedenster Produkte. Ziel ist, dass Produkte, Verfahren und Dienstleistungen in allen wirtschaftlichen Sektoren sozialen, ökonomischen und ökologisch nachhaltigen Kriterien genügt."[43] The understanding of the bioeconomy concept in media discourse is thus generally close to the official political discourse.

Few texts engage in a critical discussion. Examples thereof are the aforementioned *DIE ZEIT* portrait of BRAIN AG[44] and critical comments on policy developments.[45] As noticed with regard to the bioeconomy definition, despite the overall lack of critical engagement, there are incidences of more thorough discussions about the topic.

Although only some texts directly consider what the concept is supposed to entail and what challenges lie ahead, we found that different visions are in fact employed in all texts. According to the coding scheme, all three types of bioeconomy visions—bio-resource, bio-technology, and bio-ecology—were identified in the media texts, in some cases within the same text. Thus, visions on the bioeconomy are primarily presented implicitly without critical engagement.

3.3.1 Research Organizations and Scientists

The most active stakeholder group in media discourse, research organizations and scientists, is mainly presented in the context of bio-resource and bio-technology visions. Most notably, articles centering on this group highlight how a knowledge-based bioeconomy that rests on increasing efficiency and technological innovation holds potential for strengthening German competitiveness. For example, in a 2019 article on a lecture event at the Forschungszentrum Jülich, the representation of the bioeconomy revolves around the optimization of substances and processes, e.g. in the use of food products, and how this approach can create opportunities for a region characterized by the discontinuation of the coal industry.[46] This also applies to the numerous reports on recent research projects and outcomes. Parts of this argument are ethical considerations regarding genome editing and animal testing, which by

[42] Holdinghausen 2015, "What's not to discover-mode" (own translation).

[43] Werner 2015, "The bioeconomy is based on utilizing biological resources such as renewable raw materials for creating diverse products. Objective is for products, processes, and services across all economic sectors to fulfill the criteria of social, economic, and ecological sustainability" (own translation).

[44] Cf. Grefe (2015).

[45] Cf. e.g. Holdinghausen (2019c).

[46] Cf. Drogowski (2019).

these stakeholders are primarily justified using their value for human welfare or economic competitiveness. For example, a 2011 Der Spiegel article critically discusses the limits on genetic engineering in Germany and the consequential loss of knowledge and industry,[47] and *DIE ZEIT* offers a debate between two scientific perspectives on genetic engineering.[48] In general, however, genetic engineering plays only a relatively minor role in the overall bioeconomy discourse: it is mentioned in less than 30% of the media texts, and in those that do, most references do not go beyond the mere mentioning of genetic engineering as an example of bioeconomic activities. Overall, research organizations and scientists strongly drive the bio-technology vision.

Moreover, there is an equally strong second thematic focus of understanding the bioeconomy fundamentally as a challenge of substituting the resource base. In the 2019 article on the lecture event at the Forschungszentrum Jülich referred to in the preceding paragraph, for example, shifting away from fossil fuels is presented as the central challenge, and the bioeconomy is understood as an improved, more sustainable way of managing natural resources.[49] Here, the bioeconomy discourse is directly connected to concerns about climate change and resource scarcity. To provide another example, in a 2016 local newspaper, the development of raw materials for the future ("Zukunftsrohstoffe") is discussed.[50] The bio-resource vision is also mirrored in the already mentioned bioeconomy products lists, since these lists highlight that products known in the bioeconomy are produced with different raw materials.[51] Here, both the bio-technology and the bio-resource visions are combined by arguing that the need for (bio)technological innovation is primarily tied to global challenges of environmental degradation and resource depletion. In line with this argument, texts revolving around this stakeholder group often argue explicitly for increasing research funding, e.g. in a 2014 interview with the president of the Fraunhofer-Gesellschaft, Reimund Neugebauer.[52] In general, the dominant stakeholder group of research organizations and scientists provides bio-resource and bio-technology visions, focusing strongly on potentials for growth and competitiveness as well as the key roles of biomass and technologies.

Although the analysis shows that the sustainability paradigm was also referred to, only few texts addressed sustainability more thoroughly. A 2010 article on the recently founded research association BioSC serves as an example. In this text, sustainable biomass production is understood to rely on integrated processes.[53] However, the term "sustainable" is usually employed as an attribute without further explanation or mentioned solely with regard to material flows and their optimization.

---

[47] Cf. Schwägerl (2011).

[48] Cf. Grefe (2016).

[49] Cf. Drogowski (2019).

[50] Cf. Bayerische Gemeindezeitung (2016).

[51] Cf. e.g. Holdinghausen (2015).

[52] Cf. Sentker (2014).

[53] Cf. VDI nachrichten (2010).

Although close to 40% of all texts refer to sustainability, barely any discuss what specifically defines a sustainable bioeconomy. Thus, ecological aspects as conceptualized by the bio-ecology vision are largely missing from the research organizations and scientists' discourse presented in the media, and are outweighed by the focus on bio-resource and bio-technology.

3.3.2 Governmental Organizations and Politicians

Governmental organizations and politicians were found to be the second-most prominent stakeholder group. These stakeholders also strongly support the idea of the bioeconomy as a biomass-based growth concept, in line with the bio-resource vision. Here, again, the bioeconomy is primarily described as a concept to replace the resource base in reaction to resource scarcity and supply insecurities. For example, in the context of the 2015 meeting of agricultural ministers in Berlin, the bioeconomy is conceptualized as the utilization of biological resources with the goal of providing global food security and economic prosperity, specifically in developing countries.[54] Additionally, with regard to Germany, the potential for strengthening national competitiveness and fostering jobs in rural areas is highlighted, e.g. in a 2013 article on a visit by a Bündnis90/Die Grünen party delegation to BRAIN AG.[55] The same argument is brought forward, e.g. by UK agricultural minister Elizabeth Truss in a 2016 article on the Brexit.[56] However, especially in Germany the connection to rural areas is frequently drawn, highlighting the bioeconomy as an alternative concept of prosperity for areas of declining mining and large industry.[57] Thus, the governmental organizations and politicians strongly drive a bio-resource vision based on the swapping of raw materials and the prospect of growth, especially for rural areas.

Moreover, this stakeholder group also employs the bio-technology vision by tying it to the bio-resource perspective. For example, in the aforementioned article on the 2015 agricultural summit, the pressures on resource availability are considered to argue for innovative technologies and the optimization of processes as the bioeconomy's cornerstone.[58] Altogether, governmental organizations and politicians hold visions very similar to the perspective from research organizations and scientists.

Mirroring the results of the previous sub-section, sustainability was not discussed in depth. Similar to the stakeholder group of research organizations and scientists, the bio-ecology vision is a marginal part of governmental institutions and politicians' discourse. Thus, due to the dominant role of these two groups, ecological aspects play only a minor role in the overall public debate. Instead, media texts are

[54] Cf. Werner (2015).

[55] Cf. Mynewsdesk (2013).

[56] Cf. Bongardt (2016).

[57] Cf. e.g. Unfried (2011).

[58] Cf. Werner (2015).

primarily dominated by bio-resource and bio-technology visions, which are sometimes combined to the argument that resource pressures and economic needs are best met with a bioeconomy based on research and innovation.

3.3.3 Farmers and Biotechnology Companies

As discussed in Sect. 3.2, stakeholders from the producing and processing stages of the bioeconomy are primarily considered passively by other stakeholders who include them in their vision. To provide an example, governmental organizations often call for stronger coordination between research and business to realize the economic growth potential of the bioeconomy.[59] Of the few key actors actively involved, farmers and bio-tech companies were identified as most prominent. Thus, this section focuses specifically on those groups.

Farmers and farming associations' perspectives revolve around the idea of utilizing and converting biomass, while balancing demands for more sustainable practices and food security. Focus lies on the availability and cultivation regulations of biomass in the context of global food security.[60] In this regard, also issues of land grabbing and land consumption are mentioned, but they are mainly discussed in terms of resource scarcity and economic losses instead of ecological concerns.[61] Especially the idea is stressed that farming must be economically profitable in order to enable farmers to provide adequate food supply, e.g. by the president of the World Farmers' Association, Dr. Evelyn Nguleka.[62] Thus, farmers and farming associations primarily discuss the bioeconomy around questions of global production and trade of biological raw materials, following the argument of the bio-resource vision.

Again, the bio-resource vision also ties in with the bio-technology vision, since innovation is understood as the only measure to balance the conflicting demands for biomass. For example, the Food and Agriculture Organization's (FAO) general director Jacques Diouf is quoted on that innovative technology has to enable farmers to provide both food security to the population and raw materials to industry.[63] Although considerations of the bio-resource vision occur most frequently, the bio-technology vision is also regularly used in combination.

In turn, industry stakeholders and specifically biotechnology companies are found to primarily employ the bio-technology vision. They focus on efficiency and innovation, highlighting the central role of technologies for the bioeconomy transformation. For example, BRAIN AG CEO Holger Zinke is quoted in a 2013 article, stating that a resource-poor country such as Germany is required to invest in innovation to facilitate industry growth and thus value creation.[64] Generally, the

[59] Cf. e.g. ibid.

[60] Cf. e.g. Holdinghausen (2010).

[61] Cf. e.g. Hirn and Müller (2011).

[62] Cf. Werner (2015).

[63] Cf. Hirn and Müller (2011).

[64] Cf. Business Wire Deutsch (2012).

bioeconomy is portrayed to generate jobs, e.g. in the chemical industry.[65] In line with the biotechnology vision as employed by research organizations and scientists, this stakeholder group is also often represented in support of genetic engineering, sometimes combined with a call for an open societal debate.[66] Thus, from the group of industrial stakeholders, especially biotechnology companies provide a vision of the bioeconomy that follows the lines of thinking summarized in the bio-technology vision.

As stated earlier, this focus is readily combined with the bio-resource vision's focus on substituting fossil for biological resources. In this context, this idea is rather implicit in the numerous examples of biotechnology products which are presented to be able to substitute previously unsustainable products.[67] Altogether, we found that most stakeholders included in the media discourse provide a combination of bio-resource and bio-technology visions.

3.3.4 Environment, Development, and Consumer Associations

In contrast to the aforementioned stakeholder groups, general public actors are found to provide a minor, but diverse perspective on the German bioeconomy. As mentioned in Sect. 3.2, specific environment, development, or consumer associations offer critical voices on bioeconomy development and design. Overall, a stronger focus is set on the sustainability principle. For example, the think tank Denkhaus Bremen is quoted criticizing the lack of specific criteria on how the bioeconomy can be shaped in a sustainable way.[68] Moreover, in a 2015 article on the Bioeconomy Summit of that year, Roman Herre of the development aid association FIAN Germany states that current bioeconomy plans to support the use of biomass rather fuel resource and environmental conflicts and that economic opportunities for rural areas have not materialized.[69] Here, general public actors point out that the bioeconomy needs to include specific rules on how, not only which, natural resources are used, especially with regard to social, economic, and ecological challenges.

While the call for economic opportunities reflects the discourses of the stakeholders discussed above, in some aspects societal stakeholders voice visions which are diametrically opposed. In a 2019 statement by the Aktionsforum Bioökonomie, the commercialization of bio-ecological knowledge is criticized.[70] Focus here lies on sustainable practices and environmental protection. Moreover, instead of discussing primarily the production side of the bioeconomy, few societal stakeholders highlight also, e.g., the necessity to change individual consumption

[65] Cf. VDI nachrichten (2013).

[66] Cf. e.g. Holdinghausen (2019b).

[67] Cf. e.g. Grefe (2015).

[68] Cf. Holdinghausen (2019b).

[69] Cf. Holdinghausen (2015).

[70] Cf. Holdinghausen (2019b).

behavior[71] and to introduce principles of recycling and re-use.[72] These instances imply the idea of sufficiency, but the concept itself is only referred to in three texts. Also, this stakeholder group provides the most explicit rejection of genetic engineering.[73] An additional prominent point of criticism is lacking transparency and participation.[74] Of all stakeholder discourses analyzed, positions by societal stakeholders can be identified to follow most closely the bio-ecology vision. However, since they are seldom represented in media texts, their impact on the overall public discourse seems limited.

Overall, the discourse analysis revealed that visions of the German bioeconomy predominantly revolve around the principle of substituting biomass for fossil fuels and the idea of a central role for technology and innovation in pursuit of economic prosperity. The discussion is relatively superficial, and critical voices are limited to societal actors and only few considerations within the texts. Interestingly, despite the bioeconomy's often referred to local growth potential, the discourse mainly occurs against the background of global challenges, which results in the dominance of political and scientific actors.

4 Discussion

Given the clear dominance of stakeholders and visions concerned with biomass production and biotechnologies as discussed above, two observations shall be further elaborated in the following discussion. First, the analysis revealed a dominance of scientific actors in media discourse on bioeconomy. To some degree, this is certainly owed to the relative newness of the concept, which is still in the early implementation phase. However, this is also reflected in the strong thematic focus on research funding and cooperation, and technology transfer and commercialization—and thus, the bio-technology vision. Here, our analysis shows that currently the public discourse is relatively close to the elite-driven technocratic discourse in political strategies and expert statements,[75] because often the same stakeholders prevail in both realms. For the bioeconomy transformation in Germany, this implies that change processes are primarily directed at supporting knowledge-based corporate activities.

Moreover, the technology focus is regularly reasoned for by referencing global challenges of resource availability, connecting to the bio-resource vision. Interesting and unsuspected was also the regular consideration of supply security against the background of the bio-resource vision. Though developed countries generally suffer rather from the mismanagement of natural resources than from an actual physical

[71] Cf. Holdinghausen (2019c).

[72] Cf. Holdinghausen (2019a).

[73] Cf. Holdinghausen (2019c).

[74] Cf. e.g. Holdinghausen (2019b).

[75] Cf. McCormick and Kautto (2013), Loiseau et al. (2016) and Vivien et al. (2019).

scarcity, German media texts regularly refer to concerns about food and energy insecurities. The securitization of natural resources is known to be a successful rhetorical means in national policymaking[76] and is likely used here as well to argue for the development of a bioeconomy. These findings support assumptions made previously, according to which the bioeconomy discourse is strongly driven in a top-down manner that diverges from visions held by other stakeholders who are not as strongly involved.[77]

This first observation connects to the second one, namely a considerable cleavage between a growth-based technology focus and a sustainability-based behavioral focus. Individual texts and text segments by societal stakeholders reveal that they often hold visions that deviate substantially from the predominant discourse, focusing on environmental concerns and issues of consumption instead. Here we found a cleavage between the predominant, technology- and resource-oriented visions of established stakeholders, and alternative sustainability-focused visions of individual societal stakeholders.

However, this cleavage is not expressed in a critical debate. Stakeholders such as social and environmental think tanks and citizen organizations receive little coverage compared to politicians and scientists. Since this picture resembles so closely what has been found with regard to policy documents, we assume that the broader society has not yet fully engaged with the topic. Even though there is an apparent awareness that citizens and consumers are essential for implementing the bioeconomy, these stakeholders are barely engaged in the media discourse. Instead, the generally superficial and less critical debate on the bioeconomy's goals demonstrates that the transformation towards a bioeconomy indeed suffers from a lack of participation.

Summing up these aspects, we argue that the German public discourse on the transformation towards a bioeconomy, so far, has been primarily driven by established stakeholders with established, production-oriented visions. Societal actors and consumer-oriented, alternative sustainability visions receive little attention, but the lack of participation is recognized.

5 Conclusions

The German natural resources management system is undergoing profound changes. Two crucial change processes are the impending structural change in lignite mining regions, and the implementation of a bioeconomy, e.g. in the Rheinische Revier, the largest remaining lignite deposit in Europe. A multitude of stakeholders is affected by and shapes these processes. Therefore, a thorough analysis of stakeholder dynamics is required. Analyzing media texts offers a better understanding of which stakeholders and stakeholder visions specifically form the public perception of the bioeconomy, which may partially frame the direction it evolves into.

[76] Cf. Fischhendler and Katz (2012) and Fischhendler and Nathan (2014).

[77] Cf. e.g. Stern et al. (2018); Hausknost et al. (2017).

In this chapter, we conducted a computer-assisted, comprehensive media-based discourse analysis in order to identify and characterize stakeholders and visions which are publicly considered with regard to the bioeconomy transformation in Germany. We found that media discourse on the bioeconomy was strongly dominated by research and government stakeholders. These actors highlighted visions of the bioeconomy as a growth concept for supporting the use of biomass and biotechnologies against the background of global challenges of resource depletion. Thus, they provide visions which are categorized as bio-resource and bio-technology visions. Of the stakeholders from the production and processing stages of the bioeconomy, only farming associations and biotechnology companies were actively involved in the discourse, generally supporting these visions. Stakeholders from the general public were often called upon, but seldom expressed their own visions, which were characterized by more diverse themes and arguments, e.g. by a stronger focus on sustainability concerns and a consumption-oriented perspective. Also, media texts more often refer to global and national levels than regional, with the exception of some coverage on the transformation towards a bioeconomy in the Rheinische Revier. Overall, the media discourse on the German bioeconomy was surprisingly coherent and consistent with political strategies and expert statements.

We pursued an exploratory approach to stakeholder visions of the bioeconomy in Germany in order to analyze publicly portrayed perceptions of the topic. The results provide a useful basis for a better understanding of how media coverage shapes the transformation towards a bioeconomy in Germany, and partially also the direction into which it may drive. However, this analysis also poses several new questions for further research.

Among the observations from the analysis of media discourses on the bioeconomy, the most surprising was how predictable results were. This observation requires an in-depth comparison in order to clarify whether the resemblances are rhetorical or substantial in nature, and to explain the seeming lack of a critical public engagement with the bioeconomy. A more inductive approach might reveal more clearly where understandings of bioeconomy diverge between research and the broader public. Here, participatory approaches such as stakeholder workshops appear especially relevant—not only with regard to the often-mentioned need for more acceptance, but especially for developing a more balanced vision for a bioeconomy which is not dominated by technological fixes.

Also, the analysis poses further questions on stakeholder dynamics in the transformation towards a bioeconomy. We observed differences between stakeholder visions in time and place, i.e. an increasing public attention and variations between individual federal states and mining regions. Also, certain stakeholder groups appeared to play key roles in shaping overall public perception, while other stakeholders are barely or only passively referred to. Consequently, further research needs to engage with these differences in order to analyze how stakeholder visions spread between actors and across regions and time, and whether and why certain stakeholder groups or discourses, such as structural change, remain separate.

Further research is required on cleavages between stakeholder visions, and how these might affect a successful implementation of a bioeconomy. In this chapter, we postulated the relevance of stakeholders for transformation processes and found that different stakeholders pursue different visions. This leads us to question, whether different stakeholder coalitions can be identified, and in how far these may shape the transformation towards a bioeconomy.

References

Aachener Zeitung. (2011). Gemeinsam für nachhaltige Entwicklung; Neue Strukturen für die Region. Wirtschaftsminister Harry Voigtsberger begrüßt das Engagement des Forschungszentrums. In *Aachener Zeitung*, 21.

Aachener Zeitung. (2013). Nicht noch mal die Zeche für Kohle zahlen; Stand die Region nach dem Ende der Steinkohle-Ära plötzlich vor dem Strukturwandel, soll dies bei der Braunkohle nicht passieren. In *Aachener Zeitung*, 10.

Allgemeine Fleischer Zeitung. (2016). Hohenheim fördert kleine Produzenten. In *Allgemeine Fleischer Zeitung*, 12, 4.

Bayerische Gemeindezeitung. (2016). Straubing sucht den Plan B, wenn das Erdöl ausgeht. In: *Bayerische Gemeindezeitung*, 8, 13.

Bäumel-Schachtner, M. (2016). So wertvoll sind nachwachsende Rohstoffe; Forum zum Thema Bioökonomie Minister Brunner: Verbraucher müssen überzeugt werden. In *Passauer Neue Presse*.

Bongardt, K. (2016). Forscher warnen vor Brexit; Rückzug Großbritanniens aus der EU wäre für Studium und Forschung verheerend. *Agrarzeitung Ernährungsdienst, 23*, 9.

Bugge, M., Hansen, T., & Klitkou, A. (2016). What is the bioeconomy? A review of the literature. *Sustainability, 8*(7), 691.

Business Wire Deutsch. (2012). BRAIN beschließt Finanzierungsrunde von 60 Mio. Euro zur Ausweitung der Buy and Build Strategie; Eingeworbenes Kapital unterstützt BRAIN auf dem Weg zu voll integriertem Industrieunternehmen. In *Business Wire Deutsch*. Retrieved June 22, 2020, from, https://www.businesswire.com/news/home/20121127005652/de/

CITplus. (2012). Achema im Zentrum der Bioökonomie – Bioökonomie im Zentrum der Achema; Prof. Dr. Kurt Wagemann, Geschäftsführer der Dechema Gesellschaft für Chemische Technik und Biotechnologie e.V. In: *CITplus* 8.

Commission "Wachstum, Strukturwandel und Beschäftigung". (2019). *Abschlussbericht*. Retrieved June 16, 2020, from, https://www.bmwi.de/Redaktion/DE/Downloads/A/abschlussbericht-kommission-wachstum-strukturwandel-und-beschaeftigung.pdf?__blob=publicationFile&v=4

Drogowski, J. (2019). Die Zukunft nach der Braunkohle; Kommunale und regionale Perspektiven im Forschungszentrum Jülich aufgezeigt: Nachhaltigkeit und zirkuläre Bioökonomie. In *Aachener Nachrichten*, 17.

Editorial Nature Climate Change. (2016). The role of society in energy transitions. *Nature Climate Change, 6*, 539.

Editorial Nature Energy. (2016). That human touch. *Nature Energy, 1*, 16069.

Federal Ministry of Education and Research. (2010). *Nationale Forschungsstrategie BioÖkonomie 2030. Unser Weg zu einer bio-basierten Wirtschaft*. Retrieved June 16, 2020, from, https://www.bmbf.de/upload_filestore/pub/Nationale_Forschungsstrategie_Biooekonomie_2030.pdf

Federal Ministry of Education and Research. (2014). *Bioökonomie als gesellschaftlicher Wandel. Konzept zur Förderung sozial- und wirtschaftswissenschaftlicher Forschung für die Bioökonomie*. Retreived June 16, 2020, from, https://www.bmbf.de/upload_filestore/pub/Biooekonomie_als_gesellschaftlicher_Wandel.pdf

Federal Ministry for the Environment, Nature Conservation and Nuclear Safety. (2019). *Kohleausstieg – Kommission "Wachstum, Strukturwandel und Beschäftigung"*. Retrieved September 5, 2019, from, www.bmu.de/WS5037

Federal Ministry of Food and Agriculture. (2013). *Nationale Politikstrategie Bioökonomie. Nachwachsende Ressourcen und biotechnologische Verfahren als Basis für Ernährung, Industrie und Energie*. Retrieved June 16, 2020, from, http://www.bio-step.eu/fileadmin/BioSTEP/Bio_strategies/Nationale_Politikstrategie_Biooekonomie.pdf

Fischer, W., Hake, J.-F., Kuckshinrichs, W., Schröder, T., & Venghaus, S. (2016). German energy policy and the way to sustainability: Five controversial issues in the debate on the "Energiewende". *Energy, 115*, 1580–1591.

Fischhendler, I., & Katz, D. (2012). The use of "security" jargon in sustainable development discourse: Evidence from UN Commission on sustainable development. *International Environmental Agreements: Politics, Law and Economics, 13*, 321–342.

Fischhendler, I., & Nathan, D. (2014). In the name of energy security: The struggle over the exportation of Israeli natural gas. *Energy Policy, 70*, 152–162.

Georgis, A. (2016). Marokkanisches Öl und deutsches Bier; Erstmals ist ein afrikanischen Land Partner der Grünen Woche. Und die einheimischen Brauer feiern 500 Jahre Reinheitsgebot. In *Berliner Zeitung*, 3.

Gibbs, G. R. (2014). Using software in qualitative analysis. In U. Flick (Ed.), *The SAGE handbook of qualitative data analysis* (pp. 277–294). SAGE Publishing.

Grefe, C. (2015). Forschungsleiter Mikrobe; Pullover aus Milch, Eis aus Lupinen, Waschmittel aus Bakterien – solche Produkte sollen unter dem Label Bioökonomie die Industrie grüner machen. Ihre Vertreter treffen sich demnächst zu einem Weltgipfel in Berlin. Wie aussichtsreich sind ihre Visionen? In *Die ZEIT* 47, 45.

Grefe, C. (2016). "Das Leben bahnt sich seinen Weg". Der Biologe Andreas Weber und die Physikerin Petra Schwille streiten über das Risiko, künstliche Organismen zu erschaffen. In *Die ZEIT* 10, 33.

Hake, J.-F., Fischer, W., Venghaus, S., & Weckenbrock, C. (2015). The German Energiewende – History and status quo. *Energy, 92*, 532–546.

Hausknost, D., Schriefl, E., Lauk, C., & Kalt, G. (2017). A transition to which bioeconomy? An exploration of diverging techno-political choices. *Sustainability, 9*(4), 669.

Hempel, C., Will, S., & Zander, K. (2019). Societal perspectives on a bio-economy in Germany: An explorative study using Q methodology. *International Journal on Food System Dynamics, 10*(1), 21–37.

Hirn, W., & Müller, E. (2011). Die große Dürre. *Manager Magazine, 4*, 80.

Holdinghausen, H. (2010). Milliarden für den Rohstoffwandel; 2030 könnte ein Drittel der Industrieproduktion von Biotechnologie abhängen. Die Bundesregierung gibt daher 2,4 Milliarden Euro aus, um den Umbau der Wirtschaft voranzutreiben. In *die tageszeitung*, 8.

Holdinghausen, H. (2015). Fahrradschläuche aus Löwenzahn. Die Umstellung auf nachwachsende Rohstoffe kommt nicht so richtig voran. Umwelt- und Verbraucherschützer fordern einen grundlegenden Kurswechsel bei den offiziellen Bioökonomie-Strategien. In *die tageszeitung*, 18.

Holdinghausen, H. (2019a). Die Bioökonomie ist kein wirklicher Neubeginn. Erdöl durch Pflanzen ersetzen? Keine Lösung, sagt Josephine Koch vom Forum Umwelt und Entwicklung. Die Organisation ist eine von 18 Unterzeichnern einer Erklärung gegen die Bioökonomie-Strategie der Bundesregierung. In *die tageszeitung*, 9.

Holdinghausen, H. (2019b). Eine vage Vorstellung von der Bioökonomie. Die Bundesregierung legt einen Entwurf für ihre Bioökonomie-Politik vor. Umweltverbände vermissen eine Obergrenze für Naturrohstoffe, die Industrie den Begriff der Gentechnik. In *die tageszeitung*, 9.

Holdinghausen, H. (2019c). Erschreckend ideenlos. Bioökonomie könnte die Zukunft sein, wenn man es richtig macht. Doch was die Bundesregierung bisher plant, hat kein Konzept und vor allem kein Ziel. In *die tageszeitung*, 12.

Jacobs, I. (2014): Finanzierung ist das größte Sorgenkind der Uni. In *Stuttgarter Zeitung*, 20.

Kircher, M., & Schwarz, T. (2015). Zehn Jahre Bioökonomie: Bilanz und Ausblick; Rohstoff- und Energieeffizienz bestimmen Wirtschaftlichkeit und sind für die Zukunft unumgänglich. In *CHEManager*, 6. Retrieved August 8, 2019, from, https://www.chemanager-online.com/themen/konjunktur/zehn-jahre-biooekonomie-bilanz-und-ausblick.

Labo. (2012). In *Labo - Magazin für Labortechnik* 6, 19.

Labo. (2017). Grundlagenwerk; Bioökonomie. In *Labo - Magazin für Labortechnik* 10, 63.

Leipold, S., & Petit-Boix, A. (2018). The circular economy and the bio-based sector - perspectives of European and German stakeholders. *Journal of Cleaner Production, 201*, 1125–1137.

Leipprand, A., & Flachsland, C. (2018). Regime destabilization in energy transitions: The German debate on the future of coal. *Energy Research & Social Science, 40*, 190–204.

Lewandowski, I. (2018). *Bioeconomy: Shaping the transition to a sustainable, biobased economy*. Springer.

Loiseau, E., Saikku, L., Antikainen, R., Droste, N., Hansjürgens, B., Pitkänen, K., Leskinen, P., Kuikman, P., & Thomsen, M. (2016). Green economy and related concepts: An overview. *Journal of Cleaner Production, 139*, 361–371.

Maxwell, J. A., & Chmiel, M. (2014). Notes toward a theory of qualitative data analysis. In U. Flick (Ed.), *The SAGE handbook of qualitative data analysis* (pp. 21–34). SAGE Publishing.

McCormick, K., & Kautto, N. (2013). The bioeconomy in Europe: An overview. *Sustainability, 5*(6), 2589–2608.

Merkel, W., Kollmorgen, R., & Wagener, H.-J. (2019). Transformation and transition research: An introduction. In Id (Ed.), *The handbook of political, social, and economic transformation* (pp. 1–14). Oxford University Press.

Ministry of Culture and Science of the State North Rhine-Westphalia. (2012). *Eckpunkte einer Bioökonomiestrategie für Nordrhein-Westfalen*.

Moe, E. (2016). Germany: At a crossroads, or social and political consensus setting it on a course for structural change. In Id (Ed.), *Renewable energy transformation or fossil fuel backlash: Vested interests in the political economy* (pp. 137–165).

Mynewsdesk. (2013). Wirtschaftspolitischer Austausch zwischen Biotech-Unternehmen BRAIN und Bündnis 90/Die Grünen. In *Mynewsdesk*. Retrieved August 8, 2019, from, http://www.themenportal.de/wirtschaft/wirtschaftspolitischer-austausch-zwischen-biotech-unternehmen-brain-und-buendnis-90-die-gruenen-83870

Paltridge, B. (2006). *Discourse analysis: An introduction*. Continuum London.

Peltomaa, J. (2018). Drumming the barrels of Hope? Bioeconomy narratives in the media. *Sustainability, 10*(11), 4278.

Ronzheimer, M. (2018). Die Biologisierung der Wirtschaft; Mit der Bioökonomie soll die Wirtschaft nachhaltiger werden. Ziel ist die zunehmende Nutzung nachwachsender Rohstoffe, erneuerbarer Energien sowie der grünen Chemie. In *die tageszeitung*, 18.

Ronzheimer, M. (2019). Ökologisch fragwürdig; Zivilgesellschaft will kritische Debatte über Bioökonomie. In *die tageszeitung*, 18.

Schwägerl, C. (2011). *Der nächste Ausstieg. Der Spiegel, 34*, 30–32.

Sentker, A. (2014). "Darum beneiden uns andere". Ein Gespräch mit Reimund Neugebauer, dem Präsidenten der Fraunhofer-Gesellschaft, über Innovation, heimliche Weltmeister – und das Risiko NSA. In *Die ZEIT* 8, 44.

Speen, A. (2014). Neue Gespräche zur Zukunft der IRR. In *Rheinische Post*.

Stern, T., Ploll, U., Spies, R., Schwarzbauer, P., Hesser, F., & Ranacher, L. (2018). Understanding perceptions of the bioeconomy in Austria—An explorative case study. *Sustainability, 10*(11), 4142.

Unfried, P. (2011). "Wir brauchen keine Autofirmen" Robert Habeck. Seit der grüne Fraktionsvorsitzende das Bruttoinlandsprodukt durch eine neue Messung des Wohlstands ersetzt hat, ist Schleswig-Holstein plötzlich ein Musterland. Rechnet er das BIP-Kellerkind schön? In *die tageszeitung*, 4.

VDI nachrichten. (2010). Europaweit einmaliges Zentrum für Bioökonomieforschung. In *VDI nachrichten*, 2.

VDI nachrichten. (2013). "Chemie hat ihre goldene Zukunft noch vor sich." In *VDI nachrichten*, 9.
Vivien, F.-D., Nieddu, M., Befort, N., Debref, R., & Giampietro, M. (2019). The hijacking of the bioeconomy. *Ecological Economics, 159*, 189–197.
Werner, A. (2015). Die Bioökonomie als Teil der Lösung; Ernährungssicherung hat für Agrarminister Priorität - Politische Rahmenbedingungen fehlen in vielen Ländern. *Agrarzeitung Ernährungsdienst, 4*, 3.
Wohlfahrt, J., Frechaud, F., Gabrielle, B., Godard, C., Kurek, B., Loyce, C., & Therond, O. (2019). Characteristics of bioeconomy systems and sustainability issues at the territorial scale. A review. *Journal of Cleaner Production, 232*, 898–909.

Part V

Regulation and Economics

Bioeconomy and Genome Editing: A Comparison Between Germany and the Netherlands

11

Robin Siebert, Christian Herzig, and Marc Birringer

Abstract

This chapter examines the extent to which strategies for bioeconomic transformation have been developed in Germany and the Netherlands and how they differ in content. The analysis is based on national bioeconomic strategies as well as official statements and press releases published by the governments, ministries, and independent scientific advisory bodies and institutes of both countries until the end of 2019. The potential that both countries attribute to genome editing for the development of a post-fossil (agricultural) economy is the primary interest of the analysis. The interest of the analysis is also directed towards the way in which both countries have so far participated in the discourse on the legal classification of genome editing by the European Court of Justice. The legal classification plays a decisive role in the discourse, since stricter regulation can require genome edited organisms to undergo time- and cost-intensive approval procedures and labelling as genetically modified. For a better understanding of the discourse, the chapter also describes the general attitude towards biotechnologies in both countries in the past and illustrates insights into the social acceptance of the new biotechnological methods from first surveys carried out in both countries.

Keywords

National bioeconomic strategies · Genome editing · Social acceptance

R. Siebert (✉) · C. Herzig
Department Management in the International Food Industry, Faculty of Organic Agricultural Sciences, University of Kassel, Witzenhausen, Germany
e-mail: r.siebert@uni-kassel.de; herzig@uni-kassel.de

M. Birringer
Department of Nutritional, Food and Consumer Sciences, Applied Biochemistry for Nutrition and Environment, Fulda University of Applied Sciences, Fulda, Germany
e-mail: marc.birringer@oe.hs-fulda.de

D. Lanzerath et al. (eds.), *Bioeconomy and Sustainability*,
https://doi.org/10.1007/978-3-030-87402-5_11

1 Introduction

The concept "bioeconomy" describes a form of economy "where the basic building blocks for materials, chemicals and energy are derived from renewable biological resources".[1] According to the *Knowledge Based Bioeconomy* (KBBE) of the *European Commission* (EC) the transformation from open to closed material cycles and the increase in biomass production should ensure global food security for the growing world population.[2] Therefore, the establishment of a bioeconomic strategy pursues socio-economic goals, such as strengthening the economy by means of research funding and the occupation of future markets.[3] In the context of a bioeconomic transformation, a wide range of measures and approaches are being discussed,[4] including biotechnology and especially the new molecular biological techniques of *genome editing* (GE).[5] CRISPR/Cas, which was presented in 2012, is considered to be the best known method of GE. Compared to previous molecular biological methods, CRISPR/Cas is supposed to allow a fast, targeted and cost-effective modification of the gene structure.[6] The basic method of GE is the cut at a defined position on the double-stranded DNA using so-called *molecular scissors* and the subsequent mutation of a single base or entire base sequence introduced by the cell's own repair mechanisms.[7] The potentials and risks of GE for the bioeconomy are examined in this chapter from a comparative perspective between Germany and the Netherlands. These two countries were the focus of the *Bioeconomy and Modern Biotechnologies: Ethical, Legal and Social Aspects* retreat week in September 2019, which was funded by the German *Federal Ministry of Education and Research.* According to the *Fraunhofer Institute for Systems and Innovation Research* (ISI), Germany, through the *National Research Strategy BioEconomy 2030* (NRSB) formulated in 2010, is taking an "international pioneering role"[8] in bioeconomy, while according to Bosman and Rotmans, the Netherlands is more likely "a laggard rather than a frontrunner in Europe", partly because of its greater dependence on fossil fuels.

In the following, the relevance of bioeconomy on the political, economic and scientific level in Germany and the Netherlands will be explained first. The overview shows which bioeconomy strategies have been developed in the two countries, to what extent they differ from each other and what potential is ascribed to GE for the development of a post-fossil economy. Subsequently, the analysis focuses on the

[1]McCormick and Kautto (2013, 2589); cf. also Pietzsch (2017, VII).

[2]Cf. Albrecht et al. (2010).

[3]Cf. Kiresiewa et al. (2019).

[4]Cf. Lewandowski (2018).

[5]Cf. German Bioeconomy Council (2019a).

[6]Cf. Knott and Doudna (2018).

[7]Cf. Siebert et al. (2018).

[8]Fraunhofer Institute for Systems and Innovation Research (2017), II, "international pioneering role" (own translation).

way in which both countries have contributed and continue to contribute to the European discourse on the legal classification of GE by the *European Court of Justice* (ECJ). The legal assessment plays a decisive role in the discourse[9] since *genome edited* organisms have to go through time- and cost-intensive approval procedures due to regulation and have to be labelled as *genetically modified* (GMO). Regulation is not a ban, but a marketable application is made considerably more difficult or often practically impossible. Today, the majority of European consumers are sceptical about products labelled as GMOs. As a consequence of strict regulation, some scientists fear economic consequences for the European market and a blockade of innovation.

For a better classification of the discourse, the previous general attitude towards biotechnologies in both countries is also briefly presented and discussed. The findings from the comparison of the countries will be summarised and possible conclusions for the further course of the discourse will be derived. The analysis is based on official bioeconomic strategies and statements as well as press releases published by the governments, ministries, state and independent scientific advisory bodies and institutes of both countries until the end of 2019. More recent developments in the discourse from 2020 are not considered in our analysis.

2 Bioeconomy in Germany

In order to achieve the goal of a bio-based economy, German policymakers are pursuing the NRSB and the *National Policy Strategy on Bioeconomy* (NPSB), which was adopted by the Federal Cabinet in 2013. In addition to these two national strategies, the German states have developed different political strategies and funding measures at the federal level. In addition to state research funding programmes, there are also funding measures that are jointly financed by individual states and the federal governments. While some federal states have a broadly based research landscape on bioeconomy clusters and offer extensive research programmes, smaller states in particular appear to be keen to provide targeted support for the sale of individual bio-based products.[10] In Central Germany, for example the Bioeconomy cluster for the promotion of integrated material and energy use of non-food biomass for the production of materials, chemicals and products was established in 2012. The transformation process towards a bioeconomy, which began with the abandonment of lignite, was thus supported by the *Federal Ministry of Education and Research* in Central Germany with 80 million euros up to 2017 (of which 50% was provided by industry).

[9] Siebert et al. (2021).

[10] Cf. c/o BIOCOM AG (2020b).

2.1 National Research Strategy Bioeconomy 2030

Under the leadership of the *Federal Ministry of Education and Research*, a bioeconomic strategy was developed together with five other ministries, which provided a total of 2.4 billion euros for research and development up to the end of 2017. The NRSB has formulated two strategic goals with the vision of creating a "natural cycle-oriented, sustainable bio-based economy that carries the promise of global food supplies that are both ample and healthy, and of high quality products from renewable resource".[11] Firstly, Germany is to become a dynamic research and innovation location for bio-based products, energies, processes and services in international comparison. Secondly, research in Germany should also assume responsibility for feeding the growing world population and for climate, resource and environmental protection. Both goals are to be achieved with the help of sustainable agricultural production, the production of healthy and safe food and the industrial use of renewable raw materials, the expansion of biomass-based energy sources, international cooperation and cross-field measures.[12] The NRSB has thus initiated a change from technology- to mission-oriented research and development funding, which is geared towards overcoming social challenges.[13] The ISI, which evaluated the NSFB in 2017, assigns Germany an international pioneering role in the bioeconomy.[14] Taking into account a worldwide increasing use of genetic engineering and in order to achieve global food security, the NRSB proposes a "responsible handling of genetically modified plants".[15] In order to adapt crops to future requirements with regard to agriculture and forestry, a great need for innovative research approaches is also identified.[16] Under the name *Plant 2030*, the *Federal Ministry of Education and Research* is bundling specially funded research activities for applied plant research. These currently include the funding initiatives *Plant Breeding Research for the Bioeconomy*, *Plant Biotechnology for the Future* and various funding projects of the transnational programme, e.g. *PLANT-KBBE or Bioeconomy International*, in which public research institutions and companies from the plant breeding and bioeconomy sectors cooperate. The NRSB pointed out the possible potential of genome analysis methods at an early stage. Despite the promises of innovative biotechnologies, it also warned against glorifying technical progress as an end in itself. Rather, a careful analysis of ethical, legal and social aspects, a comprehensive formation of public opinion and participation as well as a strengthening of the dialogue and interaction between science, industry and the public are recognised as urgent.[17]

---

[11] Federal Ministry of Education and Research (2010, p. 3).

[12] Cf. ibid.

[13] Cf. Fraunhofer Institute for Systems and Innovation Research (2017).

[14] Cf. ibid.

[15] Federal Ministry of Education and Research (2010, p. 23).

[16] Cf. Federal Ministry of Education and Research (2020, p. 16).

[17] Cf. Federal Ministry of Education and Research (2010).

CRISPR/Cas is mentioned for the first time in the NRSB in the official announcement on 12 April 2017 for the promotion of research projects in plant research on *Crops for the future.*[18] Compared to established methods CRISPR/Cas is considered to have "enormous potential"[19] to realise significant progress, as it is seen to be easier, more precise, faster and cheaper to use. In order to support the German bioeconomy, the *Federal Ministry of Education and Research* promotes therefore the development of "forward-looking exploratory technological approaches which can markedly improve or accelerate molecular precision breeding of crops using optimized or novel CRISPR/Cas systems and other advanced genome editing techniques".[20]

2.2 National Policy Strategy on Bioeconomy

The objectives or guiding principles of the NPSB, which were developed under the leadership of the *Federal Ministry of Food and Agriculture* and are closely linked to the sustainability strategy of the Federal Government and the NRSB, are as follows: (1) Food security also has priority in the global context over the production of raw materials for industry and energy; (2) the use of pathways with a higher value-added potential is to be given preference in the further shaping of the framework conditions of the bioeconomy; (3) where possible and sensible, cascaded and coupled use of biomass should be implemented; (4) the securing and strengthening of the competitiveness of the bioeconomy in Germany and the growth potential on international markets must always be taken into account; (5) well-trained and informed specialists are indispensable for the competitiveness of the bioeconomy; (6) the opportunities and framework conditions for the use of key technologies and their transfer to economic use must be improved; (7) the bioeconomy must take account of growing societal demands on the way in which production is carried out—this applies to environmental, climate, nature and animal protection and to compliance with social standards; (8) the application of sustainability standards in producer countries, especially those with weak governance and weak institutions, must be expanded and efforts must be made to monitor compliance with them and (9) close cooperation between political, economic, scientific, environmental and social actors is necessary for the development of the bioeconomy.[21]

[18] Cf. Federal Ministry of Education and Research (2017).

[19] Ibid., 1.

[20] Ibid., 2.

[21] Cf. Federal Ministry of Food and Agriculture (2014, p. 21).

2.3 Bioeconomy Council

In order to ensure the implementation of both strategies, the *German Bioeconomy Council* (previously the *Bioeconomy Research and Technology Council*) was founded in 2009 by the *Federal Ministry of Education and Research* and the then *Federal Ministry of Food, Agriculture and Consumer Protection*. The independent and honorary body, consisting of 17 members, advises the German government on research and development in the knowledge-based bioeconomy, the creation of positive framework conditions for a bio-based economy, improved education and training in the bioeconomy and social dialogue.[22] By 2019, the advisory body, which was newly constituted at the beginning of its second term of office in 2012, had issued over 80 publications, including BÖRMEMO 07 on 16 January 2019, a statement on the regulation of GE by the European Court of Justice. In addition to the potential of GE for the breeding of high-yielding and resistant plants and the development of more environmentally friendly production methods, the *German Bioeconomy Council* includes the risk of ignorance and the danger of a European competitive disadvantage through regulation in its assessment of the ECJ ruling.[23] It also calls for a new law on genetic engineering and a constructive discourse on new biotechnologies for society as a whole. In the *German Bioeconomy Council's* recommendation for a further development of the NRSB, biotechnologies continue to make a significant contribution to the development of "high-tech solutions and products based on the fusion of biotechnologies with sensor, nano, information and cognitive technologies".[24] It goes on to say: "This was not considered six years ago. The new biotechnologies, such as CRISPR/Cas, require scientific assessment and societal discourse".[25] The expiry of the NRSB in July 2019 marked the end of the second term of office of the *German Bioeconomy Council*, which, however, recommends that the Federal Government further develop the advisory structure, in particular, the establishment of a German bioeconomy platform.[26]

2.4 National Bioeconomy Strategy

In order to build a bridge between technology, ecology and efficient management, to interlink the individual departments even more closely, to consolidate science-based foundations and to pursue sustainability in a concrete and consistent manner, the *Federal Ministry of Education and Research* and the *Federal Ministry of Food and*

[22] Cf. German Bioeconomy Council (2019a).

[23] Cf. German Bioeconomy Council (2019b).

[24] German Bioeconomy Council (2016, p. 7), "high-tech solutions and products based on the fusion of biotechnologies with sensor, nano, information and cognitive technologies" (own translation).

[25] Ibid., 7, "This was not considered in this way 6 years ago. The new biotechnologies, such as CRISPR/Cas, require scientific evaluation and societal discourse" (own translation).

[26] Cf. German Bioeconomy Council (2019c).

Agriculture are currently working on an overall strategy for the bioeconomy for the German government.[27] The aim of the strategy is to promote the sustainable design of agricultural and forestry production and the development of innovative, bio-based alternatives to existing products and processes. Cross-border cooperation was defined in advance as a new core topic. In order to identify the priorities for the follow-up programme of the NRSB and NPSB, the *Federal Ministry of Education and Research* has organised various events with representatives from science, industry, politics and civil society since 2016. In June 2019, the *Federal Ministry of Education and Research* asked selected stakeholders to submit their comments on the *National Bioeconomy Strategy* (NBS).

In the statement of the *German Forum on Environment and Development* (GFED), which consists of numerous German environmental and development associations, it was welcomed that "selectively central ideas of the sustainability debate",[28] such as biodiversity and climate protection, soil fertility, distributive justice, sufficiency and a possible conversion of the economic system are taken up in the Federal Government's draft. However, these are "insufficiently or not at all"[29] taken into account in the research funding of the intended political framework conditions. The GFED also misses in the draft an "indefinite commitment of the Federal Government to the precautionary principle and a clear statement against agro-genetic engineering in future bioeconomy research"[30] and criticises the Federal Government for research that is too open to technology with regard to GE and synthetic biology. In contrast, the draft was received comparatively positively by the German *Biotechnology Industry Association* (BIO-Deutschland) which welcomes the integration of the NBS into other initiatives of the German government such as the Sustainability Strategy or the High-Tech Strategy 2025 and the numerous mentions of biotechnology as a key technology for the bioeconomy.[31] However, with reference to the potential of new molecular biological techniques in agriculture, it is critically noted that the term *genetic engineering* is not mentioned in the entire draft: "Traditional mutation breeding is also genetic engineering as the ECJ clarified a year ago".[32] It is further argued that genetic engineering has the potential to provide answers to pressing issues such as increasing global warming and scarcity of resources. Particularly against the background of the public debates on this topic, some of which are "partly rather emotional and not very factual",[33] the

[27] Cf. Federal Ministry of Education and Research (2018).

[28] German Forum on Environment and Development (2019, p. 1), "selectively central ideas of the sustainability debate" (own translation).

[29] Ibid., 1, "insufficiently or not at all" (own translation).

[30] Ibid., "clear commitment of the Federal Government to the precautionary principle as well as a clear statement against agro-genetic engineering in future bioeconomic research" (own translation).

[31] Cf. Biotechnology Industry Association (2019).

[32] Ibid., 7, "Even traditional mutation breeding is genetic engineering, as the ECJ made clear a year ago" (own translation).

[33] Ibid. "rather emotional and not very objective" (own translation).

biotechnology industry association continues: "a mention of genetic engineering processes in a bioeconomy strategy [is] indispensable".[34]

3 Bioeconomy in the Netherlands

Focused mainly on food processing, chemicals, oil refining and electrical engineering, the Dutch economy is the sixth largest in the European Union.[35] The agricultural sector in particular, which employs only 2% of the labour force, is highly technological and produces a significant surplus for export.[36] Products from greenhouse production, e.g. tomatoes and vegetables, but also processed products such as starch, sugar and lactic acid are particularly noteworthy. In the Netherlands, bioeconomic potential can still be found in horticultural products such as ornamental plants or landscape woods and in the chemical industry, where many companies have already recognised the impending economic change and are focusing on bio-based chemicals and biopolymers.[37] At present, 50 of the 321 biotechnology companies are already generating a large part of their profits from the bioeconomy. The government sees considerable potential in GE, in particular, for the breeding of high-yielding and resistant organisms and the associated sustainable production of biomass, which in turn is of great importance for the bioeconomic transformation.[38] The Dutch Farmers' Association also has a positive attitude towards GMOs, but points to the resistance of many customers, especially in the important sales market in Germany.[39]

While in Germany the aim is to promote the bioeconomy on a broad basis, the Netherlands defines priority economic sectors in which a bioeconomic transformation should be promoted. These mainly include the five sectors of agriculture and food, chemicals, energy, life sciences and horticulture. To implement its bioeconomic strategies, the Netherlands follows a bottom-up approach, often led by industry through so-called *triple-helix cooperation* (also known as *cluster networks* in Germany) between scientific institutions, companies and the government at the state or regional level. With the aim of facilitating contacts between the individual areas and overcoming regulatory obstacles, the bio-based delta in the southwest of the Netherlands, for example has been established where a purely bio-based economy is to be established. The Dutch bioeconomy strategy now links more than 40 governmental, environmental and civil society organisations,

---

[34] Ibid. "it is essential that genetic engineering processes be mentioned in a bioeconomic strategy" (own translation).

[35] Cf. Bosman and Rotmans (2016).

[36] Cf. Netherlands Environmental Assessment Agency (2014).

[37] Cf. c/o BIOCOM AG (2020a).

[38] Cf. ibid.

[39] Cf. ibid.

employers' associations, trade unions and financial institutions.[40] Overall, the available biomass in the Netherlands theoretically covers the national demand for food, feed, transport and chemicals.[41] Even though the structures and strengths of the Dutch economy thus provide a good basis for bioeconomic transformation and the government is also setting the course for a greener economy, according to Bosman and Rotmans, the potential of biomass is not yet fully exploited due to the country's heavy dependence on fossil fuels.[42] Gas reserves in the Netherlands will probably come to a standstill over the next 15 years, making it an important driver for the bioeconomic turnaround. Its increasingly costly production has already led to considerable earthquakes and accidents in the north of the Netherlands, further increasing the pressure for transformation towards a bioeconomy.[43]

3.1 Werkgroep Businessplan Bioeconomy

In its *Werkgroep Businessplan Bioeconomy* (WBBE), the Netherlands formulates the goal of becoming a future centre of excellence of a global bioeconomy on the basis of renewable raw materials. By 2050, the Netherlands aims to be one of the world's leading countries in the bioeconomy, focusing on its own strengths: "A highly developed [bioeconomy] uses green resources firstly in the production of food and feed and only afterwards (or simultaneously in the case of waste products) for chemicals, materials and energy".[44] While the share of renewable energies was still 5.6% in 2014, their share in energy production is to increase to 14.5% by 2020 and to at least 40% by 2040. It is also assumed that one in three technical students will be employed in the bioeconomy by 2030. In a version of the WBBE updated in 2018, the *Dutch Ministry of Economics and Climate Policy* formulates the following eight pillars for a bioeconomic transformation: (1) the use of resources within the planetary boundaries; (2) stopping climate change; (3) greater opportunities for new jobs and businesses; (4) sustainable resource management; (5) the establishment of a stable and predictable legal framework; (6) greater cooperation in the value chain; (7) a long-term research and innovation agenda and (8) a regional strategy for the efficient use of existing biomass and agricultural land.[45]

The presentation of the content of bioeconomic strategies is now followed by an analysis of the participation of both countries in the discourse on GE. The general social attitude towards biotechnology and genetic engineering will also be discussed.

---

[40] Cf. Elyse (2015).

[41] Cf. Netherlands Enterprise Agency (2017).

[42] Cf. Bosman and Rotmans (2016).

[43] Cf. ibid.

[44] Ibid., 2.

[45] Cf. Ministry of Economic Affairs and Climate Policy (2018).

4 Participation of both Countries in the Discourse on GE

Differences in the way the topic of GE is discussed in Germany and the Netherlands exist mainly on a political and social level. Since the development of GE, the government in the Netherlands has been more active in the discourse on GE than the German government. While the German government has hardly taken an official position on GE, the Dutch government has published numerous statements and press articles.[46] Especially with regard to the ruling of the ECJ published in July 2018, the Netherlands tried to express its support for the use of GE in advance with various statements at a national and European level on regulation in the discourse. They also tried to exert increased pressure on the EC in order to obtain a ruling by the ECJ as soon as possible. In Europe, the Dutch government, which traditionally has a liberal attitude towards biotechnology, was the first to seek discussion with other EU member states in order to discuss a possible interpretation of the regulations on GMOs:

> [. . .] the Dutch authorities see no need to await the Court's rulings on the interpretation of European legislation for initiating a policy debate on New Plant Breeding Techniques. To the contrary, the Dutch authorities see a pressing need to address the underlying issues at stake in the short term, which includes making the implementation of the Directive more workable in view of ongoing technical and scientific developments.[47]

The attempt to obtain an exemption for GE in accordance with the Deliberate Release Directive 2001/18/EC and the reform of the current Genetic Engineering Act met with a positive response among the European member states. The active and official influence of the Netherlands on the discourse on GE is contrasted by the German Federal Government's attitude, which is hardly noticeable in public. In the period prior to the ECJ ruling, there is hardly any information about the German government's position on GE. Clues to the position of the *Federal Ministry of Food and Agriculture* can be found, inter alia, in isolated interviews with Federal Minister Julia Klöckner according to ECJ case law. She commented on the regulation as follows: "I think it is factually wrong to lump classical green genetic engineering together with CRISPR/Cas".[48] In order to start a broader discussion on GE in Germany, as in the Netherlands, Carina Konrad (Free Democratic Party/Germany) called on the German government to respond to the appeals from science and research: "She [Ms. Klöckner] now has to recognise the signals and act".[49]

The fact that the Netherlands and Germany could come closer together in the future with regard to their positions on GE was demonstrated at the meeting of the EU Council of Agriculture Ministers in May 2019, where, at the request of the

---

[46] Cf. Ministry of Infrastructure and Water Management (2017); Smit (2018).

[47] Ministry of Infrastructure and Water Management (2017, 2).

[48] Herrmann (2018), "To lump classical green genetic engineering together with CRISPR/Cas is, in my opinion, factually incorrect" (own translation).

[49] Karberg (2019), "It must now recognise the signals and act" (own translation).

Netherlands, the revision of the current genetic engineering law was also on the agenda, in addition to a discussion on a common agricultural policy.[50] According to the Dutch proposal for discussion, organisms obtained by GE should be distinguished from those obtained by classical genetic engineering. After Estonia had initially endorsed this proposal, 12 other Member States (Belgium, Cyprus, Finland, France, Germany, Greece, Italy, Portugal, Slovenia, Spain, Sweden and the United Kingdom) followed suit. According to the State Secretary of the *Federal Ministry of Food and Agriculture*, Hermann Onko Aeikens, Germany is generally open to a discussion on genetic engineering law.[51] If this is taken into account against the background of the nationwide ban on the cultivation of genetic engineering agreed in the coalition agreement and the, sometimes conflicting, positions of the Minister of Agriculture Julia Klöckner and the Minister of the Environment Svenja Schulze, it is questionable, according to the biochemist and molecular biologist Prof. Dr. Klaus-Dieter Jany, whether and to what extent Germany will actively support an adjustment of the genetic engineering legislation at the EC in the future.[52] While Ms. Schulze rejects CRISPR/Cas as a breeding method and advocates regulation by the ECJ, Ms. Klöckner advocates plant breeding that is open to new technologies.[53] Mr. Aeikens also emphasises that the Federal Government in Germany will only make a new push at a European level on GE when the Christian Democratic Union, Christian Social Union and Social Democratic Party agree on a common line.[54] While German politics thus still seem to wait and see, the *National Academy of Sciences Leopoldina*, the *Union of the German Academies of Sciences and Humanities* and the *German Research Foundation* formulated for the first time concrete proposals for the reform of the current genetic engineering guidelines in a joint position paper from 4 December 2019. Entitled *Towards a scientifically based, differentiated regulation of genome-modified plants in the EU,* the paper recommends that the EC amend European genetic engineering law in a timely manner, discuss each breeding method in a differentiated manner and thus create a "long-term perspective for appropriate regulatory management of new breeding technologies".[55] The opinion considers the new molecular breeding methods to be an important contribution to the bioeconomy in the coming years, in order to make "agriculture more productive, less pesticide-intensive and more climate-adapted through traits such as drought and heat tolerance".[56] Further indications that the pressure on politicians will increase fundamentally and that genetic engineering law will have to be reconsidered or revised are provided by the European citizens'

[50] Cf. Council of the European Union (2018).

[51] Cf. Jany (2019).

[52] Cf. ibid.

[53] Awater-Esper (2019).

[54] Cf. Agra Europe (2019).

[55] German National Academy of Sciences Leopoldina, Union of the German Academies of Sciences and Humanities and German Research Foundation (2019, 76).

[56] Ibid., 54.

initiative *Grow Scientific Progress* (GSP), which is seeking to reform current plant breeding legislation in the EU.[57] The initiative, which is made up of German and Dutch students from Wageningen University, aims to achieve a more liberal legal approach to the new molecular biological methods and could promote a future rapprochement between both countries with regard to biotechnology in agriculture.

On the political, economic, scientific and social level, applications of biotechnology have so far been discussed more optimistically in the Netherlands than in Germany.[58] A 2010 Eurobarometer survey showed that 53% of the Dutch population expects biotechnology to have a positive impact on our future lives, while only a quarter of those surveyed expect negative effects.[59] When asked specifically about genetic engineering, 57% of the Dutch population rejected the use of foreign genes and 46% rejected the use of related genes. In comparison, 69% and 47% of the German population rejected the use of foreign or related genes. Furthermore, according to the *Federal Ministry of the Environment, Nature Conservation and Nuclear Safety* and the *Federal Agency for Nature Conservation*, in 2009 85% of Germans considered a ban on genetic engineering in agriculture to be very or rather important.[60] Whether there is a connection between the development of GE and the recently declining number of those who agree unconditionally ("very important") to a ban on genetic engineering in agriculture cannot be conclusively answered here. One of the first more comprehensive surveys of public attitudes to GE in Germany was conducted by the Federal Institute for Risk Assessment in September 2019. With the aim of obtaining a more differentiated consumer opinion on the application of GE in the field of nutrition and human health, 20 interested German consumers were brought together with representatives from politics, science, industry and civil society. In a process lasting several days, a comparatively balanced mood regarding the potential and risks of GE was identified.[61] The 20 participants formulated the following demands on German politics: (1) retention of the precautionary principle; (2) freedom of choice for consumers; (3) freedom of information and transparency; (4) priority of social aspects over economic interests; (5) reform of patent law: no patent protection on living organisms; (6) liability regulations for unexpected damage by the producer and (7) labelling of genetically modified food. The survey shows that scepticism towards biotechnology still exists in German society. However, there seems to be a certain willingness to overcome this scepticism if decision makers openly communicate the risks.

[57] Cf. Grow Scientific Progress (2019).

[58] Cf. Durant et al. (1998).

[59] Cf. European Commission (2010).

[60] Cf. Federal Ministry for the Environment, Nature Conservation and Nuclear Safety and Federal Agency für Nature Conservation (2017).

[61] Cf. German Federal Institute for Risk Assessment (2019).

5 Conclusion

In this chapter, the development of the bioeconomy in Germany and the Netherlands was described, what importance is currently attributed to GE in this context, and how it might develop in the future. Germany is an international pioneer in the bioeconomy and, through its bioeconomy strategies, has a basic political, scientific and economic structure that forms an important basis for bioeconomic transformation. The *German Bioeconomy Council*, which is an important advisory body supporting the development of the bioeconomy in Germany, should be highlighted in this context. The Netherlands, too, has set itself high targets for 2050, which have been advanced by industry and others. GE, and in particular CRISPR/Cas, is seen as having great potential for the bioeconomy in both countries, which is also referred to in the bioeconomy strategies. Overall, the Netherlands has so far been more active or publicly perceived in the discourse on GE than Germany. Independently of each other, the Dutch government and the *German Bioeconomy Council* are in favour of a new European genetic engineering law, which should be adapted to the state of scientific knowledge and should examine each GE procedure individually.

It remains to be seen what relevance CRISPR/Cas and other similarly invasive genetic engineering methods will have in the future in the German bioeconomic strategy. It must be considered that GE is on the one hand seen as having the potential to trigger significant innovations as a "biological revolution"[62]; on the other hand, there are critical voices, such as that of the German Member of the Bundestag Harald Ebner (*Alliance 90/The Greens*), who sees CRISPR/Cas as a "one-dimensional technical apparent solution"[63] that does not provide sustainable answers to complex problems such as food security and even reduces the social acceptance of the bioeconomy. The statements of the GFED and BIO-Deutschland on the draft of the Federal Government on the hitherto unpublished NBS also show expectations of the role of GE in the future German bioeconomy that are difficult to reconcile. However, the discussion on GE in Germany also gives the impression that the social and political mood with regard to the application of biotechnology in agriculture is no longer exclusively critical or hostile and that uniform positions, which in the past could be found among actors such as the German party Alliance 90/The Greens or organic agriculture, seem to be softening.

For both countries, much will ultimately depend on whether the European Directive 2001/18/EC is amended and the Genetic Engineering Act is reformed. Whether or not majorities can be won at the political level will depend to a large extent on the new EC under the direction of Ms. Ursula von der Leyen. Ms. Petra Bosch of the *European Seed Association* explains: "I think, in the most positive way, a new regulation would be possible by the end of this legislative term of the new Commission and Parliament".[64] In the event of a (partial) deregulation of GE, it

---

[62] German Bioeconomy Council (2019b, p. 2), "biological revolution" (own translation).

[63] Denkhausbremen (2018), "one-dimensional technical sham solution" (own translation).

[64] Haas (2019), "I think that the most positive way forward would be to have a new regulation in place by the end of this legislative period of the new Commission and Parliament" (own translation).

seems likely that both countries will make greater use of the new techniques as a tool for bioeconomic transformation. Whether the new EC will wait until the end of its legislative period in 2024 is difficult to predict at this point in time.

References

Agra Europe. (2019). *Aeikens zu Klimawandel: Landwirtschaft kann nicht mit Erleichterungen rechnen*. Retrieved June 29, 2020, from, https://www.topagrar.com/acker/news/aeikens-zu-klimawandel-landwirtschaft-kann-nicht-mit-erleichterungen-rechnen-10324643.html

Albrecht, J., Carrez, D., Cunningham, P., Daroda, L., Mancia, R., Máthé, L., Raschka, A., Carus, M., & Piotrowski, S. (2010). *The knowledge based bio-economy (KBBE) in Europe: Achievements and challenges. Full report*. Retrieved June 23, 2020, from, http://cleverconsult.eu/clever3/wp-content/uploads/2015/02/KBBE_2020_BE_presidency.pdf

Awater-Esper, S. (2019). *Schulze stellt sich in der Pflanzenzucht gegen Klöckner*. Retrieved November 28, 2019, from, https://www.topagrar.com/management-und-politik/news/schulze-stellt-sich-in-der-pflanzenzucht-gegen-kloeckner-11535128.html

Biotechnology Industry Association (Bio-Deutschland). (2019). *Stellungnahme der BIO Deutschland zum Referentenentwurf für eine Nationale Bioökonomiestrategie*. Retrieved November 28, 2019, from, https://www.biodeutschland.org/de/positionspapiere/stellungnahme-der-bio-deutschland-zum-referentenentwurf-fuer-eine-nationale-biooekonomiestrategie.html?file=files/content/medien/positionspapiere/2019/BIO_Deutschland_Stellungnahme_Biooekonomiestrategie.pdf

Bosman, R., & Rotmans, J. (2016). Transition governance towards a bioeconomy: A comparison of Finland and the Netherlands. *Sustainability, 8*(10), 1017.

c/o BIOCOM AG. (2020a). *Niederlande*. Retrieved January 14, 2020, from, https://biooekonomie.de/en/node/5979

c/o BIOCOM AG (2020b). *Bioökonomie in den Bundesländern*. Retrieved January 14, 2020, from, https://biooekonomie.de/biooekonomie-deutschland

Council of the European Union. (2018). *Outcome of the Council Meeting. 3689th Council Meeting. Agriculture and Fisheries*. Retrieved November 14, 2019, from, https://www.consilium.europa.eu/media/39365/st09271-en19.pdf

Denkhausbremen. (2018). *Harald Ebner, Grüne: Lobbymacht der Wirtschaft bei Bioökonomie ausgleichen*. Retrieved November 28, 2010, from, https://denkhausbremen.de/harald-ebner-lobbymacht-der-wirtschaft-bei-biooekonomie-ausgleichen/

Durant, J., Bauer, M., & Gaskell, G. (1998). *Biotechnology in the public sphere. A European sourcebook*. National Museum of Science.

Elyse, M. (2015). *The bioeconomy: A primer*. Retrieved December 6, 2019, from, https://www.tni.org/files/publication-downloads/tni_primer_the_bioeconomy.pdf

European Commission. (2010). *Biotechnologie. Bericht. Eurobarometer Spezial 328/73.1*. Retrieved November 18, 2019, from, http://ec.europa.eu/public_opinion/archives/ebs/ebs_341_de.pdf

Federal Ministry of Education and Research. (2010). *National research strategy bioeconomy 2030. Our route towards a biobased economy*. Retrieved October 29, 2019, from, http://biotech2030.ru/wp-content/uploads/docs/int/bioeconomy_2030_germany.pdf

Federal Ministry of Education and Research. (2017). *Regulations governing the funding of plant research projects on "crops for the future" under the national research strategy bioeconomy 2030*. Retrieved November 30, 2019, from, https://www.ptj.de/lw_resource/datapool/systemfiles/cbox/2238/live/lw_bekdoc/crops-for-the-future_call-for-proposals.pdf

Federal Ministry of Education and Research. (2018). *Mit vereinter Kraft für die Bioökonomie*. Pressemitteilung: 096/2018. Retrieved December 1, 2019, from, https://www.bmbf.de/de/mit-vereinter-kraft-fuer-die-biooekonomie-5782.html

Federal Ministry of Education and Research. (2020). *Forschung für eine biobasierte Wirtschaft. Erfolge und Herausforderungen für die Bioökonomie in Deutschland.* Retrieved January 14, 2020, from, https://www.bmbf.de/upload_filestore/pub/Forschung_fuer_eine_biobasierte_Wirtschaft.pdf

Federal Ministry for the Environment, Nature Conservation and Nuclear Safety; Federal Agency für Nature Conservation. (2017). *Naturbewusstsein 2017. Bevölkerungsumfrage zu Natur und biologischer Vielfalt.* Retrieved December 2, 2019, from, https://www.bmu.de/fileadmin/Daten_BMU/Pools/Broschueren/naturbewusstseinsstudie_2017_de_bf.pdf

Federal Ministry of Food and Agriculture. (2014). *Nationale Politikstrategie Bioökonomie. Nachwachsende Ressourcen und biotechnologische Verfahren als Basis für Ernährung, Industrie und Energie.* Retrieved October 29, 2019, from, https://biooekonomie.de/sites/default/files/publications/npsb_0.pdf

Fraunhofer Institute for Systems and Innovation Research (ISI). (2017). *Evaluation der "Nationalen Forschungsstrategie BioÖkonomie 2030". Wirksamkeit der Initiativen des BMBF – Erfolg der geförderten Vorhaben – Empfehlungen zur strategischen Weiterentwicklung – Zusammenfassung.* Retrieved November 11, 2019, from, http://www.isi.fraunhofer.de/isi-wAssets/docs/t/de/publikationen/Evaluation_NFSB_Kurzbericht.pdf

German Bioeconomy Council. (2016). *Weiterentwicklung der "Nationalen Forschungsstrategie Bioökonomie 2030".* Retrieved November 16, 2019, from, https://biooekonomierat.de/fileadmin/Publikationen/empfehlungen/181116_Ratsempfelungen_fu__r_die_Weiterentwicklung_der_Forschungsstrategie_final.pdf

German Bioeconomy Council. (2019a). *Was ist der Bioökonomierat?* Retrieved October 29, 2019, from, https://biooekonomierat.de/biooekonomierat/

German Bioeconomy Council. (2019b). *Genome Editing: Europa benötigt ein neues Gentechnikrecht.* Retrieved November 12, 2019, from, https://biooekonomierat.de/fileadmin/Publikationen/berichte/BOERMEMO_07_final.pdf

German Bioeconomy Council. (2019c). *Pressemitteilung: Bioökonomie deutlich vorangebracht. Bioökonomierat der Bundesregierung verabschiedet sich im Sommer 2019.* Retrieved November 14, 2019, from, https://biooekonomierat.de/fileadmin/Pressemitteilungen/PM_BO__R_Ru__ckblick_FINAL_neu_2.pdf

German Federal Institute for Risk Assessment. (2019). *Verbrauchervotum. Ergebnis der BfR-Verbraucherkonferenz "Genome Editing im Bereich Ernährung und menschliche Gesundheit" 2019.* Retrieved December 14, 2019, from, https://mobil.bfr.bund.de/cm/343/verbrauchervotum-genome-editing.pdf

German Forum on Environment and Development. (2019). *Stellungnahme an die Bundesregierung zum Entwurf einer Nationalen Bioökonomiestrategie.* Retrieved December 3, 2019, from, https://www.forumue.de/wp-content/uploads/2019/07/Stellungnahme-Bio%C3%B6konomie.pdf

German National Academy of Sciences Leopoldina; Union of the German Academies of Sciences and Humanities; German Research Foundation. (2019). *Towards a scientifically justified, differentiated regulation of genome edited plants in the EU.* Retrieved December 15, 2019, from, https://www.leopoldina.org/uploads/tx_leopublication/2019_Stellungnahme_Genomeditierte_Pflanzen_web.pdf

Grow Scientific Progress. (2019). *Die Bürgerinitiative.* Retrieved December 4, 2019, from, https://www.growscientificprogress.org/?lang=de

Haas, L. (2019). *Neuordnung der Gentechnik-Regeln. Die EU-Kommission steht unter Zugzwang.* Retrieved December 10, 2019, from, https://www.deutschlandfunk.de/neuordnung-der-gentechnik-regeln-die-eu-kommission-steht.676.de.html?dram:article_id=465239

Herrmann, J. (2018). *Klöckner will gegen Einschränkungen neuer Gentechnik angehen.* Retrieved November 25, 2019, from, https://de.reuters.com/article/deutschland-agrar-gentechnik-idDEKCN1LL240

Jany, K.-D. (2019). *EU-Agrarministerrat: 14 Mitgliedstaaten befürworten ein einheitliches Vorgehen beim Gentechnikrecht*. Retrieved November 11, 2019, from, https://www.biotech-gm-food.com/aktuelles/eu-agrarrat-befuerwortet-aenderungen-im-gentechnikgesetz

Karberg, S. (2019). *FDP will neue Gentechnik-Gesetze*. Retrieved December 4, 2019, from, https://www.tagesspiegel.de/wissen/neue-regeln-fuer-genome-editing-gefordert-fdp-will-neue-gentechnik-gesetze/24164974.html

Kiresiewa, Z., Hasenheit, M., Wolff, F., Möller, M., Gesang, B., Schröder, P. (2019). *Bioökonomiekonzepte und DiskursanalyseTeilbericht (AP1) des Projekts „Nachhaltige Ressourcennutzung – Anforderungen an eine nachhaltige Bioökonomie aus der Agenda 2030/ SDG-Umsetzung"*. Retrieved November 25, 2019, from, https://www.umweltbundesamt.de/sites/default/files/medien/1410/publikationen/2019-07-18_texte_78-2019_sdgbiooekonomie.pdf

Knott, G. J., & Doudna, J. A. (2018). CRISPR-Cas guides the future of genetic engineering. *Science, 361*(6405), 866–869.

Lewandowski, I. (ed.) (2018). *Bioeconomy. Shaping the transition to a sustainable, biobased economy.*

McCormick, K., & Kautto, N. (2013). The bioeconomy in Europe: An overview. *Sustainability, 5* (6), 2589–2608.

Ministry of Economic Affairs and Climate Policy. (2018). *The position of the bioeconomy in the Netherlands*. Retrieved December 1, 2019, from, https://www.government.nl/binaries/government/documents/leaflets/2018/04/01/the-position-of-the-bioeconomy-in-the-netherlands/WEB_113751_brochure+ENG+Update.pdf

Ministry of Infrastructure and Water Management. (2017). *proposal for discussion on actions to improve the exemption mechanism for genetically modified plants under directive 2001/18/EC*. Ministerie van Infrastructuur en Waterstaat. Retrieved November 25, 2019, from, https://www.eerstekamer.nl/overig/20170913/proposal_for_discussion_on_actions/document

Netherlands Enterprise Agency. (2017). *The Bioeconomy in the Netherlands. Opportunities for collaboration*. Retrieved June 30, 2020, from, http://news.nost.jp/wordpress_en_3.4.1/wp-content/uploads/2018/11/1.1-Kees-Kwant-Netherlands-Enterprise-Agency-RVO.pdf

Netherlands Environmental Assessment Agency. (2014). *Green gains – in search of opportunities for the Dutch economy*. Retrieved December 1, 2019, from, https://www.pbl.nl/sites/default/files/downloads/PBL-2014-Green-gains-1262.pdf

Pietzsch, J. (2017). *Bioökonomie für Einsteiger*. Scientific advisor: U. Schurr.

Siebert, R., Richter, I., Herzig, C., & Birringer, M. (2018). Genome Editing für die Land- und Ernährungswirtschaft – Potenziale und Risiken. *Ernährungs-Umschau, 2018*(11), M639–M647.

Siebert, R., Herzig, C., & Birringer, M. (2021). Strategic framing of genome editing in agriculture: An analysis of the debate in Germany in the run-up to the European court of justice ruling. *Agriculture and Human Values*. https://doi.org/10.1007/s10460-021-10274-2

Smit, P. H. (2018). *Landbouwminister Schouten zet deur op kier voor genetische modificatie*. Retrieved December 5, 2019, from, https://www.volkskrant.nl/nieuws-achtergrond/landbouwminister-schouten-zet-deur-op-kier-voor-genetische-modificatie~bc9fbf22/?referer=https%3A%2F%2Fwww.dutchnews.nl%2Fnews%2F2018%2F10%2Fdutch-farm-minister-opens-door-to-gene-editing-crops%2F

Monitoring and Measuring Bioeconomy 12

Maximilian Kardung

Abstract

Many actors have a stake in and different definitions of bioeconomy. Even though sustainability is very prominent in many definitions, bioeconomy is not inherently sustainable. However, bioeconomy can contribute to the three pillars of sustainability: economic development, social development, and environmental protection. To assess bioeconomy's contribution to sustainable development, the first step is to measure its size and most important impacts. The economic literature on sustainable development has established sustaining human well-being as the target value, whereby one strand analyses the current generations' well-being and another strand the intergenerational well-being. Kenneth J. Arrow et al.'s contribution of the year 2012 considers the welfare of all future generations in the calculation of the current generations' well-being. They define intergenerational well-being as the discounted flow of current and future generations' well-being, and use a discount rate and the term felicity, which both have been criticized. Furthermore, it may be criticized that, here, felicity is based on the consumption of capital assets. Any felicity that is not based on consumption, for example, the value of "doing nothing," is not included. This leads us to the question of whether we should quantify such aspects for a measure, which tries to capture comprehensive wealth. The question for the bioeconomy must be whether nowadays' definition of bioeconomy is useful in promoting sustainable development. If sustainability is the main target, progress toward it should be measured to allow policymakers a fair assessment. This is not an easy task and comes with new issues and decisions that must be made, including ethical ones.

M. Kardung (✉)
Agricultural Economics and Rural Policy Group (AEP), Wageningen University, Wageningen, The Netherlands
e-mail: maximilian.kardung@wur.nl

D. Lanzerath et al. (eds.), *Bioeconomy and Sustainability*,
https://doi.org/10.1007/978-3-030-87402-5_12

Keywords

Bioeconomy · Sustainability · Measure

1 Introduction

Bioeconomy is the part of the economy that either produces biological resources from land and sea or transforms them into food, materials, and energy. Bioeconomy is crucial for the well-being of humankind because it includes food production and a variety of other economic sectors. However, the food industry is discussed controversially on a regular basis. The introduction of genetically modified crops is disputed and many consumers seem to have a negative attitude to GMO including food. Furthermore, cropland that is used for energy and other nonfood crops competes with the use for food crops, and therefore there is an inevitable competition. The range of bioeconomy is also much broader than just the agri-food sector as it is part of other economic sectors. In general, the use of all animals, plants, microorganisms, and derived biomass within economic applications are considered to be bioeconomic applications, as well as all sectors and ecosystems that rely on them.[1]

Many public, private, and civil society actors have a stake in the bioeconomy because it spans over such a wide part of the economy and society. Actors include, among others, scientists, policymakers, civil society, NGOs, or industry associations and many of them have different notions and definitions of bioeconomy. For example, the definition by the Royal Society of Biology from the United Kingdom emphasizes the use of biology to enhance economic activities.[2] In recent years, there has been a tendency to define bioeconomy in a broader sense and include everything that is in some way related to biological resources. The Global Bioeconomy Summit, which has established itself as a place for all stakeholders to meet and discuss bioeconomic developments, defines bioeconomy as "[. . .] the production, utilization, and conservation of biological resources, including related knowledge, science, technology, and innovation, to provide information, products, processes, and services across all economic sectors aiming toward a sustainable economy."[3] This is a very inclusive definition, which does not single out economic production of or with biological resources, but even includes conservation and utilization of them in other ways. The European Commission, which has been a key promoter of bioeconomy from the past to the present, defines bioeconomy in its 2018 Bioeconomy Strategy Update as follows:

"The bioeconomy covers all sectors and systems that rely on biological resources (animals, plants, micro-organisms and derived biomass, including organic waste),

[1] Cf. European Commission (2018).

[2] Cf. Royal Society of Biology (2016).

[3] Global Bioeconomy Summit (2018, p. 2).

their functions and principles. It includes and interlinks: land and marine ecosystems and the services they provide; all primary production sectors that use and produce biological resources (agriculture, forestry, fisheries, and aquaculture); and all economic and industrial sectors that use biological resources and processes to produce food, feed, bio-based products, energy, and services."[4]

This definition expands the European Commission's previous one from the 2012 European Union's Bioeconomy Strategy by including a wider array of products, sectors, and value chains. Furthermore, the strategy stresses, like the Global Bioeconomy Summit's definition, the importance of sustainability.

Even though sustainability is very prominent in its definition, bioeconomy is not inherently sustainable, which is already evident in its cornerstones, agriculture, forestry, and fisheries. Biological resources are exploited in a non-sustainable way too often. In agriculture, a higher level of intensification has increased the availability of food for many people in developed and developing countries, but that comes at the cost of environmental degradation and poor animal welfare.[5] In forestry, demand for timber and agricultural land has led to deforestation and an unsustainable level of cuttings in forests around the world. In fisheries, exploitation of the global fish stocks is done in non-sustainable ways, especially in the high seas where they are accessible by anyone. This is not only affecting future generations adversely but also global inequality because the wealthiest countries also have the main share of the global catch.[6] Therefore, sustainability must be the main target for any promotion of bioeconomy.

2 Bioeconomy and Sustainability

Sustainability is a notion that was popularized by the Brundtland Report in 1987, in which sustainable development was defined as "[...] development that meets the needs of the present without compromising the ability of future generations to meet their own needs."[7] The 2005 World Summit on Social Development further worked out three pillars of sustainability: economic development, social development, and environmental protection. Bioeconomy can contribute to all three pillars, but its largest contribution is likely toward environmental protection. However, bioeconomy is not sustainable by default and therefore this contribution must be carefully assessed.

To assess bioeconomy's contribution to sustainable development, we have to measure its size and most important impacts. This should be done with the purpose to depict a comprehensive picture of the different aspects of bioeconomy and answer some of the most pressing related questions for our society:

[4] European Commission (2018, p. 4).

[5] Cf. Garnett et al. (2013).

[6] Cf. Sumaila et al. (2015).

[7] World Commission on Environment and Development (1987).

- How many people are employed in bioeconomy sectors and how much value do they add to an economy?
- How much investment takes place and how many innovations are developed and succeed?
- How much food is available and how stable are food prices?
- How many natural resources are used and how many pollutants are emitted?

Indicators are a quantitative or qualitative measure, which can be used to find answers to these questions. They can be very specific and measure only a certain chunk or aspect of bioeconomy and therefore also answer only specific questions. One of these more specific questions is whether different types of biomass production are environmentally sustainable. So far, there has been not even a consensus on what "sustainable" means in this context and only a mixture of voluntary standards and regulations is used.[8] Examples for specific indicators are the ratio fellings and estimated maximum sustainable level of cuttings in forests or the nitrogen and phosphorus use for agricultural production. On the other hand, one indicator can capture many different aspects and help to answer several questions at the same time. For example, it can measure the sustainability of the whole bioeconomy at once.

The increased awareness and demand for sustainable development led to the need for measuring sustainability. Consequently, tools for all kinds of sustainability-related issues on different levels were developed:

- Individual: People can measure their own environmental footprint.[9]
- Corporate: Businesses measure, monitor, and report their sustainability performance, usually on a voluntary basis. Initiatives and standards have been created to promote consistency and comparability.[10]
- Country: 193 countries of the UN General Assembly adopted the 2030 Development Agenda with 17 Sustainable Development Goals (SDGs), each with a list of targets that are measured with indicators to monitor the progress.

However, a remaining issue is to define when a society can be considered sustainable. The SDGs and their corresponding targets, 169 in total, have been agreed upon by all UN member states and are an outcome of political negotiations. The high number of targets has been criticized because "[...] trying to prioritize 169 things looks very similar to prioritizing nothing."[11] It is not clear whether it is necessary or enough to reach all SDGs to achieve sustainable development. For some targets, it is also difficult to see what is pledged and how to monitor the

[8]Cf. Bosch et al. (2015).

[9]E.g. World Wide Fund for Nature (WWF) environmental footprint calculator at https://footprint.wwf.org.uk/#/

[10]E.g. UN Global Compact at https://www.unglobalcompact.org/

[11]Lomborg (2015).

progress. Take, for example, SDG Target 4.7 and the corresponding proposed indicator:

Target 4.7: "[B]y 2030 ensure all learners acquire knowledge and skills needed to promote sustainable development, including among others through education for sustainable development and sustainable lifestyles, human rights, gender equality, promotion of a culture of peace and non-violence, global citizenship, and appreciation of cultural diversity and of culture's contribution to sustainable development."[12]

Proposed indicator to be developed: "Percentage of girls and boys who achieve proficiency across a broad range of learning outcomes, including in literacy and in mathematics by the end of lower secondary schooling cycle (based on credibly established national benchmarks)."[13]

This target and its indicators are still very vague and use several buzzwords. Of course, they must be adaptable to and implementable in a wide range of countries, but still, they have to be meaningful.

Scientists have dealt with this issue extensively and scientific literature offers several suggestions for measuring sustainability. One example is the Ecological Footprint by Mathis Wackernagel and William Rees, which measures the land area a given population can use while being sustainable.[14] Another example is the UN's Human Development Index, a measure for the average achievements in a country in three basic dimensions.[15] The Environmentally Adjusted Net Domestic Product integrates environmental aspects into the traditional Gross Domestic Product.[16] Among economists, the World Banks's measure of *genuine savings* as well as the approach of Kenneth J. Arrow, Partha Dasgupta, and Karl-Göran Mäler on *inclusive wealth* and *genuine investment* led to much discussion. These three concepts are presented in more detail in the next chapter.

3 Measuring Comprehensive Wealth

The economic literature on sustainable development has established sustaining human well-being as the target value. Furthermore, one strand of the literature analyses the *current generations' well-being* and another strand the *intergenerational well-being*.[17] For the *current generations' well-being*, the question of whether development is sustainable is addressed by assessing whether the current generation is operating in a way that allows future generations to achieve a level of well-being at least as high as the current one.[18] Well-being, therefore, has to develop grow or stay constant over time. Future generations' well-being is not explicitly taken into

[12]Cf. Sustainable Development Solutions Network (2015, p. 46).

[13]Cf. ibid.

[14]Wackernagel and Rees (1996).

[15]Cf. Sagar and Najam (1998).

[16]Cf. Böhringer and Jochem (2007).

[17]Cf. Pezzey (1992).

[18]Cf. Arrow et al. (2012).

account. Contrary, *intergenerational well-being* considers the potential welfare of future generations in calculating the well-being of the current generation as described below.

The concepts of *genuine savings* as well as *inclusive wealth* and *genuine investment* establish a measure of sustainable development over time. The genuine savings rate is calculated by subtracting resource depletion and environmental degradation from the traditional net savings while adding investment in human capital.[19] Inclusive wealth and genuine investment work in a comparable way as a society's inclusive wealth are calculated by adding up the shadow value of the economy's marketed consumption goods, leisure, various health services, and consumption services by nature. Genuine investment is measured by evaluating changes in the economy's set of capital assets weighted at shadow prices. Hence, positive genuine investment is used as an indicator of sustainable development.

Arrow et al.'s contribution considers the welfare of all future generations in the calculation of the *current generations' well-being* and therefore is part of the *intergenerational well-being* strand of economic literate. Here, a society's economic development is considered sustainable at a given point of time if its wealth, termed *comprehensive wealth*, is constant or increasing at that time. They measure levels and changes in human capital, natural capital, reproducible capital, and health capital. Moreover, they incorporate the effects of technological change and changes in institutional quality in the calculation of the evolution of comprehensive wealth over time.

Arrow et al. define intergenerational well-being $V(t)$ as the discounted flow of current and future generations' utility U. Accordingly, δ denotes the felicity discount rate. Continuous time is denoted by s and t, $s \geq t$ (underlying assumptions: closed economy, infinite time horizon, constant population, changes in time varying factors are exogenous). Consequently, intergenerational well-being $V(t)$ is formalized thus[20]:

$$V(t) = \int_t^\infty \left[U(\underline{C}(s))e^{-\delta(s-t)} \right] ds, \delta \geq 0 \tag{12.1}$$

Intergenerational well-being V at time t is calculated by taking the integral of the flows of felicities of current and future generations. The integral has infinity as the upper limit, which means the researchers set no temporal limit for future generations' well-being that is considered. However, the discount rate is a limiting factor here. The discount rate is multiplied by the difference of s and t, and therefore the well-being of generations further in the future is valued less and less. Utility U of a specific generation at time s is derived through consumption C of the economy's stock of capital assets K, including manufactured goods, services provided by nature, health services, and many more. The term $U(C(s))$ is interpreted as felicity at date s. The use of the term felicity has been criticized with regard to whether this is even

[19] Cf. Hamilton and Clemens (1999).

[20] Arrow et al. (2012, p. 322).

quantifiable. Felicity is generally considered to be a subjective feeling and individuals would experience felicity in such different ways that it cannot be measured, especially not on an economy-wide level. The term well-being, which is also used by Arrow et al., seems to be more appropriate.

Arrow et al. then define sustainability as non-declining intergenerational well-being over time $dV/dt \geq 0$. Genuine investment is determined as a measure of changes in well-being, where well-being is a function of its determinants, namely the economy's stock of capital assets K and time t: $V(t) = V(\underline{K}(t), t)$. $r(t)$ $(=\partial V/\partial t)$ denotes the shadow price of time at t, and $p_i(t)$ $(\equiv \partial V(t)/\partial K_i(t)$, for all i) the shadow price of the ith capital asset at time t. By letting $Q_i(t)$ equal $\Delta K_i(t)/\Delta t$, genuine investment is[21]:

$$\Delta V(t) = r(t)\Delta t + \sum p_i(t)Q_i(t)\Delta t \tag{12.2}$$

Equation (12.2) shows that the changes in an economy's set of capital assets weighted at shadow prices, including time, equals the change in well-being. Looking at Eq. (12.2) in more detail shows that positive genuine investment increases well-being, while negative genuine investment decreases intergenerational well-being. Hence, positive genuine investment facilitates sustainable development.

4 The Value of Doing Nothing

Concerning the approach to measuring the felicity of human beings, it may be criticized that, here, felicity is based on the consumption of capital assets. The assets come from diverse sources (i.e., marketed consumption goods, leisure, various health services, and consumption services supplied by nature) and therefore it is already a more holistic approach than many others before in economics. However, this framework does not include any felicity that is not based on consumption, for example, the value of "doing nothing." Can and should we quantify such aspects and include an indicator, which tries to capture *comprehensive* wealth?

Doing nothing may be beneficial for mental health and prevent work-related stress and occupational burnout—and obviously is a way how to let nature do things on its own. However, this should be indirectly captured, at least to a certain degree, in the consumption of health services. Doing nothing might also be of high value for creative processes, which again is a value that should be measured already in comprehensive wealth.

---

[21] Arrow et al. (2012, p. 325).

5 The Use of a Discount Rate

Another conflict and major point of discussion from an ethical perspective is the use and level of the felicity (or social) discount rate. On the one hand, the discount rate represents an economic phenomenon, namely the expected rate of return for alternative uses of capital to future generations. And on the other hand, it represents the equity concerns between generations,[22] which makes it to a large part an ethical matter. On an individual level, discounting our utility is quite straightforward and uncontroversial. It is empirically proven that people have a clear time preference for immediate benefits over future benefits and value them higher.[23] Therefore, discounting future benefits is a common practice. However, if you consider not a single individuals' benefit, but the welfare between individuals or even between generations, it becomes a more controversial thing to do. A social discount rate is used in major studies on the economics of climate change.[24] Here, a controversial debate on the ethics of the discount rate is taking place, as intergenerational equity plays an important role. The policy decisions of the current generation have a large effect on future generations.[25]

6 Conclusion and Discussion

Bioeconomy consists of a range of economic sectors of different sizes and histories, which will subsist with or without nowadays a given definition of bioeconomy. Similarly, the foundation for these sectors, ecosystems, and their functions and services will continue to exist. Therefore, the question must be whether nowadays' *definition* of bioeconomy, with the objective to improve all sectors in an integrated and coherent way with the dedicated public and private initiatives, is useful in promoting sustainable development.

If sustainability is the main target, progress toward it should be measured and monitored to allow policymakers an assessment of the development. This is not an easy task and comes with new issues and decisions that must be made, including ethical ones. Our indicator to measure the sustainability of bioeconomy follows the approach by Arrow et al., which is not free of criticism. In this approach, the question if we are on a path of sustainable development is answered by looking at the utility derived from the consumption of a wide range of goods by the current generation as well as all future generations. Therefore, felicity derived from nonconsumption, e.g., the value of doing nothing is not explicitly measured. The question of whether we can and should include such aspects in the indicator is an interesting topic for future work.

[22] Cf. Arrow (1999).

[23] Cf. Thaler (1981).

[24] Cf. Stern (2007) and Garnaut (2008).

[25] Cf. Lewandowsky et al. (2017).

Furthermore, the utility derived from consumption by future generations is valued less, which, to a large degree, is an ethical matter. The use and level of the discount rate must be well thought through. Ethicists could make an important contribution to that. Policymakers, who use the indicator to guide the decisions, must be aware of the discounting and its implications.

References

Arrow, K. J. (1999). Inter-generational equity and the rate of discount in long-term social investment. In M. R. Sertel (Ed.), *Contemporary economic issues. Volume 4. Economic behaviour and design.* International Economic Association Series (pp. 89–102).

Arrow, K. J., Dasgupta, P., Goulder, L. H., Mumford, K. J., & Oleson, K. (2012). Sustainability and the measurement of wealth. *Environment and Development Economics*, 17(3), 317–353. Retrieved October 3, 2019, from, https://www.cambridge.org/core/product/identifier/S1355770X12000137/type/journal_article.

Böhringer, C., & Jochem, P. E. P. (2007). Measuring the immeasurable – A survey of sustainability indices. *Ecological Economics, 63*(1), 1–8.

Bosch, R., van de Pol, M., & Philp, J. (2015). Policy: Define biomass sustainability. *Nature, 523*(7562), 526–527.

European Commission (EC). (2018). *A sustainable bioeconomy for Europe: Strengthening the connection between economy, society and the environment. Updated bioeconomy strategy.*

Garnaut, R. (2008). *The Garnaut climate change review: Final report*, Cambridge University Press: Port Melbourne, Vic.

Garnett, T., Appleby, M. C., Balmford, A., Bateman, I. J., Benton, T. G., Bloomer, P., Burlingame, B., Dawkins, M., Dolan, L., Fraser, D., Herrero, M., Hoffmann, I., Smith, P., Thornton, P. K., Toulmin, C., Vermeulen, S. J., & Godfray, H. C. J. (2013). Sustainable intensification in agriculture: Premises and policies. *Science, 341*(6141), 33–34.

Global Bioeconomy Summit. (2018). *Communiqué. Global bioeconomy summit 2018. Innovation in the global bioeconomy for sustainable and inclusive transformation and wellbeing.* Retrieved July 9, 2020, from, https://gbs2018.com/fileadmin/gbs2018/Downloads/GBS_2018_Communique.pdf

Hamilton, K., & Clemens, M. (1999). Genuine savings rates in developing countries. *The World Bank Economic Review, 13*(2), 333–356.

Lewandowsky, S., Freeman, M. C., & Mann, M. E. (2017). Harnessing the uncertainty monster: Putting quantitative constraints on the intergenerational social discount rate. *Global and Planetary Change, 156*, 155–166.

Lomborg, B. (2015). The U.N. chose way too many new development goals. In *TIME*. Retrieved October 30, 2019. from, http://time.com/4052109/un-sustainable-development-goals/

Pezzey, J. (1992). *Sustainable development concepts. An economic analysis.* World Bank Environment Paper No. 2. Edited by The World Bank.

Royal Society of Biology. (2016). *Response from the royal society of biology to the royal society call for views on education research.* Retrieved July 13, 2020, from, https://www.rsb.org.uk/images/RSB_response_to_Royal_Society_Call_for_evidence_on_education_research_27.10.2016.pdf

Sagar, A. D., & Najam, A. (1998). The human development index: A critical review. *Ecological Economics, 25*(3), 249–264.

Stern, N. (2007). *The economics of climate change: The stern review.* Cambridge University Press: New York.

Sumaila, U. R., Lam, V. W. Y., Miller, D. D., Teh, L., Watson, R. A., Zeller, D., Cheung, W. W. L., Côté, I. M., Rogers, A. D., Roberts, C., Sala, E., & Pauly, D. (2015). Winners and losers in a world where the high seas is closed to fishing. *Scientific Reports, 5*(1), 8481.

Sustainable Development Solutions Network (SDSN). (2015). *Indicators and a monitoring framework for the sustainable development goals. Launching a data revolution for the SDGs.* Retrieved July 13, 2020, from, https://irp-cdn.multiscreensite.com/be6d1d56/files/uploaded/150612-FINAL-SDSN-Indicator-Report1.pdf

Thaler, R. (1981). Some empirical evidence on dynamic inconsistency. *Economics Letters, 8*(3), 201–207.

Wackernagel, M., & Rees, W. E. (1996). *Our ecological footprint: Reducing human impact on the earth.* New Society Publishers: Philadelphia.

World Commission on Environment and Development (WCED). (1987). *Report – Our common future*. Retrieved July 7, 2020, from, http://www.un-documents.net/our-common-future.pdf

Resource Sufficiency in a Sustainable Bioeconomy: A Predator–Prey Perspective

13

Lioudmila Chatalova

Abstract

The article elaborates on the role of sufficient resource consumption in a plant-based bioeconomy. It draws on the Lotka–Volterra model of population dynamics to illustrate the effect of alternative resource-use strategies—sufficiency, economic efficiency, and ecological consistency—and their combinations on the socioeconomic sustainability. The basic organization of the bioeconomy is mathematically defined by a set of three coupled differential equations, which describe growth dynamics of mutually dependent subsystems, namely bio-based industries, plant biomass, and arable land, in analogy to predator–prey interactions in an ecosystem. Sufficiency and efficiency in resource use are introduced into the model through variations in the values of individual growth coefficients. Consistency, by contrast, is already implied in system's mutualism, which internalizes social costs of resource overexploitation. The results demonstrate that even significant improvements in economic efficiency, either through higher factor productivity (technological advance) or higher biomass availability (biotechnological innovations), only stimulate resource overexploitation, unavoidable taking the system on an unsustainable path. However, the pressure from a moderate consumption of bio-based products (consumers' sufficiency) that reduces the amount of land under biomass crops has a system-stabilizing effect. If complemented by producers' sufficiency through deliberate degrowth or lower intensity of resource consumption, the swings in economic dynamics might be further reduced. These observations inform bioeconomy policies, which rely primarily on technological innovations. It is concluded that in a sustainable economy, seeking to align its economic interests with the carrying capacity of

L. Chatalova (✉)
Leibniz-Institute of Agricultural Development in Transition Economies (IAMO), Halle (Saale), Germany
e-mail: chatalova@iamo.de

D. Lanzerath et al. (eds.), *Bioeconomy and Sustainability*,
https://doi.org/10.1007/978-3-030-87402-5_13

the environment, sufficiency presents a critical element of the negative-feedback mechanism that regulates resource consumption.

Keywords

Lotka–Volterra model · Resource sufficiency · System's self-regulation

1 Introduction

In essence, a bio-based economy is a promise of an ecological and social modernization of industrial societies increasingly experiencing their environmental boundaries and negative rebound effects.[1] High hopes attached to it build on the assumption that breakthroughs in biotechnology and the move away from finite fossil resources will create the necessary momentum to support long-term sustainable development or even growth.[2] Although this understanding of bioeconomy acknowledges the need for a fundamental rethinking of modern lifestyles and economic models, specific strategies and actions continue to be vague or contradictory.

Academic and practitioners' debates on the bioeconomy tend to deal with technological improvements in input and throughput efficiency.[3] The essential questions regarding the growth-orientation of business models as well as the availability of non-fossil resources, their sustainability, risks, and social acceptancy continue to take a backseat in these debates. Already today, industrial countries depend on foreign biomass, while the global hunger for resources keeps growing.[4] A further increase in biomass demand raises reasonable doubts over the transformative capacity of bioeconomy.[5]

An accompanying but less prominent topic in these debates is that of sufficiency. In the sustainability context, sufficiency stands for a deliberate reduction in resource and energy consumption through both individual and collective efforts.[6] It implies modes of economic activity and lifestyles that satisfy human needs while avoiding both excess and scarcity.[7] Although discussed as a desirable or necessary complement to resource efficiency and consistency, sufficiency remains marginalized as a regressive and rather impractical idea.[8] In view of the aggravating resource-population dilemma it might however become imperative in the near future.

[1] Cf. Organisation for Economic Co-operation and Development (2009); European Commission (2018).

[2] Cf. Backhouse et al. (2017).

[3] Cf. European Bioplastics (2016) and Philp (2018).

[4] Cf. Gent (2018).

[5] Cf. World Development Movement and Transnational Institute (2012).

[6] Cf. Linz (2004).

[7] Cf. Princen (2005).

[8] Cf. Alcott (2008) and Schmid et al. (2012).

In what follows, this possibility is addressed by drawing on the much referenced model of the coexistence of predators and preys in an ecosystem, mathematically described by the Lotka–Volterra equations of population dynamics. The main tenet of this model is that in order to maintain itself, a system of metabolically interdependent subsystems (or populations) requires a self-regulating negative-feedback mechanism.[9] Accordingly, the bioeconomy is modeled as a system of two populations of predators (bio-based industries and plant biomass) feeding, respectively, on two populations of preys (plant biomass and arable land). Sufficiency in resource use is introduced into the model by varying the predator–prey rules (also referred to as the energy rules). Its contribution to the sustainability of the system is then compared against resource efficiency. It is argued that in a sustainable economy, seeking to align its economic interests with the carrying capacity of the environment, sufficiency presents a necessary element of self-regulation by means of a decrease in function, that is, of a negative-feedback loop.

The paper first outlines the concept of sufficiency, while pointing out the role that sufficiency might play in a strategic vision of a bio-based economy. It then develops a stylized model of a biomass-based economy and specifies alternative scenarios of the biomass use. The discussion of the main findings summarizes the analysis, followed by conclusions and implications for bioeconomy policies and research.

2 Sufficiency in a Sustainable Bioeconomy

2.1 The Concept of Sufficiency

Although the term sufficiency has been used mainly in the environmental context, it was economists who introduced it into the sustainability debate in the early 1970s. Studies warning against the irreversible consequences of unlimited economic growth made the case for limits on material throughput. Ernst Schumacher (1973) was among the first who put forward the idea of scaling down the use of natural resources and technologies as a means to counter growing unsustainability. In his influential book "Small is beautiful: economics as if people mattered" (1973), he advocates "enoughness" as the principle guiding the harmonization of economic interests, human needs, and ecosystem boundaries. In the same vein, Herman Daly (1974) emphasizes a critical difference between quantitative growth and qualitative development, which enables a "steady state economy" to develop without compromising the future.

Despite its large public response, the debate on the limits of growth did not affect development agendas that followed in the 1980s, such as the Brundtland Report.[10] Nonetheless, it raised awareness of the fact that sustainable development is in

[9] Cf. Lotka (1956 [1925]).

[10] Cf. World Commission on Environment and Development (1987), Annexe 2, The Commission's Mandate.

Table 13.1 Characteristics of alternative resource-use strategies

	Efficiency	Consistency	Sufficiency
Type of resource-use parameter	Technological	Techno-environmental	Social
Estimated capacity to reduce resource flows within 20 years	10–20%	50–80%	10–40%
Likelihood of rebound effects	High	Medium to low	Low
Barriers to adoption/ implementation	Low due to obvious economic benefits	Rather high due to additional investment costs and opportunistic behavior of competitors	Very high due to needed changes in collective and individual behavior on the part of consumers and producers
Complementary role in a strategy mix	Granting extra time for implementation of consistency and sufficiency	Aligning resource use to regrowth rates of renewable resources and to possibilities of reuse in a cyclic economy	Regulating excesses in resource use through deliberate reduction in production and consumption (negative feedback)

Source: Based on Huber (2000) and Institute for Futures Studies and Technology Assessment (2018)

essence the question of curbing quantitative economic growth and consumerist lifestyles and, hence, of rethinking the economic paradigm that supports them. In fact, the very foundations of neoclassical economics are based on the assumption of resource scarcity.[11] This however does not make economy environment-sensitive; instead, it defines a further condition under which the market responds to unlimited human wants. From this perspective, economy's contribution to sustainable development can only be defined in terms of improved resource efficiency. The acceptance of sufficiency as a benefit and its implementation as a business principle requires therefore a debate that goes beyond the blame-game and points out perspectives on the production and consumption side.[12]

Later economic and environmental literature develops the notion of sufficiency in this sense by moving the focus away from the limits of economic growth and towards the limits of the environmental carrying capacity. This literature addresses sufficiency not as a single approach, but as a complement to resource efficiency and eco-consistency.[13] Economic efficiency and ecological consistency are technological parameters of resource use, though to different degrees (Table 13.1). They stand

[11] Cf. Samuelson and Nordhaus (2001).

[12] Cf. Elgin (1981).

[13] Cf. Sachs (1993).

for improvements in the monetary relation between output per unit of input and minimizing industrial throughput, not questioning however the economic growth aim. Being critical to profitability and competitiveness, efficiency improvements are the easiest to implement, whereas consistency requires a reorganization of production.

Sufficiency, by contrast, is rather a social principle. It relies on social and cultural innovations to achieve deliberate changes in lifestyles and collective commitment to lower environmental impact. In this understanding, sufficiency does not oppose technological improvements in resource use, but complements them by suggesting a normative frame of reference for their realization.[14] Improvements in resource efficiency, in turn, may grant extra time for adopting less popular but more powerful strategies of consistency and sufficiency.

In its today's understanding, the notion of sufficiency is multifaceted. It reflects along with concerns for environmental overshoot and international and intergenerational justice the pursuit of the good life, thus covering the debate on the upper and the lower threshold of sustainable material wealth.[15] While the argument of the maximum (or upper) carrying capacity of the ecosystem seems to prevail, the lower threshold's advocates find support from the multidisciplinary research that questions a steady positive correlation between the level of material wealth and the quality of life.[16] This variety of sufficiency aspects has been recently discussed in terms of the distinction between the (rather objective definition of) quantitative reductions in growth and the (rather subjective understanding of) qualitative improvements in wealth.[17] At the core of this intricate debate is the idea that sustainable development requires a negative-feedback regulating excesses in production and consumption.

2.2 Sustainable Biomass Use Without Sufficiency?

In substance, the concept of sufficiency and the idea of a negative feedback are reflected in the expectations towards a bioeconomy. With its emphasis on renewable biomass, integrated production cycles, curbing energy, and resource consumption as well as reducing waste, a bio-based economy is supposed to become more sensitive to the biophysical environment, on which it critically depends.[18] Supported by digital and biotech innovations and accompanying improvements in economic efficiency and ecological consistency, an increasing use of renewable biomass is expected to act as a carbon sink and address manifold sustainability concerns.[19] Great hopes are also linked to its potential to generate new revenue streams for

[14] Cf. Huber (2000).

[15] Cf. Schneidewind and Zahrnt (2014), Fourie and Rid (2016) and Spengler (2018).

[16] Cf. Alexander (2012).

[17] Cf. Spengler (2018).

[18] Cf. German Bioeconomy Council (2016).

[19] Cf. European Commission (2011).

farmers, create new business opportunities across the sectors, and to stimulate economic growth at large.[20]

While the benefits of a gradual substitution of finite fossil resources by renewable ones seem intuitive, the ability of plant biomass to meet the projected demand of industries and growing population[21] is debatable. The uncontested fact is that the Earth Overshoot Day keeps moving up earlier in the year. Although inventory and economic estimations of available biomass differ widely, many studies underline the limited rates of regrowth of biomass, along with land-use conflicts from its production on a larger scale.[22] The efforts to address the overuse of resources, including renewable ones, by a circular economy are likewise limited in their capacity due to thermodynamic, chemical, or physical properties of composite materials and the recycling process itself.[23] Such a regenerative economy is also not immune to rebound effects or risk traps and, if it will continue to be driven by the economic imperative of growth, may even aggravate those effects.[24] Not least, renewability of biological resources does not automatically imply their environmental, social, or economic sustainability.[25]

To address the growth-sustainability trade-off in bioeconomic systems, sufficiency as a principle of feedback regulation of resource consumption should be revisited. In fact, today's expectations for biomass fall on fertile ground of a growing recognition of tangible impacts of resource overuse and the acknowledgement of the individual and collective responsibility for those impacts. Political efforts, voluntary initiatives among industries (e.g., for reduction of plastic bag use), and global social movements for climate action (e.g., Fridays for Future and Scientists for Future) are evidences of a changing stance on resource use. Given this development, which lowers the barriers for acceptance and adoption of moderate resource consumption, sufficiency may be analyzed as a realistic business and social principle that complements technological enhancements in efficiency and consistency concerning biomass use.

3 Bioeconomy as a Predator–Prey Relationship

3.1 Self-Regulating Systems

Like many other concepts of sustainable economic development—such as bionics, cradle-to-cradle, ecological modernization, or industrial ecology—bioeconomy relies on technological innovations. Unlike those other concepts, the bioeconomy

[20] Cf. Scarlat et al. (2015).

[21] Cf. European Commission (2018).

[22] Cf. Gronowska et al. (2009); Environment Agency (2009) and Piotrowski et al. (2015).

[23] Cf. Daly (1979) and Ayres (1999).

[24] Cf. Beck (1992) and Cooper (2007).

[25] Cf. ibid.

is explicitly supposed to transform the entire socioeconomic system by making the economy responsive to social and ecological sustainability concerns. Such a holistic vision of the sustainability-oriented transformation hinges on "the open systems character of the economy".[26] An open system regulates itself in a continuous exchange of resources with its external environment.[27] This view of living systems assumes that all development processes, from evolutionary to psychological to cultural ones, follow the universal laws of thermodynamics, according to which animate and inanimate nature is an energy conversion system.[28]

Mathematically, a self-regulating system was defined independently by Alfred Lotka (1925) and Vito Volterra (1931) in their models of fluctuating dynamics in chemical concentrations and among fish populations. The probably most known illustration of their model is given by the coexistence of three species, such as foxes, feeding on rabbits, and those rabbits, feeding on grass. The survival prospects of each species are described by a set of differential equations as a function of their own biological growth and reproduction rates. The latter in turn are functions of predator's consumption of its prey (attack rates) and the ability of the latter to avoid being eaten (defense rates). Consequently, each prey serves as the direct environment upon which its predator metabolically depends. Any misbalances in predation and defense rates, if not compensated, would necessarily cause threats to existence of some if not all species.

With its focus on the metabolic interdependencies, the Lotka–Volterra model raises awareness not only of limited resources but also of the indirect dependencies within the system, where foxes, although not feeding directly on grass, are dependent on its growth rate and vice versa. A clear advantage of this biological model is its formal simplicity. It reduces complex population interdependencies to their conceptual, qualitative core, where energy rules (where energy is an optimization parameter and does not necessarily relate to physical energy) are the only necessary descriptor of population dynamics and ecological consistency an element of model design. We use these properties to model the basic organization of a biomass-based economy in terms of predator–prey (or human–nature) interactions and to comparatively analyze the role of sufficiency and efficiency in its stability.

3.2 The Model: Bioeconomy as a Self-Regulating Predator–Prey Interaction

In its simplest and stylized form, a biomass-based economy can be modeled as a system of two subsystems (populations or species in biological terms): the entire industrial sector X_1 (predator) that uses or processes all types of food and non-food biomass X_2 (prey). Each subsystem grows proportionally to its own size. The

[26] Kapp (1985), 143. This formulation can also be found in several other passages in this book.

[27] Cf. Scott (1998).

[28] Cf. Ostwaldt (2017 [1912]).

interrelations within the system are defined by a set of differential equations, with X' denoting the rate of change in subsystems' size with respect to time:

$$\begin{cases} X_1' = (\beta_1 X_2 - \alpha_1)X_1 \\ X_2' = (\alpha_2 - \beta_2 X_1)X_2 \end{cases} \tag{13.1}$$

The negative growth coefficient α_1 describes the energy condition of the bio-based industry that—in biological terms—can survive only by feeding on biomass. It reflects the part of fixed operating costs that are incurred even if the production is paused or not running at its full capacity. The coefficients of attack β_1 measures the reproduction rate of industries per one unit of biomass used. Accordingly, the gains from using biomass are described by the product of the coefficient of attack and the interaction between the subsystems X_2X_1. The natural growth rate of biomass is described by the positive coefficient α_2. The growth of biomass can be negatively affected by its limited ability to avoid exploitation by the industry, denoted by the negative coefficient of defense β_2. In economic contexts, a high absolute value of β_2 can be due to low environmental standards that might otherwise prevent or reduce the discriminatory behavior of biomass processors.

The model further shows that in equilibrium, where X_1' and X_2' are zero, an amount of biomass that ensures stability of the modeled bioeconomic system $\left(X_2^*\right)$ is determined by the willingness of the bio-industry to reduce the size of its own needs:

$$X_2^* = \alpha_1/\beta_1 \tag{13.2}$$

A deliberate degrowth (a higher negative α_1) can be reached by accepting higher wear and tear rates, or by reducing the intensity of resource consumption (a lower β_1), or by both. To show that the same holds for more complex trophic interactions, the model is extended by a third subsystem, arable land,X_3:

$$\begin{cases} X_1' = (\beta_1 X_2 - \alpha_1)X_1 \\ X_2' = (\beta_2 X_3 - \alpha_2 - \gamma_2 X_1)X_2 \\ X_3' = (\alpha_3 - \gamma_3 X_3 - \beta_3 X_2)X_3 \end{cases} \tag{13.3}$$

The coefficient α_2 is now negative, implying that biomass can now be produced only by exploiting other resources, which is land. The coefficients of defense γ_2 and β_3 measure the ability of the populations biomass and land to avoid exploitation by biomass processors and producers correspondingly through, for example, environmental standards, policies, or bargaining powers. The negative coefficient γ_3 limits the growth of land resources proportionally to their own amount. This restriction accounts for the limited size and limited regenerative capacity of land that has no explicit prey in the model. In practice, land growth is limited by a number of factors such as changing climate and soil conditions, urbanization, or land conservation. The equilibrium amounts of all three subsystems are then:

$$\begin{cases} X_1^* = (\alpha_3\beta_1\beta_2 - \alpha_2\beta_1\gamma_3 - \alpha_1\beta_2\beta_3)/\beta_1\gamma_2\gamma_3 \\ X_2^* = \alpha_1/\beta_1 \\ X_3^* = (\alpha_3\beta_1 - \alpha_1\beta_3)/\beta_1\gamma_3 \end{cases} \tag{13.4}$$

Noteworthy, the amount of land is not contingent on the growth rates of its direct predator, biomass, which—just as for two populations—depends solely on the relative resource hunger (i.e., growth to consumption ratio) of the processing industry. The latter, by contrast, relies on the availability of biomass and arable land. Hence, if industrial demand for biomass grows slower than biomass and the growth of the latter does not exceed the regeneration rate of land ($[\alpha_3\beta_2 - \alpha_2\gamma_3]\beta_1 > \alpha_1\beta_2\beta_3$), then the system is able to maintain its carrying capacity within the changing conditions of resource availability and consumption habits—in other words, it is sustainable.

4 Illustration for Different Scenarios of Resource Consumption

4.1 Scenarios

The Lotka–Volterra model with three populations is used to study the role of sufficiency in bioeconomic systems. The sustainability of the industry-biomass-land system, in terms of model stability, is explored by varying its energy rules in different scenarios (Table 13.2).

Scenario 1 is the reference case for a bioeconomy that regulates its resource use through efficiency improvements only. Here, the growth of biomass processing industries is not limited by the amount of available land. This assumption stands for the mode of economic organization that disregards environmental concerns. Alternatively or additionally, it can also stand for progressive uncoupling of biomass production from land use. Mathematically, this can be reflected by an immediate regrowth of biomass. An improvement in economic efficiency of resource use is addressed in three ways, namely by variation in (a) factor productivity dX_1/dX_2 (technological improvements); (b) the amount of available biomass (biotechnological improvements of its energy content or (re)growing rate); and (c) by the combination of (a) and (b).

Table 13.2 Scenario overview

1 no limits on land use *(no sufficiency)*			2 limits on land use *(consumers' sufficiency)*		
1a higher factor efficiency	1b faster biomass regrowth	1c combination of 1a and 1b	2a degrowth/ *producers' sufficiency*	2b higher factor efficiency	2c combination of 2a and 2b
				2b+ faster biomass regrowth	2c+ faster biomass regrowth

Scenario 2 accounts for sufficiency on the part of consumers. It reflects a modified consumer behavior that (via changed demands, policies or other social projects/ values) limits the amount of arable land used for biomass production. In this scenario, the following effects are studied: (a) resource sufficiency on the part of bio-industries (by deliberate reduction of production scales); (b) factor productivity increase; and (c) combination of both options. Subscenarios 2b and 2c are further extended (to 2b+ and 2c+) by accounting for improvements in plant (re)growth rates.

Because the stability of population dynamics is determined by the relative size of populations and their metabolic balance, studying the effect of changes in the interaction (energy) rules of interest does not require real absolute parameter values. For the reference scenarios, it is assumed that the initial population of biomass, measured in any arbitrary energy units, is twice as large as the population of processing industries (80 and 40). Industries grow at the rate of 5%, while their efficiency of energy conversion from exploiting one unit of biomass is 20%. For biomass it is, respectively, 4% and 4%. The initial size of land is assumed to be 16 times its direct predator (biomass) and further limited in the second scenario by its own regeneration rate. NetLogo open source software[29] is used to graphically illustrate the system's behavior in response to variations in the energy rules in a replicable way. The system is assumed to be stable (*or* sustainable) if it can maintain all three populations over at least 1000 iterations (time steps). This number is selected so as to observe more than two cyclical adjustments of the system to resource overconsumption. The system collapses if the number of biomass processors gets to zero.

4.2 Results

Figure 13.1 summarizes results for scenario 1 where resource use is regulated by efficiency improvements only and land for biomass production is not limited. The results show population dynamics in terms of their energy levels over time. For reasons of space, the size of the population *land* is divided by four.

In reference scenario 1, the modeled bioeconomic system collapses after a bit more than 200 iterations. Even a significant increase in biomass productivity by 50% (1a) does not stabilize the system. The technologically stimulated industrial growth exceeds the level that biomass supply can sustain so that its overexploitation leads to a faster but lower industry peak. A biotechnological advance that substantially improves the availability of biomass alone (1b) or in combination with a higher biomass productivity (1c) increases the size of bio-based industries, but only slightly postpones the system's collapse. In scenario 2, the system's breakdown is addressed by introduction of two types of resource sufficiency (Fig. 13.2).

Reference scenario 2 presents the case where availability of arable land for biomass production is limited by 20% compared to scenario 1. This restriction,

[29] Cf. Wilensky (1999).

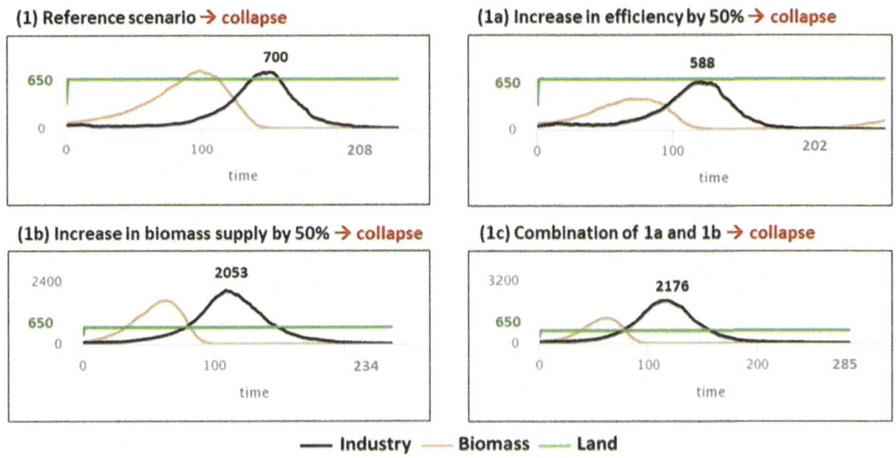

Fig. 13.1 Scenario 1—no sufficiency in resource use

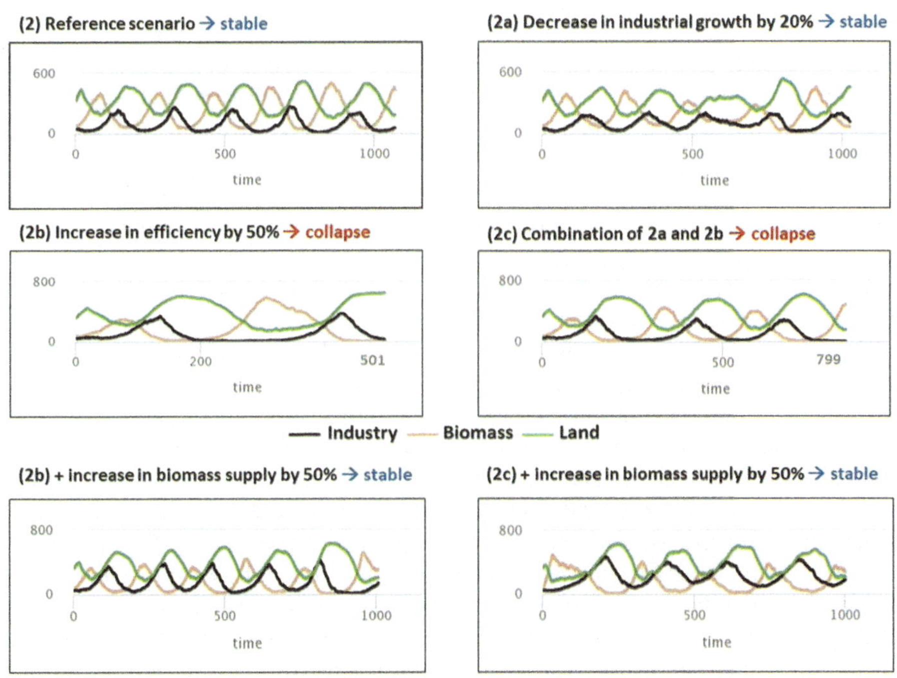

Fig. 13.2 Scenario 2—sufficiency in consumption and production

reflecting changes in consumer behavior towards sufficient lifestyles, allows the system to maintain all its elements over a longer time horizon. Sufficiency in industrial biomass use through deliberate downscaling of industrial growth

(2a) further contributes to the system's sustainability by reducing the population swings. The same or comparable effect might be possible through the use of alternative resources outside the modeled system, given, however, that the implications on the extended system's environment are rendered irrelevant. Counter to intuition, a significant improvement in biomass productivity (2b and 2c) destabilizes these sustainable dynamics, although much later compared to scenario 1. However, biotechnological advances leading to higher biomass availability (2b+ and 2c+) may reestablish sustainable population dynamics and increase the level of industrial operations.

4.3 Discussion of Findings

Other than many studies that use Lotka–Volterra equations to describe precarious human–nature interactions in high-consumption economies,[30] the present analysis is focused not on the carrying capacity of the natural environment, but on the role of the negative feedback in the socioeconomic sustainability. Its findings, although sketchy and rather preliminary, highlight the role of a moderate consumption of resources (both renewable and not) as an essential element of such a feedback mechanism. A bioeconomy, viewed of as a system of metabolically interdependent populations, was shown to be able to maintain itself in a longer run by regulating the size of its needs.

Within the given set of assumptions, this self-regulation can be achieved in three ways: by reducing consumers' demand, the scale of resource-intensive production, or by both. Our findings demonstrate that the pressure from moderate consumption (that reduces the availability of land for biomass production) may already trigger a system-stabilizing effect. If complemented by producers' sufficiency, economic volatilities in terms of swings in economic activities might be reduced. In contrast, if industrial resource consumption is driven by the logic of growth and utility maximization alone (with resource consumption rates as large as or larger than regeneration rates of land on which biomass is produced), then the system exhibits population collapse.

The variation of energy rules in our model reveals a very limited capacity of managing resource use through efficiency improvements. This observation has important implications for the bioeconomy, given its overreliance on technological innovations and the substitution of fossil resources by renewable ones. It has been displayed that improvements in resource productivity by enhanced production methods or higher biomass availability create incentives to produce more, which may overstrain direct and indirect resources. Yet, the findings from scenario 2 also show that both resource use options, efficiency and sufficiency, have a role to play as it comes to ensure system stability.[31] If complemented by sufficiency at least on the

[30] Cf. Cohen (1998) and Motesharrei et al. (2014).

[31] Cf. Huber (2000) and Institute for Futures Studies and Technology Assessment (2018).

consumers' side, a higher energy yield of available resources may lead to a sustainable economic dynamic.

The magnitude of observed negative efficiency effects is partly due to the model's assumptions. Because of the closed metabolic interdependencies in the Lotka–Volterra model, and hence of implied resource consistency, any undesired effects of economic activities cannot be externalized. In integrated production systems with closed loop supply chains and resource cascades, such effects might therefore become especially pronounced. The metabolic character of the model highlights a further important implication for the bioeconomy, which is the role of indirect dependences. While the direct dependence of bio-industries on biomass is evident, indirect dependences on land and other factors that enable biomass supply are much more difficult to fully account for and consequently easier to ignore. Blinding out indirect dependences, effects and their repercussions allows continuing business as usual, "oblivious to the catastrophic trajectory"[32] that may take the bioeconomy to its failure.

Our finding of the essential role of a moderate resource use on the part of industries and consumers for sustainable self-regulation of high-consumption societies is in line with studies and statistics showing that wealthier societies or social classes within one society deliberately reproduce at a lower rate to keep their wealth level. A move towards a society that lives better with less requires, however, more than changes in economic imperatives. It requires a higher appreciation of common goods and further social and behavioral innovations. Taking sufficient predation in an ecosystem as a role model for human–nature interactions may support this process.

5 Conclusion

Predation has been a popular and persistent metaphor for describing resource use in a capitalistic economy. Competition for resources and fight for survival in the market are viewed as underlying parallels to predator–prey behavior.[33] Although catchy, this metaphor oversimplifies predation as being driven by voracity and aggressiveness. Yet, in natural environment, predators as a rule do not overconsume their source of feed, while an unbalanced growth of predators is necessarily regulated by a decline in preys.[34]

The paper proposed to revisit the model of metabolically interdependent predator–prey dynamics to address resource use in a sustainable bioeconomy. This model, its mathematical extensions, and applications to different fields—from microbial dynamics to social conflicts to stock market behavior[35]—highlight the

[32] Motesharrei et al. (2014, p. 100).

[33] Cf. Moor (1993).

[34] Cf. Volterra (1959 [1931]).

[35] Cf. Haas (1981), Motesharrei et al. (2014) and Addison et al. (2016).

vital importance of a negative-feedback mechanism for maintaining system stability. The findings illustrate that sufficiency, both on production and consumption sides, is a key element of such self-regulation. It is not only complementary to economic efficiency, but a precondition for socioeconomic sustainability. An economy not controlling the size of its needs (or "predation") in response to environmental stress inevitably puts itself on an unsustainable trajectory.

Although this conclusion applies to all types of economies, high-consumption countries should be especially alert to the consequences of dismissing sufficiency as irrelevant or impractical. Putting high expectations on (bio)technological advances and renewable resources but ignoring sufficiency in the relevant strategies and toolsets is likely to create risks and rebounds of new quality rather than address sustainability challenges.[36] Broadening acceptance of moderate consumption and downscaling industrial (over)production is a long way to go. A challenge to start with can be the reduction of food loss and waste by policy instruments and deliberate actions on the part of producers and consumers. With its huge potential for water, land, and energy savings,[37] the reduction of food loss and waste can provide a strong argument for the feasibility and tangible advantages of sufficiency.

The findings in this paper and their outlined implications point to a number of directions to be followed by further research. As the bioeconomy is supposed to bring about a comprehensive societal change, the complexity of possible interactions will be much higher than in a stylized linear predator–prey relationship. Alternative approaches such as food web or network analysis might provide a more powerful guide to dealing with complex systemic interdependences. Defining a balanced mix of resource-use strategies, for example, would require not only economic investigations into negative effects of higher resource efficiency, but also changes in behavior, expectations and imaginaries of heterogeneous actors. In this regard, ethical and philosophical studies may provide normative arguments for advocating sufficiency in societies where the fundamental human right to personal fulfillment justifies excessive consumption.[38] Not least, interdisciplinary debates on the values added from degrowth and sufficient lifestyles (that is, from sufficient resource "predation") should be stimulated to explore opportunities for a sustainable bioeconomy.

References

Addison, L. M., Bhatt, B. S., & Owen, D. R. (2016). A prey-predator model for investing in stocks. *International Journal of Pure and Applied Mathematics, 107*(2), 487–504.

Alcott, B. (2008). The sufficiency strategy: Would rich-world frugality lower environmental impact? *Ecological Economics, 64*(4), 770–786.

[36] Cf. Parrique et al. (2019).

[37] Cf. Lipinski et al. (2013).

[38] Cf. Sloterdijk (1999).

Alexander, S. (2012). The optimal material threshold: Toward an economics of sufficiency. *Real-World Economics Review, 61*, 2–21.

Ayres, R. U. (1999). The second law, the fourth law, recycling and limits to growth. *Ecological Economics, 29*(3), 473–483.

Backhouse, M., Lorenzen, K., Lühmann, M., Puder, J., Rodríguez, F., & Tittor, A. (2017). *Bioökonomie-Strategien im Vergleich. Gemeinsamkeiten, Widersprüche und Leerstellen.* Bioeconomy & Inequalities, Working Paper No. 1.

Beck, U. (1992). *Risk society: Towards a new modernity*. Sage.

Cohen, J. E. (1998). How many people can the earth support? *Bulletin of the American Academy of Arts and Sciences, 51*(4), 25–39.

Cooper, M. (2007). Life, Autopoiesis, debt: Inventing the bioeconomy. *Distinktion: Journal of Social Theory, 8*(1), 25–43.

Daly, H. E. (1979). Entropy, growth and the political economy of scarcity. In V. K. Smith (Ed.), *Scarcity and growth reconsidered* (pp. 67–94).

Daly, H. E. (1974). The economics of the steady state. *The American Economic Review, 64*(2), 15–21.

Elgin, D. (1981). *Voluntary simplicity: Towards a way of life that is outwardly simple, inwardly rich*. Duane.

Environment Agency. (2009). *Biomass: Carbon sink or carbon sinner?* Executive Summary. United Kingdom Environment Agency for England and Wales. Retrieved September 14, 2020, from, https://fcrn.org.uk/sites/default/files/Biomass__carbon_sink_or_carbon_sinner_summary_report.pdf

European Bioplastics. (2016). *Position of European bioplastics concerning bioplastics and the circular economy*. Retrieved June 18, 2020, from, https://docs.european-bioplastics.org/publications/pp/EUBP_PP_Circular_economy_package.pdf

European Commission (EC). (2011). *A roadmap for moving to a competitive low carbon economy in 2050*. Communication from the Commission to the European Parliament, the Council, the European Economic and Social Committee and the Committee of the Regions. COM(2011)112 final. Brussels.

European Commission (EC). (2018). *A sustainable bioeconomy for Europe: Strengthening the connection between economy, society and the environment. Updated bioeconomy strategy.*

Fourie, C., & Rid, A. (2016). *What is enough? Sufficiency, justice, and health*. Oxford University Press.

Gent, R. (2018). Bioökonomie: Erdöl versus Biomasse? Chancen und Risiken aus Sicht der Deutschen Industrievereinigung Biotechnologie. *Rundbrief Forum Umwelt & Entwicklung, 1*, 4–5.

German Bioeconomy Council. (2016). *Further development of the "National Research Strategy BioEconomy 2030"*. Recommendations of the German Bioeconomy Council.

Gronowska, M., Joshi, S., & MacLean, H. L. (2009). A review of U.S. and Canadian biomass supply studies. *BioResources, 4*(1), 341–369.

Haas, C. N. (1981). Application of predator-prey models to disinfection. *Journal of the Water Pollution Control Federation, 53*(3/I), 378–386.

Huber, J. (2000). Industrielle Ökologie. Konsistenz, Effizienz und Suffizienz in zyklusanalytischer Betrachtung. In R. Kreibich & U. E. Simonis (Eds.), *Global change – Globaler Wandel. Ursachenkomplexe und Lösungsansätze* (pp. 107–124).

Institute for Futures Studies and Technology Assessment. (2018). *Effizienz, Konsistenz, Suffizienz. Strategieanalytische Betrachtung für eine Green Economy*. IZT-Text 1–2018.

Kapp, K. W. (1985). *The humanization of the social sciences*. Edited by J. E. Ullmann, R. Preiswerk.

Linz, M. (2004). *Weder Mangel noch Übermaß: Über Suffizienz und Suffizienzforschung*. Edited by Wuppertal Institute for Climate, Environment and Energy. Working paper 145.

Lipinski, B., Hanson, C., Lomax, J., Kitinoja, L., Waite, R., & Searchinger, T. (2013). *Reducing food loss and waste.* Working Paper, Installment 2 of "Creating a Sustainable Food Future". Edited by World Resources Institute.

Lotka, A. J. (1956). *Elements of physical biology [1925].* Dover Publications.

Moor, J. F. (1993). Predators and prey: A new ecology of competition. *Harvard Business Review, Reprint, 93309*, 75–86.

Motesharrei, S., Rivas, J., & Kalnay, E. (2014). Human and nature dynamics (HANDY): Modeling inequality and use of resources in the collapse or sustainability of societies. *Ecological Economics, 101*, 90–102.

Organisation for Economic Co-operation and Development (OECD). (2009). *The bioeconomy to 2030: Designing a policy agenda.*

Ostwaldt, W. (2017). *Der Energetische Imperativ [1912].* Verone Publishing.

Parrique, T., Barth, J., Briens, F., Kerschner, C., Kraus-Polk, A. W., Kuokkanen, A., & Spangenberg, J. (2019). *Decoupling debunked: Evidence and arguments against green growth as a sole strategy for sustainability.* Edited by European Environmental Bureau (EEB).

Philp, J. (2018). The bioeconomy, the challenge of the century for policy makers. *New Biotechnology, 40*(A), 11–19.

Piotrowski, S., Carus, M., & Essel, R. (2015). Global bioeconomy in the conflict between biomass supply and demand. *Industrial Biotechnology, 11*(6), 308–315.

Princen, T. (2005). *The logic of sufficiency.* The MIT Press.

Sachs, W. (1993). Die vier E's: Merkposten für einen maßvollen Wirtschaftsstil. *Politische Ökologie, 11*(33), 69–72.

Samuelson, P. A., & Nordhaus, W. D. (2001). *Economics* (17th ed.). McGraw Hill Irwin.

Scarlat, N., Dallemand, J.-F., Monforti-Ferrario, F., & Nita, V. (2015). The role of biomass and bioenergy in a future bioeconomy: Policies and facts. *Environmental Development, 15*, 3–34.

Schmid, O., Padel, S., & Levidow, L. (2012). The bio-economy concept and Knowledge Base in a public goods and farmer perspective. *Bio-based and Applied Economics, 1*(1), 47–63.

Schneidewind, U., & Zahrnt, A. (2014). *The politics of sufficiency: Making it easier to live the good life.* München, oekom Verlag.

Schumacher, E. (1973). *Small is beautiful: A study of economics as if people mattered.* Blond & Briggs.

Scott, W. R. (1998). *Organizations: Rational, natural, and open systems.* Englewood Cliffs.

Sloterdijk, P. (1999). *Sphären II: Globen.* Makrosphärologie. Frankfurt/M.

Spengler, L. (2018). *Sufficiency as policy: Necessity, possibilities and limitations.* Baden-Baden.

Volterra, V. (1959). *Theory of functionals and of integral and integro-differential equations [1931].* Dover Publications.

Wilensky, U. (1999). *NetLogo.* Center for Connected Learning and Computer-Based Modeling. Northwestern University. Retrieved June 18, 2020, from, *http://ccl.northwestern.edu/netlogo/*

World Commission on Environment and Development (WCED). (1987). *Report – Our common future.* Retrieved August 17, 2020, from, http://www.un-documents.net/our-common-future.pdf

World Development Movement (WDM) and Transnational Institute (TNI). (2012). *Bio-economies: The EU's real 'green economy' agenda?* Retrieved September 7, 2020, from, https://www.tni.org/files/download/wdm-tni__bio-economy_.pdf

Agriculture in the Bioeconomy: Economics and Policies

14

Justus Wesseler and Maximilian Kardung

Abstract

The bioeconomy is high on the policy agenda as several countries have published related strategies. We consider five major reasons why we should take the bioeconomy seriously from an economic and policy perspective, namely (1) Advances in biological sciences; (2) An increase in horizontal and vertical integration in agricultural supply chains; (3) An increase in inter- and intra-industry trade; (4) Advances in information and communication technologies; (5) An increase in globalisation.

Developing the bioeconomy promises great opportunities, but it requires some knowledge about sustainable development for examining whether or not the bioeconomy lives up to its promises. New technologies often have impacts on human health and/or the environment that may not be covered in the prices of traded goods and services. They often also include irreversibility effects. Those effects have to be evaluated differently than effects considered reversible. Further, uncertainties about future benefits and costs need to be considered as well. This is one reason why the precautionary approach has been mentioned in many regulations. An extended cost–benefit analysis based on real option theory allows

The contribution is a revised and updated version based on Inaugural lecture upon taking up the post of Professor of Agricultural Economics and Rural Policy at Wageningen University (Wesseler, 2015).

J. Wesseler
Agricultural Economics and Rural Policy Group (AEP), Wageningen University, Wageningen, The Netherlands
e-mail: justus.wesseler@wur.nl

M. Kardung (✉)
Agricultural Economics and Rural Policy Group (AEP), Wageningen University, Wageningen, The Netherlands
e-mail: maximilian.kardung@wur.nl

D. Lanzerath et al. (eds.), *Bioeconomy and Sustainability*,
https://doi.org/10.1007/978-3-030-87402-5_14

evaluating new developments in the bioeconomy under uncertainty and irreversibility. The example of vitamin A enriched rice illustrates the complexity of economic, social, and political issues related to new technologies and their controversies.

Dynamic aspects of bioeconomy policies and their economic implications along the value chain are a priority for further research. Research areas include market and policies, spatial structures as well as institutional and organisational aspects of biomass use, bioeconomy business and business models, and the link with rural development.

Keywords

Bioeconomy · Economics · Policies

1 Introduction

The bioeconomy is high on the policy agenda. The Dutch government has identified the bioeconomy as one among the "top sectors". Germany has introduced the Bioeconomy Council in 2009. The United States (US) government has published a National Bioeconomy Blueprint in 2012. The European Union (EU) has produced a Bioeconomy Strategy in 2012, reviewed this strategy in 2017, and updated it in 2018.

European Commission President Ursula von der Leyen presented the European Green Deal at the end of 2019 as a growth strategy for the EU. According to von der Leyen, the European Green Deal "shows how to transform our way of living and working, of producing and consuming so that we live healthier and make our businesses innovative."[1] The European Green Deal Investment Plan aims to mobilise at least 1 trillion Euros of investments over 10 years and the bioeconomy will likely play an essential role in achieving the goals of the plan by, among other things, making sustainable use of forests and promoting rural development.[2]

But what actually is the bioeconomy? What does it have to offer? Why do we discuss this now and not 20 years ago? When we think about agriculture, did we not always have a bioeconomy? And is it not just a buzzword that may soon be forgotten? Or should we focus on bioeconomy in our research on agricultural economics and policies? And what are the questions that need to be addressed?

When we look at the various definitions of "bioeconomy" we can observe similarities and differences. Some explicitly consider public sector research and development activities, others consider only the renewable energy sector.

[1] European Commission (2019b).

[2] Cf. European Commission (2019a).

The definition used by the European Commission includes the primary sectors as well as the up- and downstream sectors: "The bioeconomy covers all sectors and systems that rely on biological resources (animals, plants, micro-organisms and derived biomass, including organic waste), their functions and principles. It includes and interlinks: land and marine ecosystems and the services they provide; all primary production sectors that use and produce biological resources (agriculture, forestry, fisheries, and aquaculture); and all economic and industrial sectors that use biological resources and processes to produce food, feed, bio-based products, energy, and services."[3]

The European Commission recognises that "the bioeconomy, currently worth EUR 2.3 trillion in turnover and accounting for 8.2% of the EU's workforce, plays a central role in addressing a number of key interlinked challenges."[4]

In addition to the sectors mentioned, we also need to consider research and development conducted in both the public and the private sector and we will come back later to why this is important.

2 Beyond the Agricultural Sector

In our opinion there are five major reasons why we should take the bioeconomy seriously from an economic and policy perspective. These reasons are:

- Advances in biological sciences.
- An increase in horizontal and vertical integration in agricultural supply chains.
- An increase in inter- and intra-industry trade.
- Advances in information and communication technologies.
- An increase in globalisation.

2.1 Advances in Biological Sciences

The development of recombinant DNA technologies in the early 1970s was the start of modern biotechnology.[5] The Bayh-Dole act of 1980 in the US, which provided universities and other forms of organisations with the right to exploit patents that had been obtained with public funding, has been seen as key factor for innovations in modern biotechnology.[6] Some of the first successful products using rDNA technologies were a vaccine for swine diarrhoea in 1982 by the Dutch company Intervet and the production of human insulin for diabetics from genetically engineered bacteria by the US company Eli Lilly. Since 1984, the Dutch Company

[3] European Commission (2018, p. 4).

[4] Ibid., 21 f.

[5] Cf. Tramper and Zhu (2011).

[6] Cf. Stevens (2004).

Gist-Borcades (now DSM) started to insert the bovine chymosin gene in yeast cells, which allows for cultivating the yeast in large fermenters to be used for cheese production. In the late 1980s, the technology was adopted by cheese producers in Switzerland, followed, respectively, by producers in the Netherlands, Germany, and France, in 1992, 1997, and 1998. Parallel, applications for enzymes produced from genetically engineered (GE) bacteria for bakery products have been introduced.[7] In recent years, the development of New Plant Breeding Technologies (NPBTs) has set a new milestone in the advances of biological sciences. NPBTs consist of a range of technologies that allow plant breeders to develop crops with desired traits (e.g., drought resistance) in a more precise and quicker way than with traditional plant breeding techniques. For bioeconomy sectors that process biomass, NPBTs can help to develop plants that are more effective and need fewer resources. However, the Court of Justice of the European Union made a crucial decision for the development of NPBTs in 2018.[8] Plants developed by using NPBTs are since then considered GMOs and have to go through a long and expensive approval process for import and processing and/or cultivation. The same holds, possibly, also for the placement of food and feed products on the EU market, including import and processing. These regulatory hurdles could hinder the development and use of NPBTs and reduce the comparative advantage for the EU bioeconomy.[9] Other countries, such as China and the US, who do not consider NPBTs as GMOs, can benefit more from advances in these techniques.

Nevertheless, a wide array of applications of modern biotechnologies is available including applications in the food and feed sector, biofuels, biomaterials, chemicals, biorefineries, and more. These advances in the natural sciences allow us to address a number of societal challenges such as climate change.

2.2 Increase in Horizontal and Vertical Integration

In addition to this drastic technological change, supply chains became increasingly vertically and horizontally integrated.[10] Looking at the agricultural sector only and not considering the increase in up- and downstream linkages through different forms of contractual arrangements may create biases in policy analysis. Contractual arrangements cause hysteresis resulting in delayed responses to changes in external factors such as in market prices. If farmers have signed up for an environmental service scheme, they may not easily change their mode of production. Horizontal integration through mergers and acquisitions in up- and downstream sectors or voluntary collaboration at farm level can change the market power of agents with economic and distributional effects along the value chain.

[7] Cf. Tramper and Zhu (2011).

[8] Cf. Court of Justice of the European Union (2018).

[9] Cf. Gocht et al. (2020).

[10] Cf. Wesseler (2014a).

2.3 Increase in Inter- and Intra-Industry Trade

A further important aspect is the increase in inter- and intra-industry trade. The volume of world merchandise exports has more than tripled since 1990.[11] The share of intra-industry trade has more than doubled since the early 1960s.[12] Since 1960 agricultural production has tripled while the agricultural trade volume has increased by a factor of six worldwide.[13] This increase in trade with all the implications this may have for countries, producers, and consumers has increased inter-linkages between international trade and agricultural production. A drought in Brazil, resulting in a decrease in soybean yields, has an effect on European animal farming as happened in 2007/08.[14]

2.4 Advances in ICT

A further important development has been the increase in information and communication technologies. Internet and phone connections are now available almost everywhere and news about major events spread rapidly around the world. The speed at which information (news) is communicated (formally and informally) and its reach is positively affected by advances in the information and communication technology (ICT) sector: especially through mobile telephony and television networks, as well as the internet (including its social media platforms).

As ICTs improve, become more affordable, and their use spreads across the world—especially in developing countries—their impact on the bioeconomy and therefore society will gain further importance.

2.5 Increase in Globalisation

A fifth important aspect which is closely related to the previous three is the increase in globalisation. According to the Levin Institute this is understood "as a process of interaction and integration among the people, companies, and governments of different nations, a process driven by international trade and investment and aided by information technology. This process has effects on the environment, on culture, on political systems, on economic development and prosperity, and on human physical well-being in societies around the world."[15] Globalisation goes beyond increase in international trade and vertical and horizontal integration. We now find food chains such as McDonald's or Burger King and food processors such as Nestlé

[11] Cf. World Trade Organization (2014).

[12] Cf. Organisation for Economic Co-operation and Development (2010).

[13] Cf. Food and Agriculture Organization of the United Nations (2013).

[14] Cf. Backus et al. (2008).

[15] Levin Institute.

or Unilever in almost every country and food retailers such as Walmart or the Schwarz Group are following closely behind.

These five issues have important potential implications for players within the bioeconomy, especially concerning activities impacting consumer affairs and the environment.

The rapid spread of imperfect information facilitated by ICTs and the immediate responses by players within vertically integrated, cross-border value-chains may be disproportionate and have undesirable outcomes for society. An example is the May 2011 outbreak of a foodborne illness in Germany caused by a Shiga-toxin producing strain of Escherichia coli (STEC) found in contaminated fenugreek seed, which was imported from Egypt in 2009 and 2010.[16] Statements made by German officials falsely implicating imported cucumbers from Spain as the bacteria's source caused financial losses mainly to Spanish farmers and participants within the vegetable value chain, causing Russia to impose temporary import bans on all vegetables from the EU,[17] strained diplomatic relations, and tainted the image of the industry. The EHEC example illustrates the increase in horizontal and vertical interlinkages between the different sectors of the bioeconomy over space and time and how this requires us to look beyond the agricultural sector if we want to assess economics and policies. The implications for the whole value chain need to be considered, as the results of our economic and policy assessment might otherwise be biased.

But what kind of economic and policy assessments are of relevance? Every researcher may have her or his own ideas, but there are also issues of general interest formulated by different stakeholders representing civil society. The Dutch Government, The EU, the Organisation for Economic Co-operation and Development, to name only a few, expect a substantial contribution of the bioeconomy to address global as well as regional challenges. To quote the European Commission: "With the world population set to approach an estimated 9 billion by 2050, against a background of finite natural resources, Europe needs renewable biological resources—not just for securing healthy food and animal feedstuffs but also for materials and other bio-based products such as bio-fuels. A strong bioeconomy will help Europe to live within its limits. The sustainable production and exploitation of biological resources will allow the production of more from less, including from waste, while limiting negative impacts on the environment and reducing the heavy dependency on fossil resources, mitigating climate change and moving Europe towards a post-petroleum society."[18] Accordingly, this includes contributions to food security, the sustainable management of natural resources, reducing dependence on non-renewable resources, mitigating and adapting to climate change, as well as creating jobs and maintaining competitiveness (Fig. 14.1).

---

[16]Cf. European Food Safety Authority (2011).

[17]Cf. Chelsom-Pill (2011).

[18]European Commission (2012, p. 4).

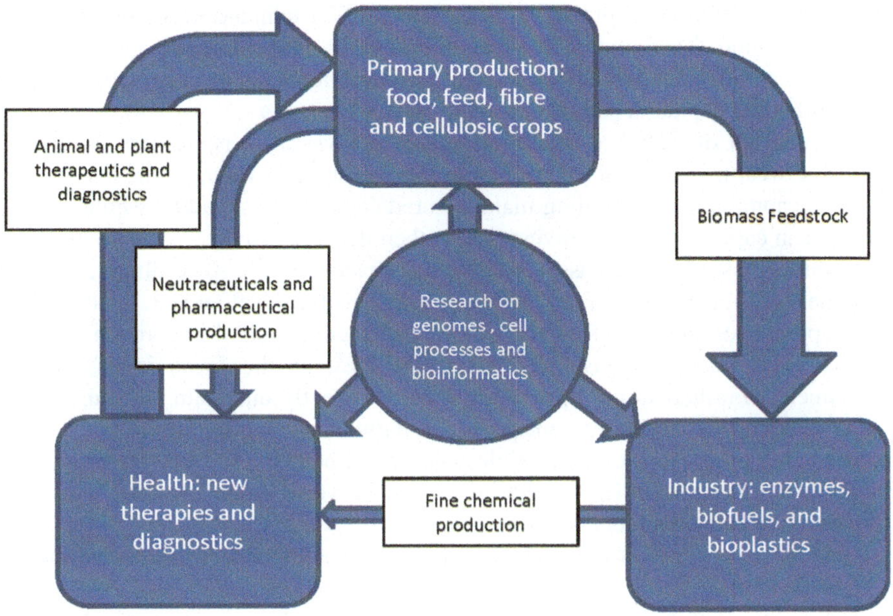

Fig. 14.1 Current and expected integration across biotechnology applications. Source: Adopted from European Commission (2012)

3 Economic Implications

In summary, developing the bioeconomy promises great opportunities for improving our well-being or equivalently sustainable development. However, it requires some knowledge about sustainable development for examining whether or not the bioeconomy lives up to its promises and leads to the question how we should define and measure sustainable development from an economic perspective.

Although there have been many attempts to measure sustainability, none has established itself so far. Examples include the Ecological Footprint (EF),[19] the UN's Human Development Index (HDI),[20] and Bhutan's Gross National Happiness Index.[21] Much discussed are the World Bank's measure of genuine savings and the approach by Kenneth Joseph Arrow et al. on inclusive wealth and genuine

[19] Cf. Wackernagel and Rees (1996).

[20] Cf. Sagar and Najam (1998).

[21] Cf. Mukherji and Sengupta (2004). Furthermore, The Netherlands Scientific Council for Government Policy published a report on the sustainability of the Dutch economy in 1987 using a dynamic multi-sector model including emissions and possibilities for emission control, cf. The Netherlands Scientific Council for Government Policy (1987).

investment.[22] Both concepts serve as measures of sustainable economic development over time. To compute the genuine savings rate, resource depletion and environmental degradation are subtracted from traditional net savings, while investment in human capital is added.[23]

The concept of inclusive wealth and genuine investment is similar: a society's inclusive wealth is determined by measuring the shadow value of the economy's stock of capital assets (including manufactured capital assets, natural capital assets, and human capital). Genuine investment is then defined as a measure of changes in the economy's set of capital assets weighted at shadow prices. Accordingly, positive genuine investment is used as an indicator of sustainable development. In contrast to other approaches, this has a forward-looking perspective. Still, the approach does not explicitly consider the existence of opportunities, as the focus is on specific investments. Further, future opportunities are inherently uncertain and this uncertainty needs to be explicitly considered, in particular when opportunities involve sunk costs or other kinds of irreversible costs and/or benefits. We will get back to the importance of opportunities, uncertainties, and irreversibilities and their relevance for sustainable development in more detail.

First we address issues related to possible negative impacts of new technologies.

3.1 Negative Impacts of New Technologies

The possibility of producing and consuming new products may have negative impacts on human health and/or the environment. Exclusion of these negative impacts from users' net-benefit assessment may warrant a restriction or ban to reduce or eliminate their negative impacts. However, if the impacts are included in the assessment and there are positive net gains, additional constraints may be unjustified from a cost–benefit perspective. Hence, it is unclear ex ante if introducing a new technology warrants additional use restrictions or even a ban, merely because of a negative health and/or environmental impact. Those have to be compared with the benefits of the new technology. Further, the impact of a new technology on human health and/or the environment may be smaller than the impacts of the technology it replaces.[24]

Following Justus Wesseler and Richard Smart, Fig. 14.2 presents the standard framework for assessing health and environmental benefits and costs of a new technology.[25] The x-axis depicts the quantity, Q, produced for either a single crop or a portfolio of crops, and refers to a specific plot, farm, or region. The y-axis represents the marginal benefit (MB) and marginal cost (MC) of producing quantity

[22] Cf. Arrow et al. (2012).

[23] Cf. Hamilton (2000).

[24] This is the link to measuring the changes in capital assets at shadow prices in the Arrow et al. (2012) approach.

[25] Cf. Wesseler and Smart (2014).

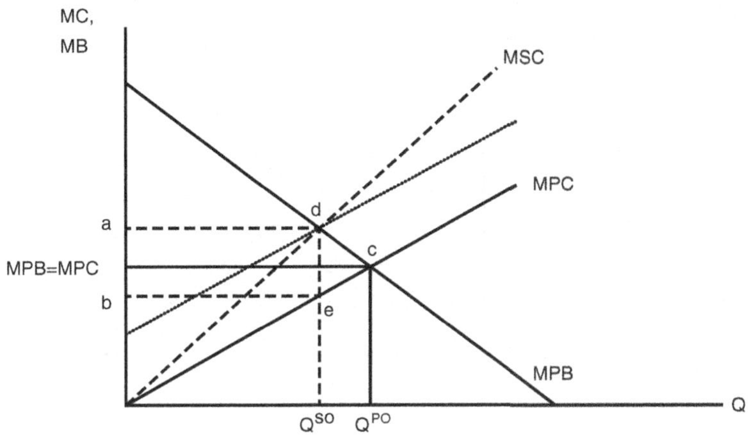

Fig. 14.2 Internalised external effects of agriculture production

Q. An increase in the production of *Q* decreases the marginal private benefit (MPB) and increases the marginal private cost (MPC). For the private producer or sector, the optimal quantity produced, *Q*, is at *c*, where MPB = MPC. If no additional benefits or costs need to be considered, *c* reflects the optimal level of production for society.

Now, let us assume that the production of *Q* bears additional costs not considered under private costs. If these costs are added to the marginal private cost, we get the marginal social cost (MSC) and the societal optimal level of production decreases from Q^{PO} to Q^{SO}. One way of reducing *Q* is by taxing production. The optimal tax rate, the Pigouvian Tax, should increase private cost so that MPC intersects with MPB at *d*. The external effects of production are then internalised and the producer pays for the extra environmental damages, equivalent to *a* minus *b*.

Figure 14.2's important message is that although producing *Q* causes external environmental damage, reducing its production to zero is suboptimal. Merely observing that producing an agricultural crop causes negative environmental impacts when regulatory policies are in place does not necessarily justify additional intervention from a cost–benefit perspective.

Regulation of environmental externalities following the Pigouvian argument has been criticised, most prominently by Ronald Coase.[26] Coase argued that observing externalities does not necessarily justify government intervention, for example via a Pigouvian tax as often argued. Stakeholders themselves should have an incentive to reduce environmental pollution. Hence, an investigation is necessary to determine if government intervention can improve the current situation of observed environmental pollution, and all institutional arrangements available to address the problem should be considered. As a reference, Coase suggests comparing the outcome of

[26] Cf. Coase (1960).

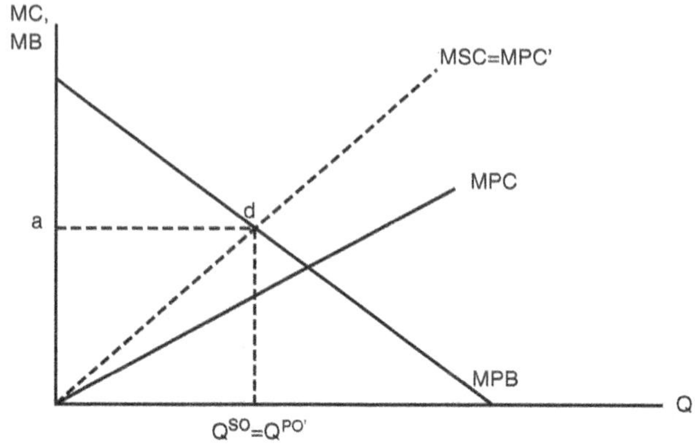

Fig. 14.3 Internalised external effects of agriculture production where MSC = MPC′ (see text for details)

alternative institutional arrangements with the existing situation.[27] An intervention by governments is warranted if a different institutional arrangement improves the outcome and requires so. The results of this reasoning are presented in Fig. 14.3, where the MPC has been adjusted by internalising the external effects of production so that the MPC is equivalent to the MSC, indicated by MPC′ = MSC.

Coase's view has been challenged by libertarians; for them, the question of government intervention depends on property rights. They argued that the courts should settle the problem of externalities. To quote Murray N. Rothbard: "We have concluded that everyone should be able to do what he likes, except if he commits an overt act of aggression against the person and property of another. Only this act should be illegal, and it should be prosecutable only in the courts under tort law, with the victim or his heirs and assigns pressing the case against the legal aggressor."[28] While ex-post liability can address a number of environmental externalities, this does not per se exclude the use of ex-ante regulations even under the libertarian view, e.g. if "everyone" freely decides to work together in a group to implement regulations imposed on members of the group. Farmers may voluntarily form a group and decide their own production standards. Further, implementing ex-post liability has its own problems due to liability avoidance, differences in wealth, and more.[29]

However, the libertarian view does not necessarily contradict the situation shown in Fig. 14.2. The expected ex-post liability cost increases MPC. Further, ex-post liability costs provide incentives for implementing ex-ante measures to reduce ex-post liability, hence increasing the MPC compared to a situation where this possibility is absent, as discussed, for example, by Volker Beckmann, Claudio Soregaroli, and Justus Wesseler for the case of coexistence.[30]

[27] Cf. Coase (2006).

[28] Rothbard (1982, p. 169).

[29] Cf. Shleifer (2010).

[30] Cf. Beckmann et al. (2010).

In conclusion, externalities create additional costs under the Pigouvian, Coasian, and libertarian views; views on measuring costs and appropriate responses differ. However, these views reach the same conclusion: the existence of externalities per se does not immediately justify government intervention and additional investigation on a case-by-case basis is needed.

3.2 The Precautionary Approach

The previous discussion fails to differentiate between different types of external costs. One of the concerns about environmental and health impacts is that they may be irreversible and/or catastrophic; this is one reason why the precautionary approach has been mentioned in many regulations of GMOs[31] and other technologies, most prominently in the Rio Declaration on Environment and Development under Principle 15: "In order to protect the environment, the precautionary approach shall be widely applied by States according to their capabilities. Where there are threats of serious or irreversible damage, lack of full scientific certainty shall not be used as a reason for postponing cost-effective measures to prevent environmental degradation."[32]

There are diverse interpretations of the precautionary approach. The one most widely held is that for a new technology, the prospect of harmful effects takes precedence over the prospect of beneficial effects. As harmful effects are potentially catastrophic, this potential cannot be excluded, and "the infinite costs of a possible catastrophic outcome necessarily outweigh even the slightest probability of its occurrence",[33] the result would be a ban of new technologies.

This line of reasoning is logically inconsistent, as pointed out by the philosopher Henk van den Belt.[34] According to Pascal's Wager: "Given a known but nonzero probability of God's existence and the infinity of the reward of an eternal life, the rational option would be to conduct one's earthly life as if God exists."[35] The contradiction is the "many gods" example: "Consider the possible existence of another deity than God, say, Odin. If Odin is jealous, he will resent our worship of God, and we will have to pay an infinite price for our mistake. Never mind that Odin's existence may not seem likely or plausible to us. It is sufficient that we cannot exclude the possibility that he exists with absolute certainty. Therefore, the very same logic of Pascal's Wager would lead us to adopt the opposite conclusion not to worship God. Pascal's argument, then, cannot be valid."[36]

[31] Cf. e.g. Council of the European Union (1999).

[32] United Nations (1993), Principle 15.

[33] Van den Belt (2003, p. 1123).

[34] Cf. van den Belt (2003).

[35] Ibid., 1124.

[36] Ibid.

In the context of new technologies, catastrophic negative effects cannot be excluded with full certainty; this interpretation of the precautionary approach is unhelpful. Van den Belt recommends a comparison of the benefits and costs of possible errors as a guideline for approval. This corresponds with recommendations by economists who suggest to "regulate until the incremental benefits from regulation are just offset by the incremental costs. In practice, however, the problem is much more difficult, in large part because of inherent problems in measuring marginal benefits and costs."[37]

A method of addressing potential environmental impacts in line with the precautionary approach, and in particular considering uncertainties and irreversible damages, is by performing an extended cost–benefit analysis. The economic literature suggests that if a new technology includes irreversible costs, the net benefits arising from the technology have to be larger than they otherwise would be.[38] The additional net benefits needed to compensate for irreversible costs are commonly calculated by using real-option models.[39]

Justus Wesseler suggests using this modelling approach for assessing new biotechnologies.[40] Because irreversible costs of GMOs are difficult to quantify, irreversible costs that are acceptable considering the net benefits of GM crop cultivation should be calculated—a threshold value Justus Wesseler, Sara Scatasta, and Eleonora Nillesen call the maximal incremental socially tolerable irreversible costs (MISTICs). This threshold level is below the threshold level ignoring uncertainties and irreversible costs. In cases where irreversible benefits are larger than irreversible costs, policies supporting the specific policy can be justified.[41]

3.3 Valuing Opportunities

Avinash K. Dixit and Robert S. Pindyck suggest the application of real option theory not only to investment problems in new technologies but also to all kinds of decision making under temporal uncertainty and irreversibility.[42] The methodology has been applied to assess a wide range of issues including the evaluation of firm investment in different sectors and in patents; the effect of subsidies and taxes on optimal investment decisions and on foreign direct investments; and more. Model applications not only include irreversible costs but also irreversible benefits, optimal abandonment, entry and exit, and uncertainty over several variables such as reversible and irreversible costs and benefits, discount rates, and others. Recent reviews by

[37] Arrow et al. (1996, p. 221).

[38] Cf. e.g. Pindyck (2000).

[39] Cf. Wesseler and Zhao (2019).

[40] Cf. Wesseler (2009).

[41] Cf. Wesseler et al. (2007).

[42] Cf. Dixit and Pindyck (1994).

Esther W. Mezey and Jon M. Conrad and Charles Perrings and William Brock discuss these applications in more detail.[43]

The advantage of real option theory is that it allows us to measure the value of future opportunities. In the most basic setting, future opportunities can be interpreted as options, where the owner of the option has the right but not the obligation to exercise the option, similar to a call option in financial markets. An option may not be exercised unless the value of the option is greater than or equal to zero. The optimal timing of exercise is important to maximise option benefit. Maximum option benefit is important to value the option. The value of the option depends on a number of important parameters including its expected return, the related uncertainty, opportunity costs, and the costs of exercise.[44]

As an illustrative example: assume a company holds the patent on a new technology, say a new seed. The patent provides opportunities for additional income if the company invests in the patented technology. These investment costs can be considered sunk costs. The value of the patent will depend on the future net benefits the technology will provide to the company. Those future net benefits are uncertain, as it is impossible to precisely predict how markets will develop. The required investments may include physical investments (green field investments) to produce the new product, or financial investments to merge and/or acquire a company that has the facilities and location to produce the new product. In addition, there may be extra costs to comply with environmental, food, and health safety standards. Real option theory tells us that although the investment opportunity yields a positive net present value (NPV), delaying the investment might be the optimal choice because losses due to irreversible investment costs can be avoided.[45]

Jim Leitzel and Erik Weisman provide an interesting contribution.[46] They argue that new government policies require sunk investments in the form of training government officials, hiring additional workers, and purchasing equipment. This argument has been picked up by David A. Hennessy and GianCarlo Moschini, although they do not explicitly refer to the contribution by Leitzel and Weisman.[47] Their paper shows that the irreversibility effect delays changes in government regulations. Johan F. M. Swinnen and Thijs Vandemoortele observe a similar effect for the case of biotechnology regulations in the EU, although they do not explicitly mention the application of a real option approach in their paper.[48]

---

[43] Cf. Perrings and Brock (2009), Mezey and Conrad (2010). Also, Dixit and Pindyck (1994), Trigeorgis (1996), Merton (1998), Smit and Trigeorgis (2004) among others offer overviews on applications and methodologies.

[44] Cf. Dixit and Pindyck (1994).

[45] Arrow and Fisher (1974) and Henry (1974) pointed this out in seminal papers in the early 1970s, cf. Wesseler (2014b).

[46] Cf. Leitzel and Weisman (1999).

[47] Cf. Hennessy and Moschini (2006).

[48] Cf. Swinnen and Vandemoortele (2010).

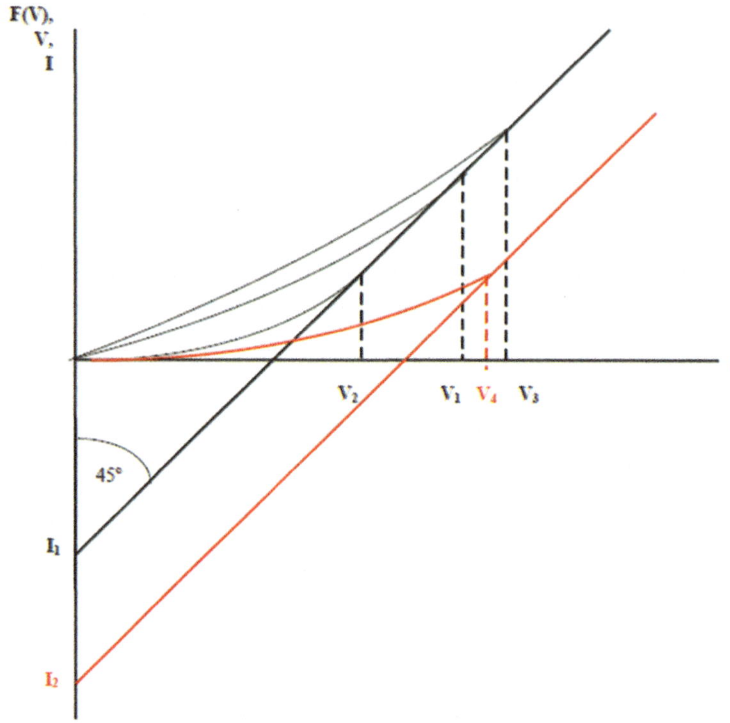

Fig. 14.4 The value of opportunities

These studies show that, with respect to government policies, an irreversibility effect exists, regulations can induce hysteresis, and this may cause additional costs. What is important to note is that without capturing the temporal dynamics—that is agents have the possibility to move from one state of nature to another state of nature and back—policy changes and their impact on the allocation of resources and resulting economic effects over time are difficult to demonstrate. These dynamic effects are difficult to capture with comparative static models.

There is one additional issue that deserves attention and that is related to the economic value of opportunities not exercised. The conceptual framework for assessing those opportunities is introduced in Fig. 14.4. The straight lines in Fig. 14.3, the so-called Marshallian lines, show the NPV of an investment opportunity. Where the straight line intersects with the horizontal axis, the NPV is zero and onwards to the right it is positive. Applying the NPV criterion as a decision rule, it would be optimal to invest—invest is used as an equivalence to saying exercising an opportunity—if the returns V of the project are equal to or greater than the irreversible investment costs I.

The value of the option to invest $F(V)$ is illustrated by the combination of the nonlinear and linear functions where the nonlinear functions smoothly match the Marshallian lines at V_1, V_2, V_3, and V_4, the real option theory's points of optimal exercise. These points are to the right of the Marshallian trigger value, implying that a greater V is needed, compared to that needed to satisfy the NPV criterion, to compensate for the irreversible investment costs. The difference is due to the so-called irreversibility effect.[49]

As a larger V is needed to induce investment, one implication is that most options will be exercised at a later point in time. The optimal threshold value for V depends on $F(V)$. Changes in V caused by changes in uncertainty, trends, opportunity costs, or a combination thereof can change $F(V)$. Further, changes in irreversible costs have an effect on $F(V)$. While Fig. 14.4 is a representation of a simple model with irreversible investment costs and uncertain returns that follow a geometric Brownian motion, more advanced models that consider entry and exit options, several stages, uncertainty about the irreversible investment costs, irreversible benefits, and more increase the complexity of possible effects. What is important to notice in the context of new developments in the bioeconomy is, that even if a project will not be exercised, the opportunity does not have a zero value. As we will observe only projects that have been exercised—values to the right of V_1, V_2, V_3, and V_4—we will miss the not directly observable real option value (ROV). Assuming a number of companies with same investment costs but different V's, the area between the $F(V)$ function and the horizontal axis to the left of the dashed lines can be used as an approximation for the ROV.[50]

If we compare the real option function for V_1, $F(V_1)$ with the real option function (in red) for V_4, $F(V_4)$, the shift is caused by an increase in the irreversible investment costs. Now, assume that the current value for V would be to the right of V_4. In this case we would not observe any effect of the policy, while the policy has substantially reduced the real option value as expressed in the changes in the area between $F(V_1)$ and the horizontal axis and to the left of V_1, and the area between $F(V_4)$, and the horizontal axis and to the left of V_4. Further, the investment trigger has moved from V_1 to V_4 resulting in a delay of exercise due to changes in investment costs. As Dixit and Pindyck and others have pointed out, the irreversibility effect can be substantial and hence so can its effect on national welfare.[51] The effects of changes in the real option value on national welfare can be expected to be substantial as well.

According to several authors, regulations of biotechnologies and GMOs in particular have unnecessarily caused a substantial increase in irreversible investment costs resulting in fewer products being developed, a concentration of the industry, reallocation of research priorities and reallocation of research and development to countries with less stringent regulations and even damaging sustainable development considering the environmental and health benefits of cultivating GE crops.[52]

---

[49] Cf. Henry (1974).

[50] This is a simplified illustration, the aggregation is much more complicated (see e.g. Mbah et al., 2010).

[51] Cf. Dixit and Pindyck (1994).

[52] Cf. Organisation for Economic Co-operation and Development (2017).

While several authors argue that regulations have negative effects on investments and reduce economic growth, others point out the positive effects of avoiding future damages. As discussed by Stefan Ambec et al., Michael Porter has argued that environmental regulations may even have positive effects on firm-level growth.[53] While the effect of regulation and in particular of environmental regulation on firm investment has been investigated,[54] less is known about the effect of regulation on research and development in the public and private sector, or about the indirect effect on sustainable development via its effect on research.

3.4 The Case of "Golden Rice"

The study by Justus Wesseler and David Zilberman on the case of vitamin A enriched rice indicates that those effects can be substantial and hence deserve attention.[55] This example illustrates the complexity of economic, social, and political issues related to new technologies and their controversies.[56]

Vitamin A enriched rice, "Golden Rice", has been developed to address vitamin A deficiency among children by using modern biotechnologies. Research has shown that one cup of "Golden Rice" per day adds enough supply of vitamin A to prevent vitamin A deficiency among children in South and Southeast Asia, has no negative health or environmental effects, and is substantially cheaper than alternative strategies. Despite the evidence, many groups are opposed to the technology with the result that until now it is not available. Greenpeace International for instance concludes: "[. . .] if introduced on a large scale, golden rice can exacerbate malnutrition and ultimately undermine food security."[57] Scientists working on the topic have been accused of being "nazi collaborators"[58]; some have lost their jobs.[59]

Wesseler and Zilberman developed a real option model to assess the health costs for India caused by not having access to the technology over one decade.[60] According to their calculations, the delay over the last 10 years has caused losses of at least 1,424,680 life years for India, not considering indirect health benefits. Our calculation also shows that the additional perceived costs by the Government of India are at least US$1.7 billion (about US$200 million annually) that would justify a delay of introduction from an economic perspective. This is a substantial amount and reflects the economic power of the opposition against the introduction of "Golden Rice". Their model explains why it is more difficult to convince regulators

[53] Cf. Ambec et al. (2014).

[54] Cf. ibid.

[55] Cf. Wesseler and Zilberman (2014).

[56] Dubock (2014) provides an excellent overview about the history of 'Golden Rice'.

[57] Greenpeace International (2012).

[58] Adams (2014).

[59] Cf. Enserink (2013).

[60] Cf. Wesseler and Zilberman (2014).

when a strong vocal opposition exists that mainly stirs uncertainty about a new technology.

The "Golden Rice" issue is an extreme case, but it shows how lobby groups can affect public opinion and off-set scientific evidence. The paper also shows how we can use economic models to assess the implications of specific policies and perhaps even generate impact. The paper has been downloaded several thousand times within a year and been widely mentioned in public media.

4 Outlook

Dynamic aspects of bioeconomy policies and their economic implications along the value chain are a priority for further research in the field of Bioeconomy Economics and Policies.

Research areas include market and policies, spatial structures as well as institutional and organisational aspects of agriculture production, agribusiness, and rural development.

Research in the field of Bioeconomy Economics and Policies should include high quality scientific economic research through economic analyses grounded in relevant and up-to-date theory, using appropriate methodologies. Research needs to create an impact in science and society by disseminating research results to other researchers, policy makers, students, and societal actors.

4.1 Market and Policies

Research focuses, among others, on impacts of the reform of Common Agricultural Policy of the EU (e.g. milk quota abolition), the effect of the bioeconomy on food prices and the effects of price volatility on farmers' income. It also includes the issues of retailer and agribusiness-driven non-tariff barriers and emerging market participants. Issues that recently gained in importance within the TTIP negotiations. The European Green Deal, as the new overarching growth strategy for the EU, has a crucial influence on the development of bioeconomy and vice versa. This growth strategy includes several components such as the European Green Deal Investment Plan, the European Industrial Strategy, the Circular Economy Action Plan, the Farm to Fork Strategy, and the Biodiversity Strategy for 2030, which all have implications for the bioeconomy.

4.2 Spatial Structures

With more and more non-agricultural residents living in rural areas in some parts of the world and depopulation of rural areas in other parts, an increasing demand and supply of multifunction goods, such as recreation, wildlife and landscape, rural areas are changing rapidly in many parts of the world. Research issues relate to the contribution of bioeconomy to regional growth, provision of green services (e.g. via contracting), landscape valuation, rural policy modelling, and land use analysis.

4.3 Institutional and Organisational Aspects

This research theme is motivated by the historically and on-going strong policy involvement in the agricultural sector, both at the national and European level. Furthermore, increasingly complex relationships within the bioeconomy demand new ways of governance. Examples of research topics are contracts between processing companies and farmers, the uptake of agri-environmental schemes (contracts between the government and farmers to provide green services) and contracts ensuring food safety and animal welfare in supply chains of food production. Regulatory hurdles such as the approval process for GMOs are an important topic for research, because they can have major influence on the development of the bioeconomy. The impacts of the Decision of the Court of Justice of the European Union on NPBT are still not clear, but the court decision likely reduces the comparative advantage of the EU bioeconomy and make it harder to reach the objectives of the European Green Deal.[61]

Research in the field of Bioeconomy Economics and Policies calls for a strong interdisciplinary approach of combining social and natural sciences.

In summary, a number of important issues emerge related to agriculture in the bioeconomy where socio-economic research can provide meaningful contributions to the scientific as well as the societal debate.

References

Adams, M. (2014). *Science for sale: The true history of silencing whistle blowers with corporate science*. In *Natural News*. Retrieved September 8, 2020, from, http://www.naturalnews.com/046097_biotech_genocide_Monsanto_collaborators_media_sellouts.html#ixzz3afUnnpFA

Ambec, S., Cohen, M. A., Elgie, S., & Lanoie, P. (2014). The porter hypothesis at 20: Can environmental regulation enhance innovation and competitiveness? *Review of Environmental Economics and Policy, 7*(1), 2–22.

Arrow, K. J., Cropper, M. L., Eads, G. C., Hahn, R. W., Lave, L. B., Noll, R. G., Portney, P. R., Russell, M., Schmalensee, R., Smith, V. K., & Stavins, R. N. (1996). Is there a role for benefit-cost analysis in environmental, health, and safety regulation? *Science, 272*(5259), 221–222.

Arrow, K. J., Dasgupta, P., Goulder, L. H., Mumford, K. J., & Oleson, K. (2012). Sustainability and the measurement of wealth. *Environment and Development Economics, 17*(3), 317–353.

Arrow, K. J., & Fisher, A. C. (1974). Environmental preservation, uncertainty and irreversibility. *The Quarterly Journal of Economics, 88*(2), 312–319.

Backus, G. B. C., Berkhout, P., Eaton, D. J. F., Franke, L., de Kleijn, A. J., Lotz, B., van Mil, E. M., Roza, P., & Uffelmann, W. (2008). *EU policy on GMOs: A quick scan of the economic consequences*.

[61] Cf. Purnhagen and Wesseler (2020).

Beckmann, V., Soregaroli, C., & Wesseler, J. (2010). Ex-ante regulation and ex-post liability under uncertainty and irreversibility: Governing the coexistence of gm crops. *Economics: The Open-Access, Open-Assessment E-Journal, 4*, 9. Retrieved September 8, 2020, from, https://poseidon01.ssrn.com/delivery.php?ID=383099002083112094100104099088089077084047001036057054066039127041058009058050053007064097068028086054009066086007120090099106083009087109015095017022022081087125006097031065091102029073&EXT=pdf

Chelsom-Pill, C. (2011). Russia lifts ban on EU vegetables. In *Deutsche Welle*. Retrieved September 21, 2020, from, http://www.dw.de/russia-lifts-ban-on-eu-vegetables/a-15182434-1

Coase, R. (1960). The problem of social cost. *The Journal of Law and Economics, 3*, 1–44.

Coase, R. (2006). The conduct of economics: The example of Fisher body and general motors. *Journal of Economics & Management Strategy, 15*(2), 255–278.

Council of the European Union. (1999). *2194th council meeting – Environment – Luxembourg, 24/25 June 1999*. C/99/203. Retrieved September 8, 2020, from, https://ec.europa.eu/commission/presscorner/detail/en/PRES_99_203

Court of Justice of the European Union (CJEU). (2018). *Judgment of the court (grand chamber) form 25. July 2018, C-528/16*, ECLI:EU:C:2018:583. Retrieved September 8, 2020, from, http://curia.europa.eu/juris/document/document.jsf?text=&docid=204387&pageIndex=0&doclang=en&mode=lst&dir=&occ=first&part=1&cid=804773

Dixit, A. K., & Pindyck, R. S. (1994). *Investment under uncertainty*.

Dubock, A. (2014). The present status of Golden Rice. *Journal of Huazhong Agricultural University, 33*(6), 69–84.

Enserink, M. (2013). Golden rice not so golden for tufts. *Science*. Retrieved September 8, 2020, from, http://news.sciencemag.org/asiapacific/2013/09/golden-rice-not-so-golden-tufts

European Commission (EC). (2012). *Innovating for sustainable growth. A bioeconomy for Europe*.

European Commission (EC). (2018). *A sustainable bioeconomy for Europe: Strengthening the connection between economy, society and the environment. Updated bioeconomy strategy*.

European Commission (EC). (2019a). *The European Green Deal*. Communication from the Commission to the European Parliament, the European Council, the Council, the European Economic and Social Committee and the Committee of the Regions. COM(2019) 640 final. Brussels.

European Commission (EC). (2019b). *The European Green Deal Sets Out How to Make Europe the First Climateneutral Continent by 2050, Boosting the Economy, Improving People's Health and Quality of Life, Caring for Nature, and Leaving No One Behind*. Press Release. Retrieved September 21, 2020, from, https://ec.europa.eu/commission/presscorner/detail/en/IP_19_6691

European Food Safety Authority (EFSA). (2011). Tracing seeds, in particular fenugreek (Trigonella foenum-graecum) seeds, in relation to the Shiga Toxin-Producing *E. coli* (STEC) O104:H4 2011 Outbreaks in Germany and France. In *EFSA Supporting Publications* 8/7, 176E.

Food and Agriculture Organization of the United Nations (FAO). (2013). *FAO statistical year book 2013 – World food and agriculture*.

Gocht, A., Consmüller, N., Thom, F., & Grethe, H. (2020). *Economic and environmental consequences of the ECJ genome editing judgement in agriculture*. Thünen Working Paper 150. Edited by Johann Heinrich von Thünen Institute.

Greenpeace International (2012). *Golden illusion: The broken promises of "golden" rice*.

Hamilton, K. (2000). Genuine saving as a sustainability indicator. In Organisation for Economic Co-operation and Development (OECD) (Ed.), *Frameworks to measure sustainable development* (pp. 65–78).

Hennessy, D. A., & Moschini, G. (2006). Regulatory actions under adjustment costs and the resolution of scientific uncertainty. *American Journal of Agricultural Economics, 88*(2), 308–323.

Henry, C. (1974). Investment decision under uncertainty: The "irreversibility effect". *American Economic Review, 64*(6), 1006–1012.

Leitzel, J., & Weisman, E. (1999). Investing in policy reform. *Journal of Institutional and Theoretical Economics, 155*(4), 696–709.

Levin Institute. (n.d.). *Globalization 101: What is Globalization?* Retrieved September 8, 2020, from, http://www.globalization101.org/what-is-globalization/

Mbah, M. L. M., Forster, G., Wesseler, J., & Gilligan, C. (2010). Economically optimal timing of crop disease control under uncertainty: An options approach. *Journal of the Royal Society Interface, 7*(51), 1421–1428.

Merton, R. C. (1998). Applications of option-pricing theory: Twenty-five years later. *American Economic Review, 88*(3), 323–349.

Mezey, E. W., & Conrad, J. M. (2010). Real options in resource economics. *Annual Review of Resource Economics, 2*(1), 33–52.

Mukherji, P. N., & Sengupta, C. (2004). *Indigeneity and universality in social science. A south Asian response.*

Organisation for Economic Co-operation and Development (OECD). (2010). *Measuring globalisation: OECD economic globalisation indicators 2010.*

Organisation for Economic Co-operation and Development (OECD). (2017). *Farm management practices to foster green growth.*

Perrings, C., & Brock, W. (2009). Irreversibility in economics. *Annual Review of Resource Economics, 1*(1), 219–238.

Pindyck, R. S. (2000). Irreversibilities and the timing of environmental policy. *Resource and Energy Economics, 22*(3), 233–259.

Purnhagen, K., & Wesseler, J. (2020). EU regulation of new plant breeding technologies and their possible economic implications for the EU AND Beyond. *Applied Economic Perspective and Policy.* Retrieved October 5, 2020, from, https://onlinelibrary.wiley.com/doi/epdf/10.1002/aepp.13084

Rothbard, M. N. (1982). Law, property rights, and air pollution. *Cato Journal, 2*(1), 55–99.

Sagar, A. D., & Najam, A. (1998). The human development index: A critical review. *Ecological Economics, 25*(3), 249–264.

Shleifer, A. (2010). Efficient regulation. In D. P. Kessler (Ed.), *Regulation vs. litigation: Perspectives from economics and law.* (pp. 27–43).

Smit, H. T. J., & Trigeorgis, L. (2004). *Strategic investment. Real options and games.*

Stevens, A. J. (2004). The enactment of Bayh – Dole. *The Journal of Technology Transfer, 29*, 93–99.

Swinnen, J. F. M., & Vandenmoortele, T. (2010). Policy gridlock or future change? The political economy dynamics of EU biotechnology regulation. *The Journal of Agrobiotechnology Management & Economics, 13*(4), 291–296.

The Netherlands Scientific Council for Government Policy. (1987). *Ruimte voor groei: kansen en bedreigingen voor de Nederlandse economie in de komende tien jaar.* Rappoorten aan de Regering 29. 's-Gravenhage.

Tramper, J., & Zhu, Y. (2011). *Modern biotechnology: Panacea or new pandora's box?*

Trigeorgis, L. (1996). *Real options. Managerial flexibility and strategy in resource allocation.*

United Nations. (1993). *Report of the United Nations conference on environment and development.* Resolutions Adopted by the Conference. A/CONF.151/26 (Vol. I).

Van den Belt, H. (2003). Debating the precautionary principle: "Guilty until proven innocent" or "innocent until proven guilty"? *Plant Physiology, 132*(3), 1122–1126.

Wackernagel, M., & Rees, W. E. (1996). *Our ecological footprint. Reducing human impact on the earth.*

Wesseler, J. (2009). The Santiello theorem of irreversible benefits. *The Journal of Agrobiotechnology Management & Economics, 12*(1), 8–13.

Wesseler, J. (2014a). Biotechnologies and Agrifood strategies: Opportunities, threads and economic implications. *Bio-based and Applied Economics, 3*(3), 187–204.

Wesseler, J. (2014b). Financial, real, and quasi options: Similarities and differences. In A. A. Pinto & D. Zilberman (Eds.), *Modeling, optimization and bioeconomy I. Contributions from ICMOD 2010 and the 5th Bioeconomy Conference 2012* (pp. 673–693).

Wesseler, J. (2015). *Agriculture in the bioeconomy: Economics and Policies*. Inaugural lecture upon taking up the post of Professor of Agricultural Economics and Rural Policy at Wageningen University on 22 January 2015. Retrieved September 15, 2020, from, https://www.wur.nl/upload_mm/6/2/2/bd6f22fa-d0ea-4c96-b4c0-b25060553443_oration%20booklet_Justus_Wesseler.pdf

Wesseler, J., Scatasta, S., & Nillesen, E. (2007). The maximum incremental social tolerable irreversible costs (MISTICs) and other benefits and costs of introducing transgenic maize in the EU-15. *Pedobiologia, 51*(3), 261–269.

Wesseler, J., & Smart, R. (2014). Environmental impacts. In J. Falck-Zepeda, K. Ludlow, & S. Smyth (Eds.), *Socio-economic considerations in biotechnology regulation* (pp. 81–95).

Wesseler, J., & Zhao, J. (2019). Real options and environmental policies: The good, the bad, and the ugly. *Annual Reviews in Resource Economics, 11*, 43–58.

Wesseler, J., & Zilberman, D. (2014). The economic power of the Golden Rice opposition. *Environment and Development Economics, 19*(6), 724–742.

World Trade Organization (WTO). (2014). *Modest trade growth anticipated for 2014 and 2015 following two year slump*. Press Release, Press/721, 14 April 2014. Retrieved September 14, 2020, from, https://www.wto.org/english/news_e/pres14_e/pr721_e.htm

Part VI

Normativity and Ethics

Bioeconomy and Ethics

15

Bart Gremmen

Abstract

Although many empirical studies have been done to elaborate the meaning of sustainability, the core of its meaning is normative. When the concept of bioeconomy emerged, the overarching concept of sustainability was used to place bioeconomy in a normative, long-term development. For example, in Germany bioeconomy is defined as a new model for industry and the economy. It involves using renewable biological resources sustainably to produce food, energy, and industrial goods. It also exploits the untapped potential stored within millions of tons of biological waste and residual materials. This definition focuses on renewable resources and biological waste. Other definitions focus on the use of biotechnology in the production of (bio-based) goods, services, or energy from biological material (or biomass) as the primary resource base. In this paper we will present an overview of the ethical issues of bioeconomy in Europe. On the one hand, we will compare the bioenergy situation in Germany and the Netherlands. We will show that the Netherlands is now switching from a bio-based economy to a circular economy. This process has been accelerated by a recent number of serious problems with the use of wood as biomass. On the other hand, we will focus on the latest developments in biotechnology and show that, next to already known ethical problems about genetic modification, CRISPR-Cas9 leads to a number of new specific ethical problems.

Keywords

Sustainability · Circular economy · CRISPR-Cas9

B. Gremmen (✉)
Department of Social Sciences, Philosophy: Ethics in Life Sciences, Wageningen University, Wageningen, The Netherlands
e-mail: bart.gremmen@wur.nl

D. Lanzerath et al. (eds.), *Bioeconomy and Sustainability*,
https://doi.org/10.1007/978-3-030-87402-5_15

1 Introduction

More than 25 years ago the concept of sustainability became a topic on academic agendas. It survived fierce discussions about its definition[1] and slowly became fashionable in the political and societal arenas of Europe. Although many empirical studies have been done to elaborate the meaning of sustainability, the core of its meaning is normative. When the concept of bioeconomy emerged, the overarching concept of sustainability was used to place bioeconomy in a normative, long-term development. For example, in Germany bioeconomy is defined as a new model for industry and the economy.[2] It involves using renewable biological resources sustainably to produce food, energy, and industrial goods. It also exploits the untapped potential stored within millions of tons of biological waste and residual materials.[3] This definition focuses on renewable resources and biological waste. Other definitions focus on the use of biotechnology in the production of (bio-based) goods, services, or energy from biological material (or biomass) as the primary resource base.[4] In this paper we will present an overview of the ethical issues of bioeconomy in Europe. On the one hand, we will compare the bioenergy situation in Germany and the Netherlands. We will show that the Netherlands is now switching from a bio-based economy to a circular economy. This process has been accelerated by a recent number of serious problems with the use of wood as biomass. On the other hand, we will focus on the latest developments in biotechnology and show that, next to already known ethical problems about genetic modification, CRISPR-Cas9 leads to a number of new specific ethical problems.

What do we mean by ethics? Ethics may be studied from several disciplinary backgrounds: law, theology, psychology, philosophy, and social science. In this paper we will study ethics from a philosophical background[5] and define it as the critical, systematic reflection on implicit and explicit moral assumptions about what we do. Traditionally, the focus is on the individual: what should/must I do? In contrast we will focus on policy making, government, and society. Our strategic vision on applied ethics is that many of the societal and scientific challenges in relation to agriculture, food, and the environment involve value conflicts, and that scientific understandings and technological solutions are often contested. In a pluralistic society, philosophy can offer proactive and constructive ways to deal with such value conflicts. Our mission as applied ethicists is to strengthen reflection on, and deliberation about, these problems and about scientific and societal responses, and thus to contribute to responsible policies and practices. The aim is to clarify the nature of values such as integrity, intrinsic value, and sustainability and explore possibilities for responsible innovation in plant and animal production systems.

---

[1] Cf. Gremmen and Jacobs (1997).

[2] Cf. German Advisory Council on Global Change (2009).

[3] Cf. Federal Ministry of Education of Research (2011).

[4] Cf. Energy Transition – The Global Energiewende (2019).

[5] Cf. Petersen and Ryberg (2007).

Our pragmatic approach starts from case material and concrete actual developments, and aims at interdisciplinarity, dialog, and collaboration.[6] In this empirical way of doing ethics, philosophical and ethical concepts are used as flexible tools that can be adapted to specific contexts. Whereas academic debates often revolve around the question of whether these sciences are benign or a threat,[7] we discuss life sciences from within.[8] This means that *the societal impact* of our work is strengthened by our bottom-up approach.

2 From Bioeconomy to Bio-Based Economy

From the mid-2000s policy makers and governments paid more and more attention to the term "Bioeconomy." They believed that the bioeconomy could help to solve the fossil fuel problem,[9] because the use of biomass can ease the transition from fossil fuels to a sustainable alternative.[10] The bioeconomy is also seen as food and non-food applications of biotechnology, especially GMOs.[11] The bioeconomy became popular with its adoption by the *European Union* (EU) and the *Organization for Economic Co-operation and Development* (OECD) as a policy agenda and framework to promote the use of biotechnology to develop new products, markets, and uses of biomass. Since then, both the EU[12] and the OECD[13] have created dedicated bioeconomy strategies, as have an increasing number of countries around the world. According to the European Commission Europe must make the transition to a post-petroleum economy.[14] In their view a greater use of renewable resources is no longer just an option, it is a necessity. As a consequence the EU has adopted a strategy "Innovating for sustainable growth: A bioeconomy for Europe".[15] In this strategy the European Commission defines the bioeconomy as "the production of renewable biological resources and the conversion of these resources and waste streams into value added products, such as food, feed, bio-based products and bioenergy. Its sectors and industries have strong innovation potential due to their use of a wide range of sciences, enabling and industrial technologies, along with local and tacit knowledge."[16] Also the importance of the bioeconomy is highlighted by calculating its economic worth. The Commission estimates that the EU's

[6] Cf. Gremmen (2002).

[7] Cf. Singer (1975) and Sandoe and Christiansen (2008).

[8] Cf. Gremmen (2007).

[9] Cf. Golembiewski et al. (2015).

[10] Cf. U.S. Energy Information Administration (2016).

[11] Cf. Heijman and Schepman (2018).

[12] Cf. European Commission (2012).

[13] Cf. Organisation for Economic Cooperation and Development (2009).

[14] Cf. European Commission (2012).

[15] Cf. ibid.

[16] Ibid, 9.

bioeconomy sectors are worth 2 trillion Euros in annual turnover and account for more than 22 million jobs and approximately 9% of the workforce.[17] The Commission is convinced that to solve the problems connected with the scarcity of non-renewable resources, global warming, and environmental pollution, the development of the bioeconomy is crucial.[18] In 2012 the USA also announced a National Bioeconomy Blueprint to encourage biological manufacturing methods.[19]

The term "bioeconomy" is a normative term because finite resources *must* be used in the most effective, efficient, and sustainable manner possible.[20] From the above we can also see a close relationship between problems that were earlier disconnected: non-renewable resources, climate change, and environmental pollution. Bioeconomy has become a normative umbrella concept with the economy at its core. It is studied from different disciplines. The result is a multifaceted concept with different definitions,[21] which makes it difficult to compare the bioeconomy on an international scale.[22] There are different views on the bioeconomy which we will use the three conceptual distinctions of.[23] Firstly, the *biotechnology view* of the bioeconomy emphasizes the importance of research into biotechnology and the commercialization of biotechnology. This view underlines understanding mechanisms and processes at the genetic, molecular, and genomic levels, and applying this understanding to creating or improving industrial processes, developing new products and services, and producing new energy. Secondly, the *bio-resource view* of the bioeconomy focuses on the role of research and development related to raw resources in the primary sector as well as establishing new value chains. Lastly, the *bio-ecology view* focuses on the importance of ecological processes that optimize the use of energy and nutrients and promote biodiversity. The first two views are largely focused on research and development in global systems. Because the bio-ecological view emphasizes the potential in regional systems, we will refer to it as the circular-economy view.

The EU strives for different goals in the transition from a fossil fuel-based to a bioeconomy.[24] It is expected that the bioeconomy will reduce its dependency on fossil fuels and achieve more sustainability as well as contribute to climate and environmental protection.[25] In bio-refineries plants are broken up into their component parts as completely as possible and converted into all kinds of materials. In this way the term "bioeconomy" is conflated with the term "bio-based economy," which

[17] Cf. ibid., 11.

[18] Cf. European Commission (2012).

[19] Cf. The White House (2012).

[20] Cf. Wield (2013).

[21] Cf. Heijman and Schepman (2018).

[22] Cf. Staffas et al. (2013).

[23] Cf. Bugge et al. (2016).

[24] Cf. European Association for Bioindustries (2011).

[25] Cf. McCormick and Kouto (2013).

refers to the *bio-resource view* of bioeconomy.[26] For example, since 2005 the Netherlands has sought to promote the creation of a bio-based economy by starting pilot power plants, a centralized organization, and supporting research (Food & Bio-based Research) being conducted. The bio-based economy is focused on biomass (crops, crop refuge as well as seaweed and algae). Bio-refining biomass results in products ranging from highest added value and lowest volume of biomass to lowest added value and highest volume of biomass: fine chemicals/medicines, food, chemicals/bioplastics, transport fuels, electricity, and heat. The main advantage of biomass is that it can produce all three types of energy carriers: electricity, heat, and fuel (liquids, solids, and gas). Another advantage is that it is easily storable, compared to energy sources such as sun or wind. However, there are two main drawbacks from an ethical point of view. The first is that, compared to energy form sun and wind, energy from biomass is not sustainable by definition. The energy of sun and wind is independent of our use. Although biomass is overall a renewable source of energy, it is also a finite source of energy: the plants that are used must definitely be replaced or regrown. It takes many years to regrow them (e.g., most tree species). The second main drawback is that the global use of biomass for fuel by rich countries may conflict with the food needs of poor countries. Other drawbacks of crops grown for fuel are: possible resource depletion and reduction of biodiversity.

Although the term "bioeconomy" may have a broad meaning,[27] the resource view is prominent in the EU. Within this view many countries focus on the use of biomass for bioenergy. In the next section we will compare two countries, Germany and the Netherlands, in their recent attempts to build a sustainable bioenergy system.

3 Bioenergy in Germany and the Netherlands

Forestry and agriculture are generally the two sources of bioenergy. Germany is, within the EU, the greatest producer of wood, and wood is by far the greatest source of bioenergy in the country.[28] It is estimated that roughly 40% of German timber production is used as a source of energy, with the rest being used as material.[29] Germany is also the leading biogas market—beginning in 2015, almost two-thirds of Europe's biogas plants were installed in Germany. In the part on biomass, we find that energy crops used in 2017 nearly 2.4 million hectares of its arable land, representing around 14% of the agricultural land in Germany. There is still room for expansion because the upper limit for bioenergy is four million hectares by 2020.[30]

[26] Cf. ibid.

[27] Cf. ibid.

[28] Cf. Energy Transition – The Global Energiewende (2019).

[29] Cf. ibid.

[30] Cf. ibid.

In contrast, the Netherlands is, within the EU, a very small producer of wood. In the past the sources of biomass for bioenergy were varied: manure, VFG (vegetable, fruit and garden waste), and roadside cuttings. In recent years, three-quarters of the wood that is burned in biomass power stations originated in the Netherlands. Most of it is pruning waste or residual wood. This flake wood is mainly from the northern provinces of the Netherlands and is left over from maintenance of forests. After fragmentation the biomass power plants buy this wood from Staatsbosbeheer and gardeners and generate steam and electricity by incineration. For example, the electricity of the medium scale power plant "Brouwer Biocentrale BV" at Balkbrug goes to the energy grid, good for 1100 households. The steam goes to neighbor Friesland Campina, which saves 5.5 million cubic meters of natural gas.[31] That wood would otherwise remain in the forest, rot, and thus emit CO_2. The Dutch government regards biomass as an important and sustainable energy source that will be needed in the coming years to achieve the climate goals and to close the gas tap in Groningen. It is expected by the government that biomass will generate more renewable energy in 2020 than solar energy or offshore wind energy. There are currently 372 power plants in the Netherlands that use wood as fuel. Of these 372,219 are operational and 153 are planned.[32]

In Germany the use of biomass for energy is not without criticism. Environmental impacts of energy crops are high on the agenda of environmental organizations. For example, the adverse effects of energy crops on the quality of groundwater and soil erosion have been debated. Also the plowing of valuable grassland to increase cultivation of corn for use in energy production has been criticized. Germany's Renewable Energy Act[33] has been developed to counteract these effects. It limits the amount of corn and grain eligible for special compensation. In addition, a set of incentives encourages the increased use of less environmentally polluting substrates.

Today, Germany uses biomass mainly of domestic origin. The challenge will be to increase biomass usage for energy without drastically increasing imports. According to the German Federal Ministry of Education and Research the expansion of biomass production for energy use must not conflict with food security, the right to food, and the protection of the environment and nature.[34] The ministry is in favor of strong sustainability criteria for biofuels and other liquid bioenergy carriers. However, we believe that enforcing these strict criteria will be a major challenge, not only in Germany, but also to prevent globally the use of biomass for energy to become the cause of increasing food prices.

Compared to Germany, the Netherlands is relatively small and most of the agricultural land is used for food crops. More than 25% of the required biomass is imported, mainly wood from the USA and wood pallets from Estland. Opponents of the use of biomass claim that forests would be cut down to meet the demand from

[31] Cf. Vries (2019).

[32] Cf. ibid.

[33] Cf. Federal Ministry of Education of Research (2011).

[34] Cf. ibid.

coal and biomass plants. According to experts this is not true.[35] They claim that all wood that comes from the USA and ends up in our biomass plants is waste wood. However, critics[36] claim that in the south of the USA, Enviva, which supplies pellets to the Netherlands, mainly uses hardwood. In short, wood from forests located in the heart of a global hotspot for biodiversity. The state of North Carolina, one of the most important pellet production centers, recently announced that burning wood for biomass is not part of the state's clean energy plan and that large-scale use of forests in North Carolina on foreign markets at national and international levels must be challenged.[37] Both in the Dutch Energy Agreement from 2013 and in the provisional Climate Agreement, there is a strong commitment to biomass. However, the Dutch government wants to know for all types of biomass how sustainable they are. Researchers are preparing an overview of the available biomass in the Netherlands, the current import and all existing applications.[38]

While the German government aims to develop the complete range of products with a focus on innovation, the Dutch government focuses on the production of electricity and heat in biomass power plants. Old coal-fired power stations are converted (in whole or in part) thanks to subsidies of many hundreds of millions of euros into biomass power stations, such as those from Uniper on the Maasvlakte and from RWE in the Eemshaven and Geertruidenberg.[39] Because biomass is more expensive than coal, there are billions of euros in subsidies available for it. Yet, there is a lot of discussion about it. Opponents are afraid that it is not just about residual products, but that more trees are disappearing than are being planted instead. In addition, the discussion is about how much biomass is available worldwide and in the Netherlands. According to Vattenfall, one of the main biomass users, every effort will be made to minimize the emission of particulate matter when building new power plants.[40] The use of filters cleans the flue gases, so that the amount of particulate matter that is already present in the air hardly increases, according to the company.[41] In addition, Vattenfall is only concerned about extracting wood from sustainably managed forests.[42] That means that they will only get their biomass from forests where more wood grows than what is extracted from it. According to the company, it is only about residual wood, from trees that are already cut down to make planks.[43]

In 2019 it became clear that relatively small biomass power plants were a danger to the health of people and the environment. The *European Academies Science*

[35] Cf. Vries (2019).

[36] Cf. Luiga and Swart (2019).

[37] Cf. ibid.

[38] Cf. Vries (2019).

[39] Cf. Mersbergen (2019).

[40] Cf. ibid.

[41] Cf. ibid.

[42] Cf. ibid.

[43] Cf. ibid.

Advisory Council (EASAC) recently stated that the method—burning wood to generate energy—is bad for the climate rather than CO_2-neutral, as the government and industry argue. In addition, a report by DNV GL consultancy, commissioned by the *Ministry of Infrastructure and Water Management*, revealed that biomass plants emit more harmful substances than plants on natural gas and even coal. The majority in the Dutch parliament want to significantly reduce the use of biomass. A motion by the political party D66 to set requirements for the emissions of smaller biomass plants is supported by the coalition partners.[44] These biomass power plants emit much more nitrogen and particulate matter than coal plants. However, they do not have to meet strict air quality requirements. The coalition parties also want no more subsidies to be given for new biomass plants. To date, more than 14 billion euros have been promised by the government. Johan Remkes' nitrogen committee recently advised the cabinet to stop subsidies for co-firing biomass.[45]

Due to these problems the government is now slowly shifting its focus from a bio-based economy to circular economy.[46] By 2050 a circular economy in the Netherlands has an interim target of reduction in the use of primary raw materials by 50% by 2030. A nation-wide Circular Economy Programme has been set up.[47] Its core concept is "Closing the loop": all raw materials and residual flows must be kept within the loop for as long as possible while keeping their quality as high as possible. Cascading and multifaceted value creation are key ambitions. For biomass and food, the key principle is to maintain soil balance, with a key focus on maintaining and boosting soil health: a key factor in sustainable production.

4 Bioeconomy as Genetic Modification

After the first half of the twentieth century, molecular biology developed ways to alter DNA in controlled ways. Genetic modification (GM) allowed the cutting and then splicing together of DNA-molecules.[48] Two methods were developed: the particle gun method and the agrobacteria method. These methods were used first in mono-cellular organisms, and subsequently applied to multi-cellular organisms. A "transgenic" mouse (a mouse containing DNA from another species) was the first GM-mammal, made in the mid-1970s. In 2015 the first GM-product to be approved for consumption in America has been the GM-salmon.[49]

Although the application of some methods and technologies is different, the basic genetics is more or less the same in plants and animals. Mutation breeding (increased mutation frequency through chemicals or radiation), for example, is not possible in

[44] Cf. Luiga and Swart (2019).

[45] Cf. ibid.

[46] Cf. Ministerie van Infrastructuur en Waterstaat (2019).

[47] Cf. ibid.

[48] Cf. Gremmen (2017).

[49] Cf. ibid.

animals, for both ethical and economic reasons, but is a common and legal method for plants. Genetic modification and marker assisted selection (MAS) have been used in plant breeding for many years. Recently genomic selection (GS), a method used in animal breeding, has become an increasingly promising method for plant breeding as well.

There are three different legal regimes of GM applications in the European Union.[50] The industrial application of GM is the first regime. Industry transfers molds, viruses, and bacteria into GMOs. This so-called contained use is allowed because it is done in sealed containers. Agriculture is the second regime. Unless there are serious arguments against it the application of GM to plants is allowed. However, the application of GM to animals is not allowed, unless there are serious arguments to do it anyway. The application of GM to humans is the third regime. This is the most severe legal regime: the genetic modification of humans is not allowed.

Although it is legal to develop and cultivate GM plants, the long and expansive regulatory road in the last decades has led to one crop in one country: corn in Spain.[51] Compared to the thousands of hectares soy, corn, rapeseed, and cotton in South- and North America, GM plants are almost non-existent in Europe. Lower price, higher quality, higher environmental value, and higher nutritional value are the potential benefits of GMOs. Higher food security, better food safety, more affordable food, higher societal health, more sustainability, and biodiversity are potential societal values of GMOs. An important problem is that these benefits and values may clash. For example, a higher food security or better price could result in a lower health or a lower food safety. It is all about setting the right priorities.

In 2010 the results of the Eurobarometer showed that people in the EU do not see benefits of GM-food.[52] Some people think that GM is probably unsafe or even harmful.[53] People in the EU have strong reservations about safety and do not see the benefits of horizontal gene transfer. They have some reservations about safety and the potential impact on the environment but accept the potential benefits of vertical gene transfer.[54] Other people have strong reservations about ethical issues, such as the use of human embryos, but consider that the science of regenerative medicine should be allowed to develop. Although strict laws are needed to alleviate concerns about ethical issues, they approve of stem cell research, transgenic animal research, and human gene therapy.[55]

The ethical arguments about GMOs are very varied in society.[56] On the critical side, some people have objections to a particular technology as such. For example in

[50] Cf. ibid.

[51] Cf. ibid.

[52] Cf. Gaskell et al. (2011).

[53] Cf. ibid.

[54] Cf. ibid.

[55] Cf. ibid.

[56] Cf. Gregorowius et al. (2012).

the case of genetic modification, this argument amounts to the claim that it is *unnatural* and therefore morally problematic. Also arguments about threatening the integrity and/or the intrinsic value of plants can be found. In another argument it is claimed that GM technology amounts to a form of *hubris* concerning man's relationship to nature (are we allowed to "play God"?).[57]

Many critics might be opposed to the different applications of GM technology and not so much opposed to GM technology as such.[58] From a consequentialist ethics perspective this means that even people who do not have an objection in principle to the technology still can be critical to its use in agriculture in general, and in food production in particular.[59] With respect to the distribution of economic benefits from its use current applications of agricultural biotechnology have also been criticized from the viewpoint of *justice*.[60] Criticism has also focused on the autonomy of consumers in deciding whether to put the products on the table and the *autonomy* of farmers in deciding whether to use the technology (e.g., are patents allowed?).[61] The *risks and uncertainties* with this new technology are emphasized by some of the critics and they argue either that there are risks to human health or the environment (e.g., sustainability and biodiversity), or that there might be such risks, and that for this reason some version of *the precautionary principle* should be applied.[62] Are these ethical issues of GM also present in more recent genetic engineering techniques like CRISPR-Cas9?

5 Genome Editing: CRISPR-Cas9

In plant breeding several new genetic engineering techniques, also referred to as genome editing, have been developed in short time. "Genome editing" can be defined as "the practice of making targeted interventions at the molecular level of DNA or RNA function, deliberately to alter the structural or functional characteristics of biological entities."[63] Because of their ability to cut and alter the DNA of any species at almost any genomic site with ease and precision, these genome editing techniques are faster, more accurate, cheaper, and more widely applicable than older techniques.[64] They have been developed to determine the site of mutation or insertion of the genes and to overcome the problem of randomness that results from mutation breeding.

[57] Cf. Comstock (2010).

[58] Cf. ibid.

[59] Cf. Jasanoff et al. (2015).

[60] Cf. ibid.

[61] Cf. Comstock (2010).

[62] Cf. ibid.

[63] Nuffield Council on Bioethics (2016, 4).

[64] Cf. Jasanoff et al. (2015).

In plant breeding CRISPR-Cas9,[65] one of the most popular genome editing techniques, has come into use in a short time. In 2012 the development of this system enabled precisely targeted alterations to DNA sequences in living cells. CRISPR-Cas9[66] is based on the "virus library" of bacteria (a natural way of bacteria to defend against phage infection) and uses RNA to locate the exact spot in a genome. It is possible to insert a new piece of DNA (in case of a cis- or transgene plant), but it also cuts the unwanted piece of DNA (i.e., point mutations).[67]

Applications of plant gene editing techniques are varied, many, and rapidly evolving, including applications that promise benefits in drought- and salt tolerance, and disease resistance. A new tomato variety that grows like a bush is one of the first cultivars. To realize the promised significant benefits of the gene editing technology, the technology needs to be firmly and fully embedded in society. For example, the "Tomelo," a variety of tomato plant geneticists from Tübingen have developed that is resistant to powdery mildew. The powdery mildew-resistant tomato has a deletion in the SlMlo1 gene. CRISPR-Cas9 enabled them to achieve this in less than 10 months, a relatively short period of time.[68] They also demonstrated that the new tomato variety is indistinguishable from naturally occurring deletion mutants and contains no foreign DNA (no natural species barrier was crossed).[69]

Although there are all kinds of material on the internet about the ethical discussion of GM, there are almost no journal papers about ethics and CRISPR-Cas9. How to refer to the new technology is an important issue: gene editing or genetic manipulation CRISPR-Cas9 or genetic modification? The use of an adequate name in the societal debate is very important.[70] The public will link the name of a new technology to an element in the name if they lack knowledge of that new technology. For example, *gen*omics has been linked to *gen*etic modification.[71] Also the aims and functions of the new technology are already described in different ways: tinkering with the genome; manipulation of DNA; repairing the genome; tools to create mutations; text processing of DNA. An inappropriate wording proves to be very hard to correct.[72]

How can we use CRISPR-Cas9 in genomes of humans, animals, and plants? First, the technology can be used to *repair* the genome (i.e., inheritable diseases), but in some cases a natural alternative to repair a genome is also possible. *Prevention* is a second use (i.e., inheritable diseases) and *improving* the genome (existing traits or new traits) is a third. Improvement by adding new traits offers endless possibilities.

---

[65] Cf. Zhang et al. (2014).

[66] Cf. Ibid.

[67] Cf. Jasanoff et al. (2015).

[68] Cf. ibid.

[69] Cf. ibid.

[70] Cf. Boersma and Gremmen (2018).

[71] Cf. ibid.

[72] Cf. Boersma et al. (2019).

The fourth way to use this technology is to *design* new genomes. This could lead, in the case of humans, to the return of earlier ethical debates about eugenics.

What are, in general, the ethical issues of CRISPR-Cas9? Since it is also possible to insert (a) gene(s) through gene editing, first of all the ethical issues of genetic modification apply. Are there any other ethical issues specific of CRISPR-Cas9? We will analyze the main technological characteristics of CRISPR-Cas9 to answer this question:

1. Compared to genetic modification it is said to be very accurate.[73] However, side effects, like off-target mutations and unexpected results, have also been reported after certain gene editing applications. How safe is CRISPR-Cas9?
2. It is said to be cheap compared to genetic modification.[74] This makes abuse by experts and companies more worthwhile.
3. It is said to be relatively fast.[75] This means that it will be difficult to exercise societal control. Because of the speed and number of innovations regulation could be slow and sometimes even implemented too late.
4. It is not by definition a transgene technique.[76] In the media a denial of its transgenic possibilities is used to make it more likeable.
5. It is said to be relatively easy.[77] This makes abuse by amateurs/terrorists more conceivable.
6. A point mutation deletion caused by CRISPR-Cas9[78] is impossible to detect. As a consequence the difference between GM and non-GM becomes undetectable, thereby blocking one of the cornerstones in the regulation of genetic modification. Therefore, it will be difficult to exercise societal regulation. It also means that it is difficult to label products developed by this technology. New transparent and responsible chains have to be developed to ensure the consumer's right on information.
7. May gene drives be used as the ultimate weapon against diseases like Malaria and the Zika-Virus? By causing mutations of chromosomes after only a few generations, gene drives are able to install new traits in every individual of a species.[79] An example is a species of mosquitos unable to carry the malaria parasite. One of the problems is that mutations can't be reversed. The long-term consequences are also unclear. For example, when we wipe out Malaria, an even more dangerous parasite could take its place. A solution may be to build in control via "safety nets": genetically decoupling of the steering parts of a gene drive. This may be understood as a kind of molecular "un-do"-button via a second

[73] Cf. Nuffield Council on Bioethics (2016).

[74] Cf. ibid.

[75] Cf. ibid.

[76] Cf. ibid.

[77] Cf. ibid.

[78] Cf. ibid.

[79] Cf. ibid.

gene drive. Specific ethical questions of gene drives are: is it a threat to biodiversity? Is it acceptable to disrupt nature? Is it acceptable to use techniques we do not yet understand and control fully? Compared to genetic modification, CRISPR-Cas9 brings about a number of extra ethical issues. Because there are relatively low costs involved and that it is relatively easy to develop innovations from an ethical point of view the possible abuse by all kinds of stakeholders is particularly worrying.

On July 25th 2018 the *Court of Justice of the European Union* (CJEU) decided that all products made by the toolbox of New Plant Breeding Technologies (NPBTs) are GMO's. NPBTs are technologies to increase and accelerate the development of new traits in plant breeding, and gene editing, especially CRISPR-Cas9, is its main technology. The content of the NPBTs toolbox has been described by two research institutes of the European Commission.[80] It contains intragenesis (technologies using transformation with genetic material restricted to the species' own gene-pool), cisgenesis, emerging techniques to induce controlled mutagenesis or insertion (ODM, Zinc Finger Nuclease technologies 1–3) and other applications such as or reverse breeding or grafting on GM-rootstocks.

There has been a long debate in European Union about the regulation and legal categorization of NPBTs. Discussions at the policy level have evolved around the question whether products from NPBTs are or should be subject to special regulation.[81] As most NPBTs could not be separated from conventional breeding techniques, some people believe that they should also not be subject to special regulation.[82] Others, highlighting the requirements of the precautionary principle, call for regulation following the regulations for GMOs.[83] Most NPBTs are subject to regular GM regulation in the EU according the CJEU judgment[84] on the mutagenesis exemption in Directive 2001/18/EC[85] (hereafter Directive). This ruling has created a regulatory system for NPBTs which is unique in the world. If organisms obtained from NPBTs are put in the same basket as GMOs, this may carry a serious risk: transferring analogous ethical problems that GMOs encountered in the past, to organisms obtained from NPBTs, while they may not address similar risks.[86]

[80] Cf. JRC and IPTS (2011).

[81] Cf. Sprink et al. (2016).

[82] Cf. New Techniques Working Group (2008).

[83] Cf. Then and Bauer-Panskus (2017).

[84] Cf. CJEU (2018): Case C-528/16.

[85] European Parliament and Council of the European Union (2001).

[86] Cf. Poortvliet et al. (2019).

6 Conclusion

In this paper we presented an overview of the ethical issues of bioeconomy in Europe. With regard to the *bio-resource view* of bioeconomy we compared the German and the Dutch bioeconomy policies. While the German government aims to develop the complete range of products with a focus on innovation, the Dutch government focuses on the production of electricity and heat in biomass power plants using mainly wood. German biomass is mainly of domestic origin and the challenge will be to increase biomass usage for energy without drastically increasing imports. Most of the Dutch biomass is wood, and the expectation is that imports will drastically increase. The Netherlands is not only too small to meet its wood needs, but also almost all agricultural land is used for food crops.[87] From the recent debates about the environmental and health impacts of the wood fueled biomass power plants we can conclude that the recent attempts to build a sustainable bioenergy system more or less failed. The debates triggered a shift in government policy from a bio-based economy to a circular economy. But is this government policy realistic? In the current situation the Dutch regional economy only provides 25% of its inputs. In case of a truly circular economy we can predict a severe reduction of agricultural production.

With regard to the biotechnology view of the bioeconomy we have first given an overview of the ethical issues of genetic modification. Many critics might be opposed to the different applications of GM technology but not to GM technology as such. Current applications of agricultural biotechnology have also been criticized from the viewpoint of the autonomy of farmers and from the viewpoint of justice. Many people[88] emphasize the *risks and uncertainties* with this new technology about human health or the environment (e.g., biodiversity and sustainability) and think that GM is probably unsafe or even harmful. Our conclusion is that the ethical arguments about GMOs in society are still very varied. This means that genetic modification will remain a controversial topic.

We also reviewed the ethical issues of the latest development in biotechnology: CRISPR-Cas9. We asked the following questions: are there any ethical issues specific to CRISPR-Cas9? How can we use CRISPR-Cas9 in genomes of humans, animals, and plants? We answered these questions by analyzing the most important technological characteristics of CRISPR-Cas9 and described seven specific ethical issues of it.

If organisms obtained from NPBTs are put in the same basket as GMOs, this may carry a serious risk: transferring analogous ethical problems that GMOs encountered in the past, to organisms obtained from CRISPR-Cas9, while they may not address similar risks. At the stage of legal interpretation (such as with the CJEU) possible consequences like those can hardly be considered. Insights from ethics on the effect of a legal definition should be taken into account in the debate on whether and how to

[87] Cf. Gremmen et al. (2019).

[88] Cf. Gaskell et al. (2011).

change this definition. To accommodate the New Plant Breeding Technologies, and CRISPR-Cas9 in particular, the existing ethical frameworks on biotechnology[89] have to be broadened. This will help stakeholders, scientists, and policymakers to understand, monitor, and evaluate, the integration of the technical, social, and ethical aspects of the modern GM-toolbox also in view of a further development of bioeconomical applications.

Putting organisms obtained from CRISPR-Cas9 in the same basket as GMOs may carry a serious risk—transferring analogous ethical problems that GMOs encountered in the past, to organisms obtained from CRISPR-Cas9, while they may not address similar risks. Possible consequences like those can hardly be considered at the stage of legal interpretation (such as with the CJEU). Rather, as discussion now unfolds whether and how to change the legal definition, insights from ethics on the effect of such a definition should be taken into account. In our view, the existing ethical frameworks on biotechnology[90] have to be broadened to accommodate the New Plant Breeding Technologies and CRISPR-Cas9 in particular. This will help scientists, stakeholders, and policymakers to understand, evaluate, and monitor the integration of the technical, social, and ethical aspects of the modern GM-toolbox also in view of a further development of bioeconomical applications.

References

Boersma, R., & Gremmen, B. (2018). Genomics? That is probably GM! The impact a name can have on the interpretation of a technology. *Life Sciences, Society and Policy, 14*(1), 8.

Boersma, R., Poortvliet, P. M., & Gremmen, B. (2019). The elephant in the room: How a technology's name affects its interpretation. *Public Understanding of Science, 28*(2), 218–233.

Bugge, M. M., Hansen, T., & Klitkou, A. (2016). What is the bioeconomy? A review of the literature. *Life Sciences, Society and Policy, 8*(7), 691.

Comstock, G. (2010). Ethics and genetically modified foods. In F. T. Gottwald, H. W. Ingensiep, & M. Meinhardt (Eds.), *Food ethics (pp.* 49–66). Springer.

Court of Justice of the European Union (CJEU). (2018). *Judgment of the Court from 25. July 2018, C-528/16,* ECLI:EU:C:2018:583. Retrieved January 8, 2019, from, http://curia.europa.eu/juris/document/document.jsf?text=&docid=204387&pageIndex=0&doclang=EN&mode=req&dir=&occ=first&part=1&cid=44391

Energy Transition – The Global Energiewende. (2019). *About us*. Retrieved June 18, 2020, from, https://energytransition.org/about-us/

European Association for Bioindustries. (2011). *Building a bio-based economy for Europe in 2020.* European Association for Bioindustries.

European Commission. (2012). *Innovating for sustainable growth: A bioeconomy for Europe.* European Commission.

European Parliament and Council of the European Union. (2001). Directive 2001/18/EC of the European Parliament and of the Council of 12. March 2001 on the Deliberate Release into the Environment of Genetically Modified Organisms and Repealing Council Directive 90/220/EEC – Commission Declaration. In *Official Journal of the European Communities* L 106/44, 17.04.2001, 1–39.

[89]Cf. Holland and Johnson (1998); Rollin (2006).

[90]Cf. Rollin (2006); Holland and Johnson (1998).

Federal Ministry of Education of Research. (2011). *National research strategy bioeconomy 2030.* Federal Ministry of Education of Research.

Gaskell, G., Allansdottir, A., Allum, N., Castro, P., Esmer, Y., Fischler, C., Jackson, J., Kronberger, N., Hampel, J., Mejlgaard, N., Quintanilha, A., Rammer, A., Revuelta, G., Stares, S., Torgersen, H., & Wager, W. (2011). The 2010 Eurobarometer on the life sciences. *Nature Biotechnology, 29*, 113–114.

German Advisory Council on Global Change. (2009). *World in transition: Future bioenergy and sustainable land use*. WBGU.

Golembiewski, B., Sick, N., & Bröring, S. (2015). The emerging research landscape on bioeconomy: What has been done so far and what is essential from a technology and innovation management perspective? *Innovative Food Science & Emerging Technologies, 29*, 308–317.

Gregorowius, D., Lindemann-Matthies, P., & Huppenbauer, M. (2012). Ethical discourse on the use of genetically modified crops: A review of academic publications in the fields of ecology and environmental ethics. *Journal of Agricultural and Environmental Ethics, 25*, 265–293.

Gremmen, B. (2002). Methodological pragmatism in bioethics: A modest proposal? In J. Keulartz, M. Korthals, M. Schermer, & T. Swierstra (Eds.), *Pragmatist ethics for a technological culture* (pp. 99–105). Springer.

Gremmen, B., Blok, V., & Bovenkerk, B. (2019). Responsible innovation for life: Five challenges agriculture offers for responsible innovation in agriculture and food, and the necessity of an ethics of innovation. *Journal of Agricultural and Environmental Ethics, 32*(5), 673–679.

Gremmen, B., & Jacobs, J. (1997). Understanding sustainability. *Man and World, 30*, 315–327.

Gremmen, H. G. J. (2007). *De zwakste schakel: over maatschappelijk verantwoorde genomics*. Inaugural speech.

Gremmen, H. G. J. (2017). *Ethics of livestock farming? Who cares?* Inaugural Lecture.

Heijman, W. J. M., & Schepman, T. (2018). Measuring the size of the Dutch bio-economy. *Visegrad Journal on Bioeconomy and Sustainable Development, 7*(2), 67–72.

Holland, A., & Johnson, A. (1998). *Animal biotechnology and ethics*. Springer.

Jasanoff, S., Hurlbut, J. B., & Krishanu, S. (2015). CRISPR democracy: Gene editing and the –need for inclusive deliberation. *Issues in Science and Technology, 32*(1), 25–32.

Joint Research Centre (JRC) and Institute for Prospective Technological Studies (IPTS). (2011). *New plant breeding techniques. State-of-the-art and prospects for commercial development.* Publications Office of the European Union.

Luiga, M., & Swart, F. (2019). Houtverbranding in biomassacentrales is totaal niet duurzaam. In *Het parool*. Retrieved July 7, 2020, from, https://www.parool.nl/columns-opinie/houtverbranding-in-biomassacentrales-is-totaal-niet-duurzaam~b4108a1a/?referer=https%3A%2F%2F02025.nl%2Fbericht%2F39314%2Fhoutverbranding-in-biomassacentrales-is-totaal-niet-duurzaam-

McCormick, K., & Kouto, N. (2013). The bioeconomy in Europe: An overview. *Sustainability, 5*(6), 2589–2608.

Mersbergen, C. van. (2019, October 31). Vattenfall: Vorloopig geen nieuwe biomassacentrales in Nederland. In *Algemeen Dagblad* Retrieved July 15, 2020, from, https://www.ad.nl/binnenland/vattenfall-voorlopig-geen-nieuwe-biomassacentrales-in-nederland~ae6cc277/

Ministerie van Infrastructuur en Waterstaat. (2019). *Aanbiedingsbrief Uitvoeringsprogramma Circulaire economie*. Retrieved June 18, 2020, from, https://www.rijksoverheid.nl/onderwerpen/circulaire-economie/documenten/kamerstukken/2019/02/08/aanbieding-uitvoeringsprogramma-circulaire-economie-2019-2023

New Techniques Working Group. (2008). *Final report*. Retrieved June 19, 2018, from, http://www.seemneliit.ee/wp-content/uploads/2011/11/esa_12.0029.pdf

Nuffield Council on Bioethics. (2016). *Genome editing. An ethical review*. Nuffield Council on Bioethics.

Organisation for Economic Cooperation and Development. (2009). *The bioeconomy to 2030: Designing a policy agenda. Main findings*. OECD.

Petersen, T., & Ryberg, J. (2007). Normative ethics: 5 questions. Automatic Press.
Poortvliet, P. M., Purnhagen, K. P., Boersma, R., & Gremmen, B. (2019). On the legal categorisation of new plant breeding technologies: Insights from communication science and ways forward. *European Journal of Risk Regulation, 10*(1), 180–186.
Rollin, B. E. (2006). *Science and ethics*. Rodopi.
Sandoe, P., & Christiansen, S. B. (2008). *Ethics of animal use*. Wiley.
Singer, P. (1975). *Animal liberation: A new ethics for our treatment of animals*. HarperCollins.
Sprink, T., Eriksson, D., Schiemann, J., & Hartung, F. (2016). Regulatory hurdles for genome editing: Process- vs. product-based approaches in different regulatory contexts. *Plant Cell Reports, 35*(7), 1493–1506.
Staffas, L., Gustavsson, M., & McCormick, K. (2013). Strategies and policies for the bioeconomy and bio-based economy: An analysis of official national approaches. *Sustainability, 5*(6), 2751–2769.
The White House. (2012). *National bioeconomy blueprint*. The White House.
Then, C., & Bauer-Panskus, A. (2017). *Playing Russian Roulette with biodiversity*. Test Biotech.
U.S. Energy Information Administration (EIA). (2016): *Annual energy outlook 2016, with projections to 2040*. U.S. Energy Information Administration (EIA).
Vries, A. de (2019). Initiatiefnemer nieuwe biomassacentrale Zwolle: 'Onderzoek over uitstoot is halve waarheid'. In *De Stentor*. Retrieved June 18, 2020, from, https://www.destentor.nl/zwolle/initiatiefnemer-nieuwe-biomassacentrale-zwolle-onderzoek-over-uitstoot-is-halve-waarheid~a13953d2/
Wield, D. (2013). Bioeconomy and the global economy: Industrial policies and bio-innovation. *Technology Analysis & Strategic Management, 25*(10), 1209–1221.
Zhang, H., Zhang, J., Wei, P., Zhang, B., Gou, F., Feng, Z., Mao, Y., Yang, L., Zhang, H., Xu, N., & Zhu, J. K. (2014). The CRISPR/Cas9 system produces specific and homozygous targeted gene editing in rice in one generation. *Plant Biotechnology Journal, 12*(6), 797–807.

Bioeconomy from the Perspective of Environmental Ethics

16

Marion Stahl

Abstract

The article discusses the field of bioeconomy primarily from the perspective of environmental ethics. Even though bioeconomy has the potential for eco-friendly production methods, it is by no means eco-friendly and sustainable per se. Against this background, it seems necessary to first deal with the foundations of environmental ethics and then to develop its "added value" for the bioeconomic context. Thus, in this contribution, fundamental positions of environmental ethics are presented in a selective manner by taking into account both anthropocentric and physiocentric approaches. Albeit the instrumental value of nature in the context of bioeconomy appears as a priority, it is nevertheless important to take sufficient account of non-instrumental values in nature. In view of the far-reaching consequences of the application of modern biotechnologies with regard to ecosystems, biodiversity and resource use, integral, physiocentric approaches as well as nature-aesthetic, anthropocentric positions of environmental ethics can provide important impulses for the complex discourse on sustainable bioeconomic action. By focusing on central topics of environmental ethics in general, a value-based view of bioeconomic fields of application is to be conveyed, which might possibly remain unnoticed in the context of a primarily technology-driven view. Finally, the elaborated environmental–ethical reflections in this article can be seen as an informative basis for developing guidelines for action and implementation strategies in the context of politics, economy and society, which allow bioeconomic strategies to be value-based and sustainable.

Keywords

Environmental ethics · Non-instrumental values · Biodiversity

---

M. Stahl (✉)
Department of Philosophy, University of Hamburg, Hamburg, Germany
e-mail: stahl.m@mail.de

D. Lanzerath et al. (eds.), *Bioeconomy and Sustainability*,
https://doi.org/10.1007/978-3-030-87402-5_16

1 Introduction

In the current debate on the goals and challenges of bioeconomy, economic, social, political and ethical discourses emphasize the enormous economic potential of bioeconomic strategies and modern biotechnologies as well as their fundamental potential to contribute significantly to improving global living standards and to a more sustainable economic development.[1] Even if the energetic use of regenerative biomass represents a more environmentally and climate-friendly alternative to the use of fossil fuels, this does not necessarily imply that a biologically based industrial and economic form is environmentally friendly and sustainable per se. Rather, bioeconomy and the application of modern biotechnologies hold both opportunities and risks with regard to environmental and social factors. Therefore, in addition to extensive risk-benefit analyses, fundamental ethical reflections are needed to investigate the possible consequences of the use of modern biotechnologies and bioeconomic concepts with regard to life, environment and social contexts.

Bioeconomic activities affect contexts of life and the environment both directly and indirectly. In this sense, bioeconomy also touches on issues of environmental ethics that can be located on the horizon of the issue of sustainability. In the National Research Strategy Bioeconomy 2030, sustainable goals and strategies are listed in the context of five central fields of action against the background of the desired bioeconomic transformation.[2] These goals relate in part to the defined "sustainable development goals" of the United Nations. In Germany, the so-called Bioeconomy Council was established in 2009, which in its advisory capacity also takes a position on fundamental ethical questions. However, discourses on environmental ethics dealing with the basic principles of ecological sustainability have so far been neglected in the bioeconomic context. If we do not want to understand the bioeconomy as a new form of "overexploitation" of nature and social and economic systems,[3] bioeconomic strategies must be defined and restricted with the help of ethical principles and guidelines for environmental ethics. Especially in view of the high complexity of bioeconomic strategies, it seems to be necessary to deal with selected approaches in environmental ethics in greater depth and to examine their relevance for the bioeconomic context.

After a brief introduction to environmental ethics, selected approaches from both, the anthropocentric and the physiocentric perspectives will be presented. This is followed by an outline of fundamental ethical fields of conflict in the context of bioeconomy. Finally, the question of the "added value" of environmental ethics for the bioeconomic context will be addressed. By focusing on central topics of

[1] Cf. Wesseler and von Braun (2017, 276 ff).

[2] On the National Research Strategy Bioeconomy 2030 and the five fields of action "securing global nutrition," "ensuring sustainable agricultural production," "producing healthy and safe foods," "using renewable resources for industry" and "developing biomass-based energy carriers," cf. Federal Ministry of Education and Research (2010, 16 ff).

[3] Cf. Grefe (2016, 302).

environmental ethics, an integral, value-based view concerning bioeconomic fields of application is to be conveyed. Such a perspective might go unnoticed in the context of a primarily research- or technology-centred view. The elaborated ethical reflections can be seen as a basis for developing guidelines for concrete instructions for action and implementation strategies in the context of politics, economy, and society, which allow bioeconomic strategies to be value-oriented and sustainable.

2 On the Foundations of Environmental Ethics

Since the 1970s, environmental ethics has established itself as a sub-area of applied ethics. Environmental ethics developed from a far-reaching discourse on the consequences of the increasing use of modern technologies and a growth-oriented economy and consumer behaviour. On the scientific level, this discourse was conducted in connection with "political ecology" and a wide-ranging debate on environmental justice and future justice.[4] Irrespective of its political origins, the "concern" for nature in view of finite resources and the consequences of new technical intervention options is also reflected in the context of current reflections on environmental ethics, starting from the question of the value of nature.[5] Thus, one of the central questions of environmental ethics is whether nature has a moral intrinsic value or is only of value in relation to human subjects.[6] The assumption of an intrinsic value of non-human beings is accompanied by the attribution of a specific moral status. Based on the ascribed moral status and intrinsic value of non-human natural beings, forms of responsibility or certain duties towards animate or inanimate nature can be formulated. In the context of the anthropocentric perspective of environmental ethics, the focus is mostly on the instrumental value of nature in terms of human needs. In the context of the physiocentric perspective, however, the morally relevant intrinsic value of nature is regarded as completely independent of the benefit for humans.[7]

Even in the early phases of the nature conservation movement and in early texts on environmental ethics, there are expressions of appreciation concerning the autonomy and the intrinsic value of nature.[8] In his essay *The Land Ethic* (1949), for example the forestry scientist and environmentalist Aldo Leopold (1887–1948) develops his own conception of environmental ethics, which includes as a central element the so-called "community concept." According to this concept, each individual is to be considered as a member of a community. A general "community concept" is what Leopold ultimately identifies as the premise of all ethics. The special feature of such land ethics, however, is that it extends the boundaries of the

[4]Cf. Ott et al. (2016, 3 ff, 127 ff).

[5]Cf. Eser (2014, 221 ff).

[6]Cf. Krebs (1997a, 7).

[7]Cf. Krebs (1997b, 337 ff).

[8]Cf. Eser (1999, 129).

"community" and also includes the "land," so that man and nature are regarded as equal members of a "biotic community." In such a view man no longer appears in the role of the conqueror and ruler of nature, but as an equal member of a "land community," which should exist in harmony with nature.[9]

The concept of nature in contemporary environmental ethics, of course, usually refers to a cultivated nature already formed by humans. In this respect, current nature conservation measures include not only the protection of largely untouched nature—the "wilderness," but above all the protection of cultural landscapes and even forms of extensive agriculture.[10] If so-called ecosystem services are threatened, and as a result, the "output" of nature seems to decrease, the question of the instrumental value of nature comes into focus. However, the question of the supposed intrinsic value of nature is pushed into the background. Bioeconomy can exacerbate this trend if, given the availability of modern biotechnologies, the idea becomes prevalent that all that is needed are suitable techniques and strategies to make unlimited use of regenerative resources. Against this background, it is also important to consider the non-instrumental value of nature in the context of the debate around bioeconomy. In the following, selected conceptions of environmental ethics from both the anthropocentric and the physiocentric perspectives will be presented.

2.1 Anthropocentric Arguments of Environmental Ethics

In an anthropocentric perspective of environmental ethics, arguments in which nature conservation is justified by functional or instrumental values for humans predominate the discussion. The so-called "basic needs" argument comes to the fore: Basic needs arguments generally assert that natural resources on which people are fundamentally dependent are threatened by technical interventions in nature. Therefore, out of self-interest and in consideration of the interests and needs of present and future generations, it seems morally imperative to implement appropriate nature conservation measures in order to preserve the natural resources necessary for mankind.[11] However, with regard to the conservation of plants, the argument would only argue for the preservation of some plants that have proven to be useful for mankind, while excluding those plants whose usefulness is not foreseeable in the long term.[12]

Despite the apparent plausibility of the argument that nature should be protected at least out of pure self-interest, the basic needs argument does not seem sufficiently convincing in every respect. Hans Jonas names one of the reasons for this in his book *Prinzip Verantwortung* (1979): Predictive knowledge usually lags behind technical knowledge, so that the "Kluft zwischen Kraft des Vorherwissens und Macht des

---

[9]Cf. Leopold (1949, 203 ff).

[10]Cf. Siep (2004, 296 f); cf. also Ott et al. (2016, 46).

[11]Cf. Krebs (1997b, 364 ff).

[12]Cf. Kallhoff (2002, 72 f).

Tuns"[13] creates a new ethical problem. In order to avert irreversible damage for present and future generations in view of the uncertainty of possible consequences of concrete technical interventions in nature, binding criteria are required. As an ethical guideline for responsible environmental action, Jonas formulated his own version of the so-called "precautionary principle." Given the indeterminable risk of modern technology, responsibility for future generations means for Jonas "Vorrang der schlechten vor der guten Prognose."[14] The "precautionary principle," which is a guiding principle of environmental policy in Germany since the early 1970s, became a component of international law since the UN Conference on Environment and Development held in Rio de Janeiro (Rio Conference) from 3 to 14 June 1992.[15]

Non-instrumental values are also cited within the framework of the anthropocentric perspective—for example in the context of natural aesthetic arguments. Aesthetic or eudaimonistic value attributions emphasize the added value of nature for human beings beyond the mere satisfaction of needs. For example the experience of aesthetic sensation during a stay in the nature or the transforming power of concrete experiences in nature. The "aesthesis argument" of the German philosopher Gernot Böhme justifies nature conservation, especially with regard to the sensual-bodily feeling of well-being conveyed by aesthetic experiences in nature. Referring to the platonic triad of "the good, the true and the beautiful," he identifies the beauty of nature as an indicator for the "good" of nature and finally makes a plea for landscape and nature conservation in urban space.[16]

The argument of aesthetic contemplation developed by the philosopher Martin Seel goes one step further: Seel ascribes a non-instrumental, aesthetic *intrinsic* value to nature, whereby the aesthetic experience of nature is considered as a central, irreplaceable element of a successful life.[17] Seel makes a clear distinction between *moral* recognition, which applies to all beings capable of suffering as sensitive subjects, and *aesthetic* recognition, which applies to concrete, experienceable states and situations in and with nature. His thesis is that *direct* recognition of nature can only exist in non-instrumental, *aesthetic* recognition.[18] As an elementary and irreplaceable component of a good life, the beauty of nature is to be regarded as a "Korrektiv individueller und kollektiver Ideale der Existenz."[19] According to Seel, the aesthetic value also has an ethical dimension, whereby these two dimensions cannot be separated.

Natural aesthetic arguments are also relevant with regard to the issue of responsibility for future generations. As the German bioethicist and philosopher Dieter

[13] Jonas (1984, 28), "gap between the power of foreknowledge and the power of action" (own translation).

[14] Ibid., 70, "the priority of bad over good prognoses" (own translation).

[15] For more information on the precautionary principle, cf. Gottschlich (2017, 195 ff).

[16] Böhme (1989, 31 ff, 56 ff).

[17] Cf. Seel (1993, 205–227).

[18] Cf. ibid., 207 ff.

[19] Seel (1991, 288), "a corrective of individual and collective ideals of existence" (own translation).

Birnbacher remarks, responsibility for future generations means more than just averting possible dangers and securing the *status quo*. According to Birnbacher, responsibility for the future also includes improving the present situation as well as the conservation of non-material, cultural and aesthetic values for future generations.[20] With regard to the bioeconomic context, natural aesthetic arguments can at least have an informative effect. They remind us that it seems to be morally imperative to preserve and create places of "natural beauty," since they represent a special aesthetic, sensual and even non-instrumental value for man, beyond the mere satisfaction of material needs.

2.2 Physiocentric Arguments of Environmental Ethics

Physiocentric approaches to environmental ethics do not primarily focus on human needs and interests when considering the ethical principles for nature conservation but on the moral intrinsic value of nature itself. Pathocentrism, biocentrism, ecocentrism and the most "radical" approach of so-called holism can be distinguished as central approaches of the physiocentric view. In the context of pathocentric concepts only all living beings capable of suffering are attributed an intrinsic value. In biocentric arguments, however, all living beings are supposed to have an intrinsic value. Ecocentrism, on the other hand, includes entire ecosystems as part of the so-called "moral community." Holistic concepts go one step further, attributing a morally relevant self-value to the natural world as a whole and to all of its individual animate and inanimate components.[21]

In the following, the holistic approach of "deep ecology" and the biocentric approach of Paul W. Taylor will be outlined first, followed by a discussion of the "added value" of physiocentric concepts in the context of bioeconomy. In the doctrine of deep ecology, a similar idea of an equal relationship between human beings and nature is advocated, as we find it in Aldo Leopold's land ethic. Deep ecology, whose founder is the Norwegian philosopher and mountaineer Arne Naess (1912–2009), can be considered as a holistic approach to environmental ethics, which also refers to natural philosophical and religious foundations. Even though deep ecology does not follow a strictly scientific approach, it has nevertheless developed a significant influence within the discourse of environmental ethics. In his essay *The Shallow and the Deep, Long-Range Ecology Movement. A Summary* (1973), Naess initially distinguishes the deep ecology movement from the so-called "shallow" or superficial ecology movement. The shallow ecology movement is concerned with a fight against environmental pollution and the overexploitation of natural resources to provide health and prosperity for people in developed countries. The deep ecology movement, however, aims at implementing a wide-ranging form

[20]Cf. Birnbacher (1988, 217 ff, 227 ff).

[21]The different concepts listed above are partly interwoven. For an overview of the various physiocentric arguments, cf. Ott et al. (2016, 12 ff); cf. also Krebs (1997b, 345 ff).

of environmental protection based on the appreciation of nature for its own sake and seeks to avoid environmental pollution and resource exploitation from the outset.[22] With regard to the relationship between man and nature, Naess cites the image of a "relational, total-field" within which all organisms are like "knots in the biospherical net or field of intrinsic relations."[23] This means that all organisms are related to each other in a constitutive way. Another central aspect of the deep-ecological doctrine is the so-called "biospherical egalitarianism," according to which all life forms are to be considered as equal: "To the ecological field-worker, the equal right to live and blossom is an intuitively clear and obvious value axiom."[24] According to the holistic argument in general, environmental protection is thus ultimately based on the unity of man and nature and on the right to self-realization of all natural entities. Apart from fundamental argumentative weaknesses, however, it remains questionable to what extent the postulate of unity of man and nature overcomes the limits of an anthropocentric view, according to the claim of deep ecology.[25]

The biocentric approach of the philosopher Paul W. Taylor is close to the deep ecology's core principle of biospherical egalitarianism. In his book *Respect for Nature* (1986), Taylor uses a teleological argument to justify moral respect for purposeful, animated nature. Four elements constitute the biocentric outlook: First, the conviction that humans are members of Earth's community in the same way as other beings. Second, all human and non-human beings are integral elements of a system of interdependence, whereby the survival and well-being of each element depend on its relationship to every other animate element. Thirdly, all living beings are regarded as teleological organisms that strive for their own "good." Since this is true of all living beings, fourthly, the assumption of a superiority of man over all other beings must be rejected.[26] The "attitude of respect" for living nature is closely related to these four principles of the biocentric outlook.

Taylor highlights two concepts that underpin the attitude of respect for nature: "the concept of the good of a being" and "the concept of inherent worth."[27] The first concept states that every individual organisms has a "good"[28] of its own, independent of other organisms and their goods. In the second concept, Taylor refers to the attribution of an intrinsic value as a precondition for being able to meet non-human, living nature with a respectful attitude: "To have the attitude of respect for nature is to regard the wild plants and animals of the Earth's natural ecosystems as possessing inherent worth."[29]

---

[22] Cf. Naess (1973, 95 ff).

[23] Ibid., 95.

[24] Ibid., 96.

[25] Cf. Krebs (1997b, 361 ff).

[26] Cf. Taylor (1986, 99 ff).

[27] Cf. ibid., 60 ff, 71 ff.

[28] Ibid., 69.

[29] Ibid., 71.

If one takes into account Taylor's rules or duties, such as those of "nonmaleficence" or "non-interference",[30] resulting from the principle of respect for animated nature, the question arises as to how far Taylor's concept allows interventions in nature at all. I think that, despite their argumentative weaknesses, integral approaches of environmental ethics can have a clarifying effect with regard to the bioeconomic context. By referring to the fragile interplay of the human-nature relationship and by pointing out complex interrelationships of ecosystems and biodiversity, these concepts direct our attention to significant aspects that are commonly not taken into account in the context of bioeconomy.[31]

3 Bioeconomy from the Perspective of Environmental Ethics

When approaching the bioeconomy from the perspective of environmental ethics, it seems important to reflect on two important issues that can only be touched upon briefly here: Firstly, it is necessary to look at the decisive environmental developments of recent decades in order to determine the relevant, environment-related starting factors for the bioeconomic context. Secondly, it is important to analyze whether and to what extent the relationship between man and nature changes in the context of bioeconomy in general and how the meaning and values attributed to nature and environment may change in detail. Thirdly, it must be questioned what "added value" integral approaches of environmental ethics can offer for the bioeconomic context beyond the purely instrumental dimension of values.

With view to the first topic, two drastic developments are to be analyzed: Firstly, the now undeniable factuality of climate change, and secondly, the problem of increasing scarcity and pollution of resources worldwide, which particularly affects the Global South. These two developments go hand in hand with already visible and perceptible effects on human life. In the light of these drastic developments that can no longer be ignored, the new dimension of "responsibility" resulting from modern technology in the sense of Hans Jonas becomes tangible. Nature in its "critical vulnerability" has itself become a "human responsibility," which requires a fundamentally changed human practice based on a thorough ethical reflection.[32]

Against this background, the technology-centred bioeconomy may even appear to be a solution for dealing with the "vulnerability" of nature: Indeed, bioeconomic strategies and modern biotechnologies offer opportunities for a more careful use of resources through targeted interventions and intelligently linked value chains. In addition, they promise to significantly reduce CO2 emissions through the energetic use of regenerative biomass and to counter climate change with innovative technical

[30] Cf. ibid., 172 ff.

[31] For a critical review of Taylor's egalitarian-biocentric approach, cf. Pinsdorf (2016, 147).

[32] Cf. Jonas (1984, 26 f).

solutions.[33] On the other hand, bioeconomic strategies themselves carry the risk of exacerbating existing environmental problems: Examples for this are the relatively large water footprint of the "green" bioeconomy or the far-reaching consequences for natural cycles caused by genetically modified plants. The bioeconomy could even contribute significantly to the increasing loss of biodiversity unless no effective compensation measures are taken.[34]

The issue of climate change and the finite nature of resources in the context of the first topic certainly comprises far-reaching ethical and social aspects that go beyond the field of environmental ethics. It should be borne in mind that the bioeconomy has the potential to influence the distribution of the biomass available worldwide decisively. This can have an impact on food security if the balance of the global food situation changes as a result.[35] In addition, the increasing "economisation" of life and the specific modification of nature in the bioeconomic context raises far-reaching ethical questions, such as how we deal with life and our environment in general and what "nature" we want.[36]

Bioeconomic strategies are increasingly located in a global context. Moreover, the relevant areas in politics, economics and research are closely intertwined. Therefore, due to their complexity and global range, bioeconomic concepts require their own "governance." The question also arises whether in the bioeconomic context separate dynamics and structures of power will develop or have already been developed.[37] Critical voices even speak of a "totalitarian approach" and a "deliberate deception" of the public with regard to the goals and benefits of bioeconomic approaches, which would only serve power politics and economic considerations.[38] One does not have to agree with this. But it is hard to deny that the bioeconomy has a far-reaching impact, both in terms of their power potential and their effects on environment.

With regard to the second topic, it is important to reflect whether and to what extent the bioeconomic action is not even accompanied by a fundamental change in meaning and values concerning the specific human-nature relationship. The instrumental value of nature as a "source," "resource", and "sink" seems to be of central importance in the bioeconomic context. However, there seems to be a shift in meaning—albeit only gradual—within the respective instrumental value. Thus, the notion of inexhaustible resource availability and unlimited economic growth may become entrenched by bioeconomy. The efficiency-oriented approach of bioeconomy may consolidate the idea that the "source" nature can be cultivated and exploited extremely effectively using suitable technologies. On the other hand, bioeconomy can

[33] Technical interventions such as the so-called "geoengineering" are also discussed in the context of the ethical debate on climate and future justice. Cf. Long (2016, 109–120).

[34] Cf. Fritsche and Rösch (2017, 182 ff).

[35] Cf. von Braun (2016, 7–20).

[36] Cf. Grefe (2016, 35).

[37] Cf. Schaper-Rinkel (2012, 155–179).

[38] Cf. Gottwald and Krätzer (2014, 42 ff, 88 ff, 97 ff).

promote a growing awareness of the limitedness and vulnerability of natural resources, since bioeconomic policies and strategies are directly dependent on renewable resources and intact ecosystems.

4 Bioeconomy, Biodiversity and the Intrinsic Value of Nature

In their fundamental reliance on natural resources, bioeconomic strategies seem to focus primarily on the instrumental value of nature. But the issue of biodiversity makes it clear that consideration of the intrinsic value of nature is also relevant to bioeconomy. It is undisputed that the intensive cultivation of energy crops by green bioeconomy is in fact a far-reaching threat to biodiversity. At the same time, green bioeconomy is largely dependent on intact ecosystems and thus also on biodiversity.

In the first place, it is important to shed light on what is meant by the complex concept of biodiversity. The definition proposed by McNeely et al. seems to be informative in this respect: "Biodiversity encompasses all species of plants, animals and microorganisms and the ecosystems and ecological processes of which they are parts. It is an umbrella term for the degree of nature's variety."[39] According to this, biodiversity does not only consist of biological species diversity, but encompasses it as one aspect—among others—of natural diversity. According to the German philosopher Konrad Ott, biodiversity simply is "der Inbegriff für die Mannigfaltigkeit des Lebendigen."[40] According to Ott, a distinction can be made between biodiversity on (a) the *genetic* level, (b) the *species* level and (c) the *ecosystem* level.[41] For the philosopher Ludwig Siep, biodiversity is basically to be regarded as part of the human heritage and as part of those goods, "an denen sich kollektives und individuelles menschliches Handeln orientieren soll."[42] At this point, I would like to refer to another important aspect that Holmes Rolston has introduced into the philosophical discourse on biodiversity. Rolston points out that species must always be considered in the context of their "niche" they inhabit or the ecosystems that are relevant to them. According to Rolston, we must not value the species itself, but the place it inhabits: "A species is what it is inseparably from the environmental niche into which it fits. [. . .] It is not merely *what* they are, but *where* they are that humans must value correctly."[43]

The assumption made by Rolston that it is primarily the location, i.e. the "niche" of the species that is worth estimating, leads directly to the following question: Does biodiversity merely have a value for us humans, or does the issue of biodiversity also refer to intrinsic values of nature? Dirk Lanzerath has pointed out that the view that

[39] McNeely et al. (1990, 17).

[40] Ott (2015, 47), "the epitome of the diversity of life" (own translation).

[41] Cf. ibid., 48.

[42] Siep (2004, 292), "to which collective and individual human action should be oriented" (own translation).

[43] Rolston (2013, 252).

nature is only there for the sake of man seems intuitively mistaken, given the natural diversity and the innumerable functions of species for other non-human living beings and their own habitats and ecosystems.[44] It is a fact that biodiversity is increasingly being reduced at the global level. Instrumental as well as non-instrumental values can be cited that considers the protection of biodiversity as "good" and necessary. A physiocentric approach that draws attention to the potential for the unfolding of species and that acknowledges the emergence of species as the result of a creative, complex process can create a deeper awareness that species are worthy of protection because of their own value.

How can these previous analyses about biodiversity and the intrinsic value of nature be applied to the context of bioeconomy? It should first be noted that through intensive agriculture of green bioeconomy, genetic engineering and the increased use of pesticides, nature is primarily seen in its function as a resource. Even if effective compensatory measures were implemented, bioeconomic strategies run the risk of fundamentally disregarding the intrinsic value of nature. This intrinsic value also includes the specific value that natural, non-human entities have for other natural, non-human entities.

On the other hand, the use of modern biotechnology also enables extremely targeted interventions that do not burden nature more than necessary. Forms of modern biotechnology based on knowledge-based strategies could prove to be both efficient in terms of their productive "output" and consistent in the sense of gentler interventions in natural systems, in contrast to conventional applications. Ideally, by means of both efficient and consistent biotechnologies, the damage to natural entities could be kept relatively low, provided that suitable strategies are developed. Awareness of the fact that the value of biodiversity may not be reduced to values that exist only for us humans can motivate the development of bioeconomic strategies that limit drastic interventions in nature from the outset. Such an approach would require strict regulation and may not be compatible with purely economic interests.

However, if bioeconomic concepts are combined with measures of effective nature conservation, both the concern to recognize the intrinsic value of nature could be taken into account, as well as the instrumental interest to largely conserve natural resources with regard to their purely functional values or "output" for human beings. In view of the threat to biodiversity posed by the green bioeconomy, measures and mandatory standards for biomass production are to be introduced, such as increasing diversification in the cultivation of energy crops, the targeted cultivation of wild plants, alternative concepts for crop rotations and concepts of cascade use and recycling management.[45]

[44] Cf. Lanzerath (2008, 193).

[45] For more information on the standards and opportunities of bioeconomic production methods, cf. Fritsche, Rösch (2017, 187 ff).

5 Concluding Remarks

Given the enormous complexity of bioeconomic applications and the far-reaching consequences of the use of modern biotechnologies with regard to ecosystems, biodiversity and natural resources, physiocentric approaches, as well as nature aesthetic approaches of environmental ethics, can provide important impulses for the complex discourse on sustainable bioeconomic strategies. In consideration of integral approaches to environmental ethics, bioeconomic strategies and modern biotechnologies should be understood as an opportunity to enable a more careful use of natural resources. Natural aesthetic experiences can convey a sense of the added value of nature for us humans. The idea of a unity of man and nature and the assumption that every living part of nature must be met with an attitude of respect due to its inherent value can be a motivation to develop sophisticated bioeconomic strategies that intervene in natural cycles in the gentlest possible way.

The promise propagated by bioeconomy to reconcile supposedly environmentally friendly technologies with economic growth is by no means guaranteed per se. The further development of sustainability standards and monitoring systems must ensure the sustainability of bioeconomic concepts in terms of both ecological and social aspects in practice. In the discussion about the risks and opportunities of the bioeconomy, it is generally important to bear in mind that the growth-oriented approach of the bioeconomy is located in a world that is endowed with finite resources and that is characterized by a complex interplay of "natural" conditions. In view of increasing consumer needs worldwide and existing global inequalities with regard to elementary basic needs, it is urgently necessary to discuss once again the issues of the "limits to growth," global justice and how the balance between economy and ecology can be achieved in a sustainable way in the global context.[46] This also includes an in-depth discussion of consumer behaviour that has become taken for granted, especially in rich countries of the global North.

The envisaged bioeconomic transformation goes hand in hand with the danger of competitive advantages being exploited at the global level. This may exacerbate the problem of global inequalities in general. Poorer countries must also be able to benefit from future developments, which is why the discourse on new biotechnologies must be conducted not only from the point of view of economic efficiency but also from that of common welfare.[47]

However, if the potential of bioeconomy is assessed primarily in terms of profits for companies and shareholders, not only the question of the intrinsic value of nature is lost sight of, but also the fundamental, teleologically oriented question of what inventions in the context of modern biotechnologies bring "good" to mankind, as the philosopher Ludwig Siep noted.[48] Since there is certainly more to this "good" than the satisfaction of elementary material needs and economic interests, namely also

[46] Cf. Meadows et al. (1972); also Daly (2007).

[47] Cf. von Braun (2016, 17).

[48] Cf. Siep (2004, 294); cf. also Gottwald and Krätzer (2014, 88 ff).

social participation and self-determination[49] as well as the enjoyment of cultural and aesthetic values, ethical debates in the bioeconomic context should take into account the aspects mentioned in this contribution.

References

Birnbacher, D. (1988). *Verantwortung für zukünftige Generationen*. Stuttgart: Reclam.

Böhme, G. (1989). *Für eine ökologische Naturästhetik*. Frankfurt / Main: Suhrkamp.

Daly, H. (2007). *Ecological economics and sustainable development. Selected essays of Herman Daly. In:* J. van den Bergh (Ed.), *Advances in Ecological Economics*. Cheltenham, UK / Northampton, MA: Edward Elgar.

Eser, U. (1999). *Der Naturschutz und das Fremde. Ökologische und normative Grundlagen der Umweltethik.* In Campus Forschung. Vol. 776. Frankfurt / Main / New York: Campus.

Eser, U. (2014). Umweltethik und Politische Ethik: Natur als Gegenstand von Interessenkonflikten. In M. Maring (Ed.), *Bereichsethiken im interdisziplinären Dialog* (pp. 220–238). Karlsruhe: KIT Scientific Publishing.

Federal Ministry of Education and Research. (2010). *Nationale Forschungsstrategie BioÖkonomie 2030. Unser Weg zu einer bio-basierten Wirtschaft*. Retrieved July 21, 2020, from, https://www.bmbf.de/upload_filestore/pub/Nationale_Forschungsstrategie_Biooekonomie_2030.pdf

Fritsche, U., & Rösch, C. (2017). Die Bedingungen einer nachhaltigen Bioökonomie. In J. Pietzsch (Ed.), *Bioökonomie für Einsteiger*. Scientific advisor: U. Schurr (pp. 177–203). Berlin / Heidelberg: Springer.

Gottschlich, D. (2017). *Kommende Nachhaltigkeit. Nachhaltige Entwicklung aus kritisch-emanzipatorischer Perspektive*. In Feminist and Critical Political Economy. Vol. 4. Baden-Baden: Nomos.

Gottwald, F.-T., & Krätzer, A. (2014). *Irrweg Bioökonomie. Kritik an einem totalitären Ansatz.* Edition unseld 51. Berlin: Suhrkamp.

Grefe, C. (2016). *Global Gardening. Bioökonomie – Neuer Raubbau oder Wirtschaftsform der Zukunft?* In Schriftenreihe der Bundeszentrale für politische Bildung (bpb). Vol. 1737. Bonn: Kunstmann.

Jonas, H. (1984). *Das Prinzip Verantwortung. Versuch einer Ethik für die technologische Zivilisation* [1979]. Frankfurt / Main: Suhrkamp.

Kallhoff, A. (2002). *Prinzipien der Pflanzenethik. Die Bewertung pflanzlichen Lebens in Biologie und Philosophie*. Frankfurt / Main: Campus.

Krebs, A. (1997a). Einleitung. In Id (Ed.), *Naturethik. Grundtexte der gegenwärtigen tier- und ökoethischen Diskussion* (pp. 7–12). Frankfurt / Main: Suhrkamp.

Krebs, A. (1997b). Naturethik im Überblick. In Id (Ed.), *Naturethik. Grundtexte der gegenwärtigen tier- und ökoethischen Diskussion* (pp. 337–379). Frankfurt / Main: Suhrkamp.

Lanzerath, D. (2008). Der Wert der Biodiversität: Ethische Aspekte. In German Reference Centre for Ethics in the Life Sciences (Ed.), *Biodiversität*. Ethik in den Biowissenschaften – Sachstandsberichte des DRZE, (Vol. 5, pp. 147–213). Freiburg / München: Alber.

Leopold, A. (1949). The land ethic. In Id (Ed.), *A sand County Almanac. And sketches here and there* (pp. 201–226). New York: Oxford University Press.

Long, J. (2016). Bringing geoengineering into the mix of climate change tools. In C. J. Preston (Ed.), *Climate justice and geoengineering. Ethics and policy in the atmospheric anthropocene* (pp. 109–120). London / New York Rowman & Littlefield.

[49] Cf. Martha C. Nussbaum's "capability approach," Nussbaum (2007, 155 ff); cf. also Nussbaum (2011, 17 ff).

McNeely, J. A., Miller, K., Mittermeier, R. A., Reid, W. V., & Werner, T. B. (1990). *Conserving the World's Biological Diversity*. Edited by International Union for the Conservation of Nature and Natural Resources; World Resources Institute; Conservation International; World Wildlife Fund-US; World Bank. Gland, CH / Washington, D.C: WRI, Conservation International, WWF-US, World Bank.

Meadows, D. H., Meadows, D. L., Randers, J., & Behrens, W. W. (1972). *The limits to growth: A report for the club of rome's project on the predicament of mankind*. New York: Universe Books.

Naess, A. (1973). The shallow and the deep, long-range ecology movement. A summary. *Inquiry: An Interdisciplinary Journal of Philosophy, 16*(1), 95–100. Oslo: University of Oslo.

Nussbaum, M. C. (2007). *Frontiers of justice. Disability, nationality, species membership*. Cambridge MA / London: Belknap (Harvard University Press).

Nussbaum, M. C. (2011). *Creating capabilities. The human development approach*. Cambridge MA / London: Belknap (Harvard University Press).

Ott, K. (2015). *Zur Dimension des Naturschutzes in einer Theorie starker Nachhaltigkeit*. In Beiträge zur Theorie und Praxis starker Nachhaltigkeit. Vol. 8. Marburg: Metropolis.

Ott, K., Dierks, J., & Voget-Kleschin, L. (2016). *Handbuch Umweltethik*. Stuttgart: Metzler.

Pinsdorf, C. (2016). *Lebensformen und Anerkennungsverhältnisse. Zur Ethik der belebten Natur*. In L. Siep & D. Sturma (Eds.), Studien zu Wissenschaft und Ethik. Vol. 7. Berlin / Boston: De Gruyter.

Rolston, H. (2013). Biodiversity. In L. Gruen, D. Jamieson, & C. Schlottmann (Eds.), *Reflecting on nature. Readings in environmental ethics and philosophy* (2nd ed., pp. 244–255). New York: Oxford University Press.

Schaper-Rinkel, P. (2012). Bio-Politische Ökonomie. Zur Zukunft des Regierens von Biotechnologien. In S. Lettow (Ed.), *Bioökonomie. Die Lebenswissenschaften und die Bewirtschaftung der Körper* (pp. 155–179). Bielefeld: Transcript.

Seel, M. (1991). *Eine Ästhetik der Natur*. Frankfurt / Main: Suhrkamp.

Seel, M. (1993). Ästhetische und moralische Anerkennung der Natur. In J. Huber & A. M. Müller (Eds.), *Raum und Verfahren* (pp. 205–227). Zürich: Stroemfeld / Roter Stern.

Siep, L. (2004). *Konkrete Ethik. Grundlagen der Natur- und Kulturethik*. Frankfurt / Main: Suhrkamp.

Taylor, P. W. (1986). *Respect for nature: A theory of environmental ethics*. In M. Cohen (Ed.), Studies in Moral, Political, and Legal Philosophy. Princeton / New Jersey: Princeton University Press.

Von Braun, J. (2016). Bioökonomie – das Konzept und die Chancen für die Agrar- und Forstforschung. In Dachverband Agrarforschung e.V. (Ed.), *Bioökonomie: Welche Bedeutung haben die Agrar- und Forstwissenschaften?* Series Agrarspectrum of the Dachverband Agrarforschung (Vol. 48, pp. 7–20). Frankfurt / Main: DLG.

Wesseler, J., & von Braun, J. (2017). Measuring the bioeconomy: Economics and policies. *Annual Review of Resource Economics, 9*(1), (pp.275–298). Palo Alto, CA.

17 Conditions for an Ethically Responsible and Sustainable Bioeconomy Based on Hans Jonas' Ethics of Responsibility

Jana Franziska Schoop

Abstract

Through its treatment of the non-human environment, humanity endangers the entirety of living nature and thus the foundation of present human life and the conditions of the possibility of future human life. A potential concept to respond to this threat seems to be offered by the bioeconomy, which is often described as an ethically responsible and sustainable economic model. But what conditions must it fulfil to meet these two attributed demands? The establishment of such normative conditions implies the necessity for the specification of the terms of ethical responsibility and sustainability. This contribution concretizes the concept of responsibility on the basis of Hans Jonas' ethics of responsibility, since the current ecological situation does not solely require an economic transformation but a fundamental philosophical reflection of the relationship between human and nature. To specify the concept of sustainability, it is shown that sustainability, as an ethical principle, can be derived from Jonas' ethics. Based on these two attributions, seven conditions are deduced, which a bioeconomy has to satisfy to be considered as actually ethically responsible and sustainable. These criteria point out that the value of nature is not exhausted in its economic usability, establish the necessity to reject economic growth as the leading paradigm of a bioeconomy in favour of a sustainable and demand-orientated instead of growth-oriented treatment of nature, postulate a reflected, cautious dealing with newly developed technologies, demand for a national and international legal regulation of the bioeconomic approach to nature and suggest that it is not the only economic transformation that must be pursued, but rather a systemic change.

J. F. Schoop (✉)
German Reference Center for Ethics in the Life Sciences (DRZE), University of Bonn, Bonn, Germany
e-mail: schoop@drze.de

D. Lanzerath et al. (eds.), *Bioeconomy and Sustainability*,
https://doi.org/10.1007/978-3-030-87402-5_17

Keywords

Ethics of responsibility · Sustainability · Bioeconomy

1 Introduction

In recent decades, human interaction with nature has become a central subject of the philosophical disciplines of ethics and applied ethics. While nature was changed by man in the pre-modern age, but not in a significant long-term, profoundly endangered way, it has become the plaything of human action at the latest since industrialization and the rapid spread and development of modern technologies. Through the far-reaching use of nature and the exceeding of the carrying capacity and planetary boundaries, only made possible by highly modern technology, mankind endangers its own basis of existence and threatens to annihilate the future of its descendants through its own actions.[1] Human life and survival are impossible without any intervention in nature, because humans as living organisms are indispensably dependent on nature. However, too extensive use of nature ultimately leads to the endangerment of the future of humanity. The human-nature relationship is therefore always characterized by this ambivalence and requires fundamental philosophical reflection in order to meet the challenges that this very ambivalence involves.

This ambivalence, as well as the threat to the future existence of humanity caused by human-technological dealings with nature, which seems to become increasingly clear and prompt new considerations on how to interact with nature, was recognized, amongst others, by the philosopher Hans Jonas as early as the last century, who dedicated his work "Das Prinzip Verantwortung. Versuch einer Ethik für die technologische Zivilisation,"[2] first published in 1979, to this topic.

This work played a decisive role in determining the beginnings of the philosophical debate on the future and nature and was often interpreted as a kind of "emergency ethics," which, as a response to the challenges of technological civilization, made securing the survival of humanity a key objective. The core of Jonasian ethics is the normative justification of responsibility of the present generations towards future humanity as well as towards the entire living nature. The resulting *ecological*

[1] At this point it can be referred to the 'Earth Overshoot Day' as an example. The global Earth Overshoot Day marks the date on which the annual demand of humanity for natural resources exceeds the amount that can be reproduced by nature within 1 year. In 2020, this day already fell on August 22. According to this, humanity uses nature on average 1.6 times faster than it can regenerate naturally—thus it can be said that 1.6 Earths would be needed to cover the global consumption of resources. Cf. Global Footprint Network 2020, on further anthropogenically caused existential risks cf. Bostrom (2018, pp. 23 ff.).

[2] In the English edition, the original German title translation is "The Imperative of Responsibility. In Search of an Ethics for the Technological Age."

imperative, the *precautionary principle* and the *precedence of the bad over the good prognosis* are still present in technical and bioethical debates today.

In view of this threat to the human basis of life, the concept of sustainability was put on the political agenda in the 1960s, which aims to ensure the lasting satisfaction of the needs of future generations through sustainable action by present generations that preserves the regenerative capacity of natural systems.

However, since both the ethics advocated by Jonas and the concept of sustainability are initially merely theoretical concepts, the question remains as to how these can be operationalized in a practical and political sense to meet the challenges of the present and, in particular, of the future and to counteract the endangerment of human existence. An approach for the implementation of these demands seems to be offered by the bioeconomy: it is understood as an economic system that turns away from the use of non-regenerative raw materials and towards the use of regenerative resources in order to make products and their manufacturing processes more sustainable and thus to counter the challenges outlined above.[3] The bioeconomy is very frequently described as a "responsible and sustainable economy." However, as often as the normative terms of sustainability and responsibility are used in scientific and political publications on the bioeconomy, it is barely explicated what is to be meant by them. At no point is it made clear how responsibility and sustainability are normatively substantiated, what concrete understanding of these terms is assumed and what demands these terms or concepts contain. This leads to the following question: what conditions must the bioeconomy fulfil to be considered ethically responsible and sustainable?

In order to answer this research question and to define the concept of responsibility as well as the concept of sustainability more precisely, it will be proceeded and argued as follows: *Firstly*, to justify the duty of responsibility towards future generations and the entire living nature, the future ethics and ethics of responsibility of Hans Jonas will be subjected to an analysis in order to provide the normative foundations for the requested conditions of the bioeconomy. Thereby, it will be clarified how Jonas justifies responsibility (Sect. 2.1), what concept of responsibility he specifically advocates (Sect. 2.2) and what requirements and principles for responsible action result from his ethics (Sect. 2.3). *Secondly*, the concept of sustainability becomes the subject of the examination. Here, the origin of the term is discussed, and a distinction is made between various concepts and terms of sustainability (Sect. 3.1). It is then shown how sustainability can be derived from Jonas' ethics of responsibility not only as a resource-economic but also as an ethical principle, and which specific sustainability concept or term follows from Jonas' ethics of future responsibility (Sect. 3.2). *Thirdly*, on the basis of these considerations, the bioeconomy becomes the object of consideration. After an concise introduction to the concept, its foundations and its objectives (Sect. 4.1), seven conditions are formulated on the basis of the previous research, by which the

---

[3] Cf. Federal Ministry of Education and Research 2010, p. 4 f.

bioeconomy must be oriented in order to be considered future-oriented, ethically responsible and sustainable in the Jonasian sense taken as a basis here (Sect. 4.2).

2 Hans Jonas' Philosophy of Responsibility

In his work "Das Prinzip Verantwortung. Versuch einer Ethik für die technologische Zivilisation" Hans Jonas embeds his ethical considerations and concepts in a comprehensive analysis of modern technology and the ecological situation. He states the justification of a new ethics to be necessary, since the developments of modern technology and the resulting "veränderte Wesen menschlichen Handelns"[4] have made it possible for mankind to endanger the existence of all human and non-human life in the present and future for the first time. This, on the one hand, by means of a *direct* threat to humans and nature in the present and in the future, for instance, through the use of nuclear energy and nuclear weapons,[5] and, on the other hand, *indirectly* by means of the overstimulation of the natural earthly load limits and planetary boundaries resulting from human interaction with nature and the associated threat to the human basis of life.[6] In view of this realization of the menace to the future life, Jonas aims the development of a complementary "Notstandsethik"[7] and, in an act of overcoming the normative concept of anthropocentrism, to establish responsibility both for the future of humanity and for nature as a whole.

2.1 Substantiation of Responsibility

Jonas substantiates the duty of responsibility essentially by three main aspects. *Firstly,* with reference to the purposefulness inherent in all being, he ascribes an absolute intrinsic value to the entire animate nature and thus also to humanity, to which a claim for existence or continued existence is immanent. Through this inherent worth and the claim contained in it, the existence of all being has absolute priority over non-existence and must therefore absolutely be protected and preserved. From the worthiness of all purposeful being, Jonas derives the duty to preserve the entire living nature in the present and future.[8]

However, this constitutes only one partial aspect and thus only a *necessary* condition of Jonas' justification of the duty of responsibility for the entire animate nature in the present and future. In order that the duty of responsibility represented

[4]Jonas (2017a, p. 13), "changed character of human action" (own translation).

[5]Cf. Werner (2016, p. 41), Jonas (1994a, p. 140), Jonas (2016, p. 173), Jonas (2017a, pp. 80, 34, 7 f., 247).

[6]Cf. Jonas (1993, p. 85), Jonas (1994a, p. 140) and Jonas (2017a, pp. 251 f.).

[7]Jonas (2017a, p. 250), "emergency ethics" (own translation).

[8]Cf. Jonas (2017a, pp. 100, 105 f., 142 f., 153–171, especially pp. 153–157); Müller (1988, pp. 35–39); Buddeberg (2011, pp. 61–71) and Werner (2016, pp. 46–49).

by Jonas can be *sufficiently* justified, two further conditions must be added to the condition of the absolute and intrinsic value of all being, which each refer to the ability of the implementation of the claim for existence or continued existence that lies in all being, to be able to satisfy the claim to preservation.

Thus Jonas substantiates the duty of responsibility *secondly* through the specific abilities of the sensation of responsibility and the takeover of responsibility, which characterize only the human being. According to Jonas, the ethical ability of responsibility arises from the specific human, ontological ability of freedom: "Der Mensch ist das einzige uns bekannte Wesen, was Verantwortung haben kann. Indem er sie haben *kann*, hat er sie."[9]

The *third* and last aspect of justification also concerns the possibility of implementing the absolute intrinsic value and the claim for existence and preservation of all being in the present and the future contained therein and consists in the extent of human possibilities of action, the dimension of power. In order for a person to be able to meet the claim for the existence of an entity, this being must be located within its scope of action and therefore succumb to its power. The greater the scope of action and thus a person's power over things in the world, the greater the extent of the things which, with their claim for existence and continued existence, have an entitlement on him to exert his power or to refrain from exercising it. Accordingly, the greater the power of human action, the higher and more far-reaching the moral demands on action become. A person's responsibility, according to Jonas, is proportional to the extent of his power: "meine Kontrolle dar*über* [impliziert] zugleich meine Verpflichtung da*für*."[10] Since modern technology, which is characterized by both a global and a temporal expansion of the consequences of action and technology,[11] gives man an extremely extensive amount of power, he is responsible for everything that is affected by his power to act. Since, through precisely these human-technological actions, the entire terrestrial biosphere of the present, but because of the remote effects and long-term consequences of modern technology even more so of the future, becomes the object of the human power of action and is thus influenced and endangered, man is responsible for this very thing.[12]

These three aspects of the substantiation of responsibility put forward by Jonas can now, taken together, be regarded as *sufficient* conditions for justifying the duty of responsibility of present generations towards the entire animate nature and thus also towards humanity in the present and in the future. Jonas also calls this duty of

---

[9] Jonas (2016, p. 165), "Man is the only being known to us who can have responsibility. By being *able* to have it, he has it" (own translation). Cf. also Jonas (1994a, p. 129), Jonas (2016, pp. 165, 171, 184), Jonas (2017a, pp. 157, 162–171, 185, 246 ff.) and Buddeberg (2011, pp. 72–75).

[10] Jonas (2017a, p. 176), "my control *over* it [implies] at the same time my obligation *for* it" (own translation).

[11] Cf. Jonas (1993, p. 85) and Jonas (2017a, pp. 72 f., 55).

[12] Cf. Jonas and Löw (1990, pp. 21, 30 ff.), Jonas (1994a, p. 133), Jonas (2016, pp. 166 f.) and Jonas (2017a, pp. 174 ff., 248).

responsibility the "Pflicht zur Zukunft,"[13] and specifically related to the preservation of humanity, he titles it as the duty "daß eine Menschheit sei."[14]

2.2 Concept of Responsibility

Now that Jonas' strategy of justifying responsibility has been outlined, a closer definition of the concept and term of responsibility underlying his ethics will be given in the following.

The concept of responsibility, which originates from Roman jurisprudence, is generally closely related to the terms "accountability" and "justification."[15] Thus, according to a common understanding, responsibility can be understood as a four-digit relational concept—as a duty which is, on the basis of a *normative foundation*, imposed on a *subject of responsibility* for an *object of responsibility*, which in turn can be demanded by an *instance of responsibility*.[16] Jonas also always understands responsibility in the context of accountability and justification of a person's actions[17] and bases his work "Das Prinzip Verantwortung" and the future ethics and ethics of responsibility developed therein on a four-digit relational concept of responsibility, whose elements will now be explicated in more detail.

The *first relatum*, the normative foundation, which indicates why the subject of responsibility must assume responsibility for the object of responsibility and justify itself to the instance of responsibility, has already been explained: the intrinsic value of all being, the human capability for the feeling and the takeover of responsibility as well as the extent of the human power of action constitute the normative basis of responsibility and thus the first relatum of Jonas' concept of responsibility.

From this first relatum, the substantiation of responsibility, it can now be seen what the *second relatum*, the subject of responsibility, consists of, because according to Jonas' premise that only man has the ability to assume responsibility, only man can be precisely this searched subject. However, Jonas concretizes this initially rather vague characterization of the subject of responsibility by identifying both individuals[18] and collectives[19] as well as the level of politics[20] as the subject of responsibility. He thus argues for individual, collective and political responsibility that is simultaneous but located at different levels.

---

[13] Jonas (2017a, p. 84), "duty to the future" (own translation).

[14] Ibid., p. 90, "that humanity exists" (own translation).

[15] Cf. Kreß (2017, p. 646).

[16] Cf. Ropohl (1993, pp. 154 ff.); Buddeberg (2011, p. 3).

[17] Cf. Jonas (2016, pp. 165 f.) and Jonas (2017a, p. 55).

[18] Cf. Jonas and Gebhardt (1994, p. 208), Jonas (1994a, p. 142), Jonas (2016, pp. 175, 181) and Jonas (2017b, p. 298).

[19] Cf. Jonas (1993), Jonas (1994b, p. 25), Jonas (2017a, pp. 26, 32), Jonas (2017b, pp. 58, 32, 275), Jonas (2017c, p. 44) and Claas (2011, p. 47).

[20] Cf. Jonas and Gebhardt (1994, p. 207) and Jonas (2017a, pp. 37, 197, 221, 263).

The *third relatum*, the object of responsibility, for which the subject is responsible on the basis of valid normative claims, can also be identified from the previously explained first relatum and the Jonasian substantiation of responsibility: "Wofür ich verantwortlich bin, sind natürlich die Folgen meines Tuns – in dem Maße, wie sie ein Sein affizieren. Also ist wirklicher Gegenstand meiner Verantwortung dies von mir affizierte *Sein* selber."[21] The object of responsibility is, therefore, that entity to which the actions of the subject of responsibility, including the consequences of those actions, relate and affect. Although human actions can also affect only one or several entities, due to the enormous spatial and temporal expansion of the remote effects and late consequences of modern technology, human actions and technology applications can also extend to the entire terrestrial biosphere in the present and future. Thus, the present and future existence of the entire terrestrial biosphere becomes the object of responsibility, which includes the entire animate nature and thus also humanity.[22]

According to Jonas, the *fourth relatum*, the instance of responsibility, is not formed by religion, a God, the divine or earthly court of justice or, for example one's own conscience, but lies outside the subject of responsibility and in being itself, in the object of responsibility. Through the absolute intrinsic value of all being, the object of responsibility has, as it were, a claim to demand accountability and justification from the subject of responsibility. But while the object of responsibility is not necessarily the entire being, the entire earthly biosphere, but can also refer to a specific being and merely individual entities, the instance of responsibility is the "Sein des Ganzen in seiner Integrität."[23] Jonas thus shifts, in a way, the instance of responsibility into the object of responsibility[24]:

> [D]ie Werthaltigkeit des Seins im Ganzen [. . .], letztlich dies Ganze [erscheint] als dasjenige nicht nur, *für* das ich jeweils partikular mit meinem Tun verantwortlich *werde*, sondern auch als das, *wovor* ich immer schon mit all meinem Tunkönnen verantwortlich *bin* – weil sein *Wert* ein Recht auf mich hat.[25]

---

[21] Jonas (2016, p. 165), "What I am responsible for, of course, are the consequences of my actions – to the extent that they affect a being. Therefore, the real object of my responsibility is this *being itself* that is affected by me" (own translation).

[22] Cf. ibid., pp. 165 f.; Jonas (2017a, pp. 26 f., 198, 248) and Buddeberg (2011, pp. 57 f.).

[23] Jonas (2016, p. 166), "being of the whole in its integrity" (own translation).

[24] Cf. Jonas (2016, pp. 165 f.), Jonas (2017a, pp. 94 f., 98 f.), Wetz (1994, p. 134), Buddeberg (2011, pp. 58 ff.) and Werner (2016, p. 49).

[25] Jonas (2016, p. 166), "The intrinsic value of being as a whole [...], ultimately this whole [appears] as that for which I *become* individually responsible with my actions, but also as that for which I have always *been* responsible with all my actions – because its *value* has a *right* to me" (own translation).

2.3 Requirements and Principles of Responsibility

How can the duty of responsibility founded by Jonas, the duty of generations living today, not to endanger the present and especially the future existence of mankind and nature be fulfilled? What exactly does this duty involve, or: how can we act responsibly, according to Jonas? In the following, it will be shown which concrete requirements and principles arise from his ethics.

He bundles the central demands of his ethics of responsibility in his categorical imperative, which functions as a "praktische *Verpflichtung* gegenüber der Nachwelt einer entfernten Zukunft und als Prinzip der Entscheidung in gegenwärtiger Aktion"[26]:

> ‚Handle so, daß die Wirkungen deiner Handlung verträglich sind mit der Permanenz echten menschlichen Lebens auf Erden'; oder negativ ausgedrückt: ‚Handle so, daß die Wirkungen deiner Handlung nicht zerstörerisch sind für die künftige Möglichkeit solchen Lebens'; oder einfach: ‚Gefährde nicht die Bedingungen für den indefiniten Fortbestand der Menschheit auf Erden'; oder, wieder positiv gewendet: ‚Schließe in deine gegenwärtige Wahl die zukünftige Integrität des Menschen als Mit-Gegenstand deines Wollens ein'.[27]

Accordingly, the imperative demands a comparison of one's own actions with their effects on the future, real, i.e. genuine or dignified human existence that must be secured. Although Jonas is primarily concerned with the "Überleben der Menschheit als Gattung",[28] he also wants the "humane Leben im sozial-kulturellen Kontext"[29] to be preserved.[30] Even though Jonas aligns the formulation of the imperative strongly towards safeguarding the future of humanity, the preservation of nature and all non-human existence is "*sine-qua-non* offenkundig mitenthalten."[31] This on the one hand for the reason that all being has an absolute intrinsic value and must therefore be preserved, but on the other hand also because of the indissoluble interweaving of man with non-human nature.[32] The latter refers to the following:

The imperative includes the *precautionary principle*, which can be seen, inter alia, from the use of the terms "possibility" and "conditions." In order not to jeopardize the possibility of a genuine, dignified human existence in the future, the

---

[26] Jonas (2017a, pp. 33 f.), "practical *obligation* to posterity of a distant future and as a principle of decision in present action" (own translation).

[27] Ibid., 36, "'Act so that the effects of your action are compatible with the permanence of genuine human life on earth'; or negatively expressed: 'Act so that the effects of your action are not destructive to the future possibility of such life'; or simply: 'Do not endanger the conditions for the indefinite survival of humanity on earth'; or, again positively expressed: 'Include the future integrity of man in your present choice as a co-object of your will'" (own translation).

[28] Schmidt (2013, p. 178), "survival of humanity as a species" (own translation).

[29] Ibid., "humane life in a socio-cultural context" (own translation).

[30] Cf. Jonas (1994a, p. 138), Jonas (2016, pp. 171 f.), Jonas (2017a, pp. 88–91, 186 f., 250) and Claas (2011, p. 54).

[31] Jonas (2017a, p. 245), "*sine-qua-non* obviously included in it" (own translation).

[32] Cf. Jonas (2017a, pp. 245 ff.), Claas (2011, p. 53) and Schmidt (2013, p. 179).

necessary conditions for it must be provided and secured "und das heißt unter anderem, diese physische Welt so zu erhalten, daß die Bedingungen für ein solches Vorhandensein intakt bleiben."[33] The precautionary principle requires that these preconditions for future existence are protected and preserved: *precaution* must be made for the biological basis of life and the conditions of existence which are essential and indispensable for human existence, as well as for the prerequisites for satisfying basic needs.[34]

According to Jonas, the comparison of consequences, long-term effects and remote effects of one's own actions or technical applications to future genuine human and non-human existence as well as the indispensable prerequisites for this, which the Jonasian imperative demands, requires the introduction and establishment of a new "Wissenschaft hypothetischer Voraussicht, eine[r] ‚vergleichende [n] Futurologie‘,"[35] which should also be integrated into political practice. The aim and purpose of this discipline, which can also be described as "*Technikfolgenabschätzung*," technology assessment, is to calculate and process the various possible consequences, the long-term impact and remote effects of already implemented or planned technology applications[36] and serves as a "Tatsachenwissenschaft von den Fernwirkungen technischer Aktion."[37]

But this futurology is faced with a problem: due to the complex world-interconnection and cause-effect relationships, a total, overall knowledge of the effects of actions and technical applications is usually hardly possible, so that predictions and forecasts can only be made with a certain degree of probability. But then how is a comparison between actions and applications of technology on the one hand and their consequences and effects on the existence of future life and the prerequisites necessary for this on the other hand possible, as the imperative postulates, if the consequences of actions and long-term effects cannot be reliably predicted? As a solution to this problem and to bridge the "Kluft zwischen Kraft des Vorherwissens und Macht des Tuns,"[38] Jonas devises the risk-averse decision rule or guiding principle "*in dubio pro malo*:"[39] if a prognosis regarding the manner, extent or probability of occurrence of an action or application of technology is tinged with uncertainty, the prophecy of doom or calamity is to be given priority over the prophecy of salvation, so as not to endanger the future human and non-human existence and the conditions necessary for it, which must be absolutely preserved.

---

[33] Jonas (2017a, p. 34), "and that means, among other things, preserving this physical world in such a way that the conditions for its existence remain intact" (own translation).

[34] Cf. ibid., pp. 186, 250, Birnbacher and Schicha (2001, pp. 24 ff.), Gottwald and Krätzer (2014, p. 130).

[35] Jonas (2017a, p. 62 f.), "science of hypothetical foresight, a 'comparativen futorology'" (own translation).

[36] Cf. Jonas (1994a, pp. 134 f.), Jonas (2016, p. 168), Jonas (2017a, pp. 62–69) and Werner (2016, p. 45).

[37] Jonas (2017a, p. 62), "factual science of the remote effects of technical action" (own translation).

[38] Ibid., p. 28, "gap between the power of foreknowledge and the force of action" (own translation).

[39] Jonas (2016, p. 175).

If the application of a technique is suspected of endangering the future of humanity and nature or, even worse, making it entirely impossible, then, according to Jonas, caution is called for, and the bad prognosis is to be given precedence over the good one.[40]

This reflects the *principle of caution*, sometimes also referred to as the *risk principle*. Thus Jonas speaks several times of caution, deliberation and restraint[41] and emphasizes that "Vorsicht [. . .] [ist] ein Gebot der Verantwortung."[42] In accordance with Jonas, caution must be exerted in dealing with high-risk technologies and techniques in order not to endanger the absolutely morally required possibility of the existence of humanity and nature in the present and future, as well as the indispensable conditions for this in any way. Thus is "Vorsicht zur höheren Tugend geworden, wohinter der Wert des Wagens zurücktritt [. . .], denn die Einsätze sind zu groß geworden für das Spiel."[43]

3 Sustainability

Just like the concept of responsibility, the concept of sustainability also requires a more detailed definition and a filling of content so that it cannot be subjected to any suspicion or reproach of empty formulas and is not accused of inflationary use on the one hand,[44] and so that conditions for an ethically responsible and sustainable bioeconomy can be specified on the other. Therefore, a condensed overview of some theoretical foundations of the concept of sustainability will be given at first.

---

[40] Cf. Jonas (2016, pp. 174 f.) and Jonas (2017a, pp. 28, 66–72, 76). This decision-making maxim raised by Jonas was and is often criticized as being too regressive, developmentally inhibiting and risky against the backdrop of the danger of inaction and renunciation, especially with regard to the ambivalence of technology concerning good and bad consequences and the possible necessity of new technologies to solve existing problems. Cf. for instance, Hirsch Hadorn (2000, pp. 233 f.); Werner (2016, p. 45); more generally Sunstein (2005) as well as Pinsdorf and Wesseler and Kardung (2022) in this volume. However, it should be added that Jonas is quite aware of this problematic of the peculiarities of technology, such as the ambivalence of its consequences, and the resulting fact that "die Befolgung einer gewonnenen ethischen Einsicht [. . .] selbst wieder zum Problem [werden kann]" (Jonas, 2017f, p. 13). That "following a gained ethical insight [...] itself [can] become a problem again" (own translation). For this, for instance, cf. Jonas (2017a, p. 323, 2017c, 2017e). If this principle of action is indeed judged to be of the aforementioned critical characteristics, then this rule must be effectively expanded and complemented by a new maxim for action which implies the duties and principles resulting from Jonas' ethics, but captures and takes the risks of renouncing the use of technology into account.

[41] Cf. Jonas (2017a, pp. 55, 71, 82, 338) and Schmidt (2013, p. 177).

[42] Jonas (2017a, p. 338), "caution [...] [is] an imperative of responsibility" (own translation).

[43] Jonas (2017d, p. 67), "caution has become a higher virtue, behind which the value of risk recedes [...], because the stakes have become too great for the game" (own translation).

[44] Cf. Birnbacher and Schicha (2001, p. 25) and Ostheimer (2013, pp. 397 ff.).

3.1 Theoretical Foundations

The origin of the concept of sustainability can be traced back to eighteenth-century forestry. In 1713, Hans Carl von Carlowitz demanded that the use of forest resources should not affect the possibility of a sustainable yield—no more trees should be felled than can regenerate naturally. According to this origin, sustainability is to be understood as a resource-economic principle, which should ensure the constant use of a resource and thus the permanent economic yield.[45] However, the concept of sustainability has changed and expanded significantly since its emergence in the eighteenth century.[46] The definition of sustainability, which is still the most widely used and accepted today, is taken from the "Brundtland Report" of 1987, whose official name is "Our Common Future" report. It states: "Dauerhafte Entwicklung ist Entwicklung, die die Bedürfnisse der Gegenwart befriedigt, ohne zu riskieren, daß künftige Generationen ihre eigenen Bedürfnisse nicht befriedigen können."[47] This definition already shows the clear future-relatedness of sustainability – responsibility for the future is even considered as the fundamental premise of sustainability as "einem Weg, der uns Menschen eine dauerhafte Existenz auf diesem Planeten eröffnet."[48] These normative foundations imply that, according to today's understanding of the term, sustainability can no longer be regarded as a mere resource-economic principle but as a normative principle.[49] However, in order to be able to develop the term and the concept of sustainability in more detail, some differentiations must be contemplated:

Firstly, sustainability as an interdisciplinary concept must always be understood in its multidimensionality. Sustainability in the sense of the Brundtland definition, which calls for ensuring the possibility of satisfying the needs of future generations in favour of the possibility of the permanent existence of humanity, is only possible if sustainability is striven for in these three interdependent dimensions, which constitute the so-called "three-dimensional model:" in the ecological, the economic and the social dimensions of sustainability, to which ecological, economic and social

[45] Cf. Hamberger (2013, pp. 431–434) and Pufé (2017, pp. 37 ff.).

[46] Detailed and systematic overviews of the stations, summits and agreements that mark this change can be found in Jischa (2005, pp. 146–151) and in Pufé (2017, pp. 36–65) for instance.

[47] Hauff (1987, p. 46), "Sustainable development is development that meets the needs of the present without compromising the ability of future generations to meet their own needs" (own translation). On a similar definition, see Pufé (2017, p. 42). For the English term "sustainable development" there are many different translation variants in German. In addition to "nachhaltige Entwicklung" and "zukunftsfähige Entwicklung" the term "dauerhafte Entwicklung" is popular, as can be seen from the definition in the German translation of the Brundtland Report. "Nachhaltigkeit" and "Dauerhaftigkeit" can be understood as synonyms in this context.

[48] Ekardt (2014, p. 28), "a path that opens up a permanent existence on this planet for us humans" (own translation). The similarity of the wording as well as the closeness of content to Jonas' imperative can be minded.

[49] Cf. Hauff (1987, p. 46), Grunwald and Kopfmüller (2006, pp. 27, 55), Hamberger (2013, p. 427), Ostheimer (2013, p. 400), Ekardt (2014, pp. 26, 28) and Pufé (2017, p. 116).

capital are assigned. The aim of each dimension is to preserve or increase its own form of capital.[50]

Secondly, these three dimensions are weighted differently depending on how sustainability is normatively substantiated. Either all dimensions are given the same weight, or they are weighted unequally, so that, for instance, one dimension is considered to have priority over the other two in order to ensure sustainability according to the aforementioned definition.[51]

Thirdly, the weighting of the dimensions is also reflected in the strength of the sustainability concept represented in each case. Thus, a distinction must be made between a weak, a strong and a balanced conception of sustainability. These positions differ in their views on the question of whether substitution, i.e. the replacement of the various types of capital between one another, is permitted or not. While representatives of the weak sustainability concept consider substitution to be permissible, since they only demand the preservation of the total amount of existing capital, advocates of the strong concept of sustainability strictly reject the substitution of goods of different forms of capital among themselves, since the preservation of each individual form of capital has to be guaranteed. In term, the middle position of balanced sustainability considers substitution to be permissible if each form of capital is not endangered by falling below critical limits.[52]

The *fourth* and last differentiation concerns the way in which the respective sustainability concept is implemented. The substantive understanding of sustainability demands a political-institutional implementation "from above," along with defined action strategies and rules for achieving sustainability goals. The procedural understanding of sustainability, however, emphasizes the relevance of individual participation as well as the engagement of civil society and calls for the implementation of sustainability strategies "from below" at individual and collective levels.[53]

In order to ensure that the concept of sustainability is not used as an inflationary empty formula in any context, as for instance, Dieter Birnbacher, Christian Schicha and Jochen Ostheimer note,[54] it is necessary to specify, on the one hand, how sustainability as a normative concept can be normatively substantiated and, on the other hand, which particular concept of sustainability follows from the specific normative justification. This need for a normative justification and substantive definition will now be met on the basis of Hans Jonas' ethics of responsibility.

[50]Cf. Pufé (2017, pp. 96 f., 99 f., 101 ff., 110–115) and Von Hauff et al. (2018, pp. 20, 22 ff.).

[51]Cf. Ekardt (2013, p. 188), Pufé (2017, pp. 97 f.) and Von Hauff et al. (2018, p. 24).

[52]Cf. Grunwald and Kopfmüller (2006, pp. 37–43), Pufé (2017, pp. 105–110), Von Hauff et al. (2018, p. 21). A detailed analysis of the different positions and their argumentative foundations can be found in Klaus Mathis' habilitation thesis, cf. Mathis (2017, pp. 166–185).

[53]Cf. Grunwald and Kopfmüller (2006, pp. 40 ff.).

[54]Cf. Birnbacher and Schicha (2001, p. 25) and Ostheimer (2013, pp. 397 ff.).

3.2 Sustainability as an Ethical Principle

If one takes another look at the Jonasian imperative with its demands and principles and compare this with the Brundtland Report's definition of sustainability, the equality of content of these demands becomes clear. They both identify the future as the dimension for which present generations are responsible; they both demand a comparison of one's own actions in terms of their impact on the possibility of a good life for future generations and the necessary precondition of the satisfaction of their needs. Both the demands of Jonas' ethics of responsibility and sustainability are aimed at "die langfristige Sicherung und Weiterentwicklung der Grundlagen der menschlichen Zivilisation angesichts der begrenzten Belastbarkeit der natürlichen Umwelt und ökonomischer und sozialer Zukunftsrisiken".[55] The Jonasian *precautionary principle* already resonates in the definition of sustainability referred to here.[56] Furthermore, his *principle of caution* has also found its way into the sustainability debate.[57] Since Jonas' imperative follows from the normative justification structures of his ethics and the principles contained therein, and since the demands of the imperative can be regarded as congruent with the demands of the standard definition of sustainability, the demands of sustainability also follow from Jonas' ethics. Thus, the concept of sustainability can be derived from Jonas' ethics of responsibility for the future and, due to its normative justification, can be understood as a principle that is by no means merely resource-economic, but rather a thoroughly ethical one.[58]

Now it remains to be clarified which specific understanding of sustainability follows from the Jonasian ethics, because depending on the respective normative substantiation of sustainability, the previously explicated differentiations are filled out differently in terms of content.[59]

According to Jonas, *firstly*, the ecological dimension and thus ecological sustainability must be prioritized and given greater weight than economic and social sustainability. In his ethics, Jonas demands the preservation of future, genuine human existence and of the indispensable preconditions for it, which, in accordance

---

[55] Grunwald and Kopfmüller (2006, p. 27), "the long-term safeguarding and further development of the foundations of human civilization in view of the limited resilience and capacity of the natural environment and economic and social future risks" (own translation).

[56] Volker Hauff, for instance, the German member of the Brundtland Commission on Environment and Development, from which the definition used here is based, emphasizes that the entire Commission was "maßgeblich von Jonas inspiriert" (Schmidt, 2013, p. 180), "significantly inspired by Jonas" (own translation).

[57] In the United Nations Rio Declaration on Environment and Development of 1992, the precautionary principle is recognized as a guiding principle and legally manifested in a separate principle (Principle 15). Cf. United Nations Conference on Environment and Development (1992), Pufé (2017, pp. 48 f.) and Huber (2018, p. 414).

[58] Jan C. Schmidt and Susanne Moser also underline this close connection between Jonas' ethics and sustainability. See Schmidt (2013, pp. 180 f.) and Moser (2016, p. 72).

[59] Cf. Ekardt (2013, p. 188).

with the *precautionary principle*, necessarily includes the preservation of the integrity of nature, since this forms the inalienable basis of human life. While a threat to this basis of life, i.e. to ecological capital, also means a threat to future human existence, a threat to economic or social capital would not have such an impact on the existence of humanity in the future, which must be imperatively preserved. Since future existence is endangered by current human-technological action and the way we treat nature, in that ecosystem functions and ecosystem services and thus the human basis of existence is massively impaired, emphasis must be placed on the preservation and protection of this very ecological dimension, whose goods are jeopardized by current action. In addition, the economic and social dimensions of sustainability are in a state of dependency on the ecological dimension, since where there is no intact nature to provide the basis for future existence, there will incontestably be no possibility of a profitable economy and no just society in the future.

Secondly, Jonas' ethics leads to a *strong ecological* sustainability concept in that no ecological capital must be substituted by economic or social capital. Since the possibility of a real existence in the future requires an intact nature with intact ecosystem functions and services as the basis of life, ecological capital must be absolutely protected. If ecological goods were substituted by the other two forms of capital, this would, from a certain degree of substitution, affect the integrity or intactness of nature as the basis of human existence and thus the possibility of a genuine human existence in the future. A further reason that can be put forward against the substitution of ecological capital, according to Jonas, is the absolute intrinsic value of nature and all its entities and components, to which a claim for existence or continued existence is immanent and which therefore cannot simply be substituted by a good of one of the other two forms of capital.

Since Jonas argues in his ethics for parallel individual, collective and political responsibility located at different levels, his ethics *thirdly* leads to an understanding of sustainability, which requires both substantive and procedural implementation. Accordingly, sustainability must be addressed and legally manifested by politics "from above" in accordance with political-institutional responsibility, but it must also be carried by the civil society and individual commitment, including changes in lifestyles, consumption and behaviour, in the sense of individual and collective responsibility "from below."[60]

In summary, it can be said that Jonas' ethics thus gives rise to a strong understanding of ecological sustainability that prioritizes the ecological dimension and simultaneously requires both substantive and procedural implementation.

[60] Cf. Jonas (2016, pp. 175 ff.) and Jonas (2017d, pp. 67–70).

4 Bioeconomy as a Responsible and Sustainable Economic Model?

But since both Hans Jonas' ethics of responsibility and the ethical principle of sustainability arising from it are initially merely theoretical concepts dedicated to safeguarding future human and non-human existence and the preconditions necessary for this, the question remains as to how these can be operationalized in a practical-political manner to counter the threat to future human existence and nature effectively. An approach to the implementation of these demands seems to be offered by the bioeconomy. But can it really be considered an ethically responsible and sustainable economic model, as which it is frequently described? In order to investigate which conditions or criteria the bioeconomy must fulfil in order to comply with these attributes and to meet the demands of responsibility and sustainability in the Jonasian sense represented here, the concept and conception of the bioeconomy must first be examined more closely.

4.1 Basics, Objectives and Definition of the Bioeconomy

The German Federal Ministry of Education and Research and the Federal Ministry of Food and Agriculture define bioeconomy as follows: "Die Bioökonomie ist die wissensbasierte Erzeugung und Nutzung nachwachsender Ressourcen, um Produkte, Verfahren und Dienstleistungen in allen wirtschaftlichen Sektoren im Rahmen eines zukunftsfähigen Wirtschaftssystems bereitzustellen."[61] The bioeconomy as an economic model is intended to provide the possibility of counteracting the global challenges of the future, also in view of the increasing world population growth, by ensuring the security of health and nutrition, the supply of water, energy and other elementary resources, as well as contributing to climate and nature protection.

This is to be enabled by a transformation of the economy away from the use of fossil, non-renewable raw materials towards the use of renewable resources. This transformation emerges from an awareness of the finite nature of non-renewable resources such as, for instance, petroleum, which currently serves as a production base in numerous fields of application.[62] Thus, the Federal Ministry of Education and Research emphasizes that "[m]it Blick auf knappe fossile Rohstoffe, Klimawandel und wachsende Weltbevölkerung nachhaltige und ressourceneffiziente Strategien gefragt [sind], um langfristig den Wohlstand moderner Gesellschaften zu garantieren.

[61] Federal Ministry of Education and Research and Federal Ministry of Food and Agriculture (2014, pp. 3 f.), "The bioeconomy is the knowledge-based production and use of renewable resources to provide products, processes and services in all economic sectors within the framework of a sustainable economic system" (own translation).

[62] Cf. Federal Ministry of Education and Research (2010, pp. 2–6) and Federal Ministry of Education and Research (2017, p. 4).

Die Bioökonomie bietet die Chance, Wirtschaftswachstum im Einklang mit Natur- und Umweltschutz zu erreichen."[63] The required transformation of the economy calls for the production and use of biological resources such as microorganisms, plants and animals to be of the highest efficiency.[64] This, in turn, necessitates extensive research and expansion of knowledge about nature, its organisms and cycles in order to make the best possible use of them. This knowledge should then serve to initiate the economic transformation of the bioeconomy through the development and application of innovative technologies and techniques. Within the bioeconomy and the exploration of its possibilities, there is a collaboration of many different scientific disciplines such as biology, economics, agricultural sciences, biotechnology and engineering, to give just a glimpse of the interdisciplinarity. As broad as the bioeconomy is in its disciplinary exploration, so are its impacts, which reach "in alle [. . .] wirtschaftlichen Sektoren."[65] The areas of application of the bioeconomy range from pharmaceuticals and the automotive industry to the energy sector and everyday consumer goods such as product packaging. Thus, Anita Krätzer and Franz-Theo Gottwald elaborate that there are essentially six technological fields of application inherent in the bioeconomy – these are biotechnology including genetic engineering, synthetic biology, precision farming, the use of biomass for energy and the production of bioplastics as well as nutrigenomics.[66]

Essential to the understanding of the bioeconomy as a knowledge-based and bio-based economic approach is the attested, fundamental claim of linking economic growth, including competitiveness and increased efficiency, with sustainability and nature conservation, so that economy and ecology do not confront each other as opposing antagonists, but rather merge harmoniously in an economic concept. Thus, the Federal Ministry of Education and Research characterizes bioeconomy as a "nachhaltige Wirtschaftsweise" that "verantwortungsvollen Umgang [. . .] mit natürlichen Ressourcen ermöglicht" and thus meets the "zentrale [n] Herausforderung[en] der Zukunft".[67] This paradigm of sustainability should be reflected in the entire bioeconomic value chain, from the participants in research to production and to the consumers, and thus requires the involvement of actors from science, business and society.[68] This suggests that the proponents of the bioeconomy

[63] Federal Ministry of Education and Research and Federal Ministry of Food and Agriculture (2014, p. 2), "in view of scarce fossil resources, climate change and a growing world population, sustainable and resource-efficient strategies are needed to guarantee the long-term prosperity of modern societies. The bioeconomy offers the opportunity to achieve economic growth in line with the protection of nature and the environment" (own translation).

[64] Cf. Gottwald and Krätzer (2014, pp. 23 f.).

[65] Federal Ministry of Education and Research and Federal Ministry of Food and Agriculture (2014, pp. 3 f.), "in all [. . .] economic sectors" (own translation).

[66] Cf. Gottwald and Krätzer (2014, pp. 27–41).

[67] Federal Ministry of Education and Research (2017, pp. 11, 5, 27), "sustainable economy" "enables the responsible use [. . .] of natural resources," "central challenge[s] of the future" (own translation).

[68] Cf. Federal Ministry of Education and Research (2017, pp. 12, 19, 26).

are striving to promote and demand the pursuit of sustainability within the economic, ecological and social dimensions to achieve sustainability. Sustainability is therefore understood in accordance with the "three-dimensional model." This is also reflected in the Federal Ministry of Education and Research's statement that "[f]ür eine lebenswerte Zukunft [. . . .] Innovationen notwendig [sind], die Umweltaspekte und wirtschaftliche sowie gesellschaftliche Belange *gleichermaßen* berücksichtigen."[69] Here it also seems to become clear that, in accordance with the aforementioned third differentiation in the context of sustainability, all three dimensions are weighted equally, and no priority is given to one over the others.

In addition to sustainability as the quintessential basis of the bioeconomy, the fundamental orientation towards the future and the emphasis on responsibility for the future can also be regarded as constitutive of this new economic concept.[70] Thus, the Federal Ministry of Education and Research emphasizes that bioeconomy makes it possible to meet the "Verpflichtung gegenüber internationalen Partnern und nachfolgenden Generationen".[71]

However, as often as the normative terms "sustainability" and "responsibility" are employed in scientific or political publications on the bioeconomy, little is explained about what is meant by them. At no point is it made clear how responsibility and sustainability are normatively justified, what concrete understanding of these terms is assumed and what demands these terms or concepts contain. For this reason, the reproach of empty formulas, which has already been raised against the concept of sustainability, can also be drawn here.[72] In order to avoid the accusation that the terms "sustainability" and "responsibility for the future" are strategically used in the context of the bioeconomy to disguise or conceal primarily economic interests and to legitimize them through the use of these two normatively shaped, positively charged terms,[73] it must be clarified how responsibility and sustainability are normatively substantiated, how these terms are filled with content depending on their justification, and which demands arise from them. Only then it can become apparent how the bioeconomy must be constituted to meet the demands of ethical responsibility and sustainability that are attributed to it.

[69] Federal Ministry of Education and Research; Research for Sustainability – FONA, emphasis added, "for a livable future [. . .] innovations [are] necessary that take into account environmental aspects as well as economic and social concerns *in equal measure*" (own translation).

[70] Cf. Federal Ministry of Education and Research (2010, pp. 3 ff., 14, 16, 19, 23, 42 ff.); Federal Ministry of Education and Research and Federal Ministry of Food and Agriculture (2014, pp. 5 f.), Gottwald and Krätzer (2014, 23) and Federal Ministry of Education and Research (2017, pp. 3, 8, 10).

[71] Federal Ministry of Education and Research (2010, p. 2), "obligation to international partners and future generations" (own translation).

[72] Cf. chap. 3.1 of this contribution as well as Gottwald and Krätzer (2014, pp. 21, 23 f.) and Grefe (2016, pp. 12 f.).

[73] Franz-Theo Gottwald and Anita Krätzer, amongst others, raise this objection of concealment, see Gottwald and Krätzer (2014, p. 27).

The previous specification and substantive concretization of these two concepts on the basis of Hans Jonas' ethics of responsibility for the future and the ethical principle of sustainability derived from it can now be applied to the subject of the bioeconomy.

4.2 Conditions for an Ethically Responsible and Sustainable Bioeconomy

Like any concept which, by virtue of its power and range, is of enormous spatial and temporal scope and has an impact on the future, the bioeconomy should be guided by the duty of responsibility towards the entire living nature in the present and in the future, the precautionary principle, the principle of caution or risk principle implied therein, and the ethical principle of sustainability that follows from this. However, in order to prevent moral overtaxation caused by these extensive duties and demands and the resulting attitude of resignation, these far-reaching obligations, which initially may seem abstract, must be concretized and transferred to the specific bioeconomic approach to nature. From these quite comprehensive and general normative claims of the Jonasian ethics, the following seven, more precise conditions result in the context of the bioeconomy, which must be met in order to be able to actually be considered as a responsible and sustainable economy, as attributed by its advocates, in the underlying Jonasian sense:

4.2.1 Condition 1: Rejection of Economic Growth as the Guiding Paradigm of the Bioeconomy in Favour of a Strong Environmental Sustainability

If bioeconomy, according to the two different interpretations of the term, is not understood as the ecologization of economy, but as the complete economization of nature[74] (which seems to be suggested by the goal of the best possible, most efficient use of biological resources) and if, as a result, the carrying capacity and planetary boundaries are further exhausted, this endangers the intactness of nature as a necessary basis of human existence and thus not only the future *genuine* and *dignified* existence but the existence in general. According to Jonas, this must be imperatively prevented. In accordance with the *precautionary principle*, the biological basis of life must be protected unconditionally in order not to endanger the possibility of future existence. However, since the economic growth aimed at by the bioeconomy, including increased efficiency and competitiveness, is based on the consistent use of all biological resources for the creation of economic profit, but since this use of nature endangers future existence, economic growth must not be regarded as the primary paradigm of a bioeconomy that wants to be perceived as

[74] For this terminological differentiation, which also has an impact on the substantive definition as well as the conceptualization and practical implementation of the bioeconomy, cf. for instance Gottwald and Krätzer (2014, p. 12); Vogt (2016, pp. 3 f.) and Pinsdorf (2022) in this volume.

sustainable and ethically responsible. The ethical principle of strong sustainability, which follows from Jonas' ethics, also demands that the ecological dimension of sustainability hast to be prioritized over the economic and social dimension in order to secure future existence and the indispensable prerequisites for it. This prohibits the substitution of ecological goods by economic goods since from a certain extent this would have a negative impact on the intactness of nature and thus on the possibility of future existence. An overexploitation of nature and an exceedance of the planetary boundaries in favour of economic profit maximization and economic growth are, in the words of Hans Jonas, due to the "simplen Wahrheit, daß begrenzte Erde und unbegrenztes [ökonomisches] Wachstum unverträglich sind und jene das letzte Wort behält"[75] therefore identified as ethically irresponsible and unsustainable. If the bioeconomy shall be sustainable and ethically responsible, the ecological dimension of sustainability must be prioritized over the economic dimension, and the primary concern must be to preserve or increase ecological capital, but not economic capital.

4.2.2 Condition 2: Technology Assessment

Since the bioeconomy is largely based on the development and application of innovative, newly technologies, such as, for instance, genetic engineering and synthetic biology,[76] their spatial and, in particular, temporal remote effects must always be examined to determine how they will affect the future real, genuine existence of humanity and nature.

This, therefore, requires, *first*, the establishment of the discipline of futurology or technology assessment in science and industry, as Jonas also calls for in his "Ethik für die technologische Zivilisation"[77] in view of the destructive potential of technology. *Secondly*, an individual decision based on a case-by-case assessment of each new technology that is to be developed and applied in the context of bioeconomic research is necessary. As Jonas has already emphasized, due to the discrepancy between the "Kraft des Vorherwissens und Macht des Tuns,"[78] assessment and evaluation of the remote effects of modern technologies can always be shaped by uncertainty and ambiguity regarding their type, extent and probability of occurrence.[79] If, in the context of the bioeconomy, a new technology involves the possibility of endangering future human existence or the preconditions necessary for it, then, in accordance with Jonas' *principle of caution* or *risk principle*, a risk-

[75] Jonas (1994a, pp. 142 f.), "simple truth that finite earth and infinite [economic] growth are incompatible and that the latter has the last word" (own translation).

[76] Cf. Gottwald and Krätzer (2014, pp. 27–41).

[77] This is the subtitle of the work "The Imperative of Responsibility," Jonas (2017a, p. 3), "Ethics for the Technological Age" (original translation).

[78] Ibid., 28, "power of foreknowledge and the force of action" (own translation).

[79] Franz-Theo Gottwald and Anita Krätzer also stress the potential unpredictability of the knowledge of the consequences of modern technologies in the context of the bioeconomy, cf. Gottwald and Krätzer (2014, pp. 29 ff.).

averse decision-making rule such as the "*in dubio pro malo*"[80] rule must be adopted in order not to endanger the future existence that must be preserved unconditionally and to meet the demands of responsibility for the future as well as the ethical principle of sustainability derived from it. If this rule is perceived as too inhibiting to development, then a new maxim has to be formulated, which includes the responsibility towards future generations as well as towards the entire living nature substantiated by Jonas.[81]

4.2.3 Condition 3: Economic, Social and Individual Transformation

The analysis of the subject of responsibility of Jonasian ethics has shown that he advocates individual and collective and political responsibility. For the understanding of the ethical principle of sustainability, which can be derived from Jonas' ethics of responsibility, it follows that sustainability must be understood in a substantive as well as in a procedural sense and therefore requires implementation at the individual, collective and political level. In order to be considered ethically responsible and sustainable, the bioeconomy must therefore not only be an economic transformation that is substantively directed by actors and stakeholders in politics and business but must also be understood as a social change that is procedurally supported by individuals through frugality and changes in their individual ways of living, consuming and acting as well as through collective social commitment.[82] This in turn requires a sensitization and awareness of individuals regarding nature and future issues, which could be manifested in the educational sector, for instance.[83] Therefore, it can be stated that bioeconomy only appears to be ethically responsible and sustainable in its entirety if it is not understood merely as a change in the economic model but as a comprehensive systemic change.

4.2.4 Condition 4: Appreciation and Acknowledgement of the Absolute Value of the Entire Animate Nature and Its Entities

The bioeconomy can only be considered ethically responsible if it acknowledges, respects and protects the *absolute value* of nature and all its components, as substantiated by Jonas. Accordingly, nature with all its organisms and life forms must not be understood as a "storehouse" that can be accessed at any time. Contrary to the Federal Ministry of Education and Research's description of "biologischen

[80] Jonas (2016, p. 175).

[81] For possible criticism of the priority of the bad over the good prognosis demanded by Jonas cf. footnote 40 in this contribution.

[82] On the value of frugality in Jonas ethics, see Jonas (2016, pp. 175 ff.) and Jonas (2017d, pp. 67–70).

[83] Markus Vogt makes a similar demand, see Vogt (2016, p. 13).

Ressourcen [. . .] als biochemische ‚Fabriken',"[84] the bioeconomy must therefore not engage the "Um*wertung* alles Lebendigen zum Rohstoff ‚Biomasse'"[85] and demote or degrade them to exploitable capital with merley instrumental-economic value, but must respect, acknowledge and protect the absolute value of the entire animate nature including all ilts living entities, which is inherent in the claim for existence or continued existence, and which is justified in the framework of the Jonasian ethics.

4.2.5 Condition 5: Demand Orientation Instead of Growth Orientation

From the recognition and protection of the absolute value of nature, emphasized in condition four, it does by no means follow that man must not use nature for the satisfaction of his needs. Rather, Jonas also underscores that encroachment and destruction of other life is part of life itself:

> Übergriff in anderes Leben ist mit der Zugehörigkeit zum Lebensbereich eo ipso gegeben, da jede Art von anderen lebt oder deren Umwelt mitbestimmt und daher die bloße, von Natur betriebene Selbsterhaltung einer jeden einen fortwährenden Eingriff in das übrige Lebensgefüge darstellt.[86]

From the combination of this legitimate, human satisfaction of needs by the use of the living nature on the one hand and the absolute value of all being that urges on further existence, on the other hand, it then results from that man must use nature for his purposes only to the extent that he actually needs it for his real existence. This indicates that if overexploitation of nature and overloading of the planetary boundaries does not result from the actual necessity of satisfying human (basic) needs and securing genuine human existence,[87] but from purely economic interests in maximizing profit through the economic exploitation of the living nature, this use of nature must be shown to be ethically irresponsible. Applied to the bioeconomy, this means, with reference to the first condition mentioned above, that in order to be

---

[84] Federal Ministry of Education and Research (2017, p. 21), "biological resources [...] as biochemical 'factories'" (own translation).

[85] Gottwald and Krätzer (2014, p. 8), emphasis added, "Re*value*ation of all living things into the feedstock 'biomass'" (own translation).

[86] Jonas (2017a, pp. 246 f.), "Encroachment into other life is given with the affiliation to the sphere of life eo ipso, since every species lives from others or co-determines their environment and therefore the mere self-preservation of each individual, which is pursued by nature, represents a continuous encroachment into the rest of the structure of life" (own translation).

[87] Here, it becomes clear that there is the need to determine which goods are necessary to be able to lead a real, genuine, humane and good existence and which go far beyond that in order to be able to distinguish between a legitimate use of nature and an illegitimate use. Although Jonas primarily aims at securing human existence and considers the question of preserving a concrete way of existence to be of secondary importance in view of the current ecological and technical situation, he also wants the "echte" and "humane Leben im sozial-kulturellen Kontext" (Schmidt, 2013, 178) to be safeguarded; "genuine" and "humane life in a socio-cultural context" (own translation).

See Jonas (1994a, p. 138), Jonas (2017a, pp. 86–91, 186 f., 250) and Birnbacher and Schicha (2001, p. 29) and Meyer (2018, pp. 40 ff.).

considered ethically responsible and sustainable, it must be understood as a *demand-oriented* rather than growth-oriented economy, which should be based on values such as sufficiency, frugality and moderation.[88]

4.2.6 Condition 6: Use of Renewable and Non-Regenerative Resources

In order for the bioeconomy to meet the demands of responsibility for the future and sustainability, the development of concrete rules for the bioeconomic use of regenerative and non-regenerative resources is necessary. Any resource that is regenerative in principle can be turned into a non-renewable resource through overuse and depletion until it is finally completely used up and no longer exists. From a certain extent, this could have a negative impact on the integrity of nature due to the complex natural system interrelationships, which, according to Jonas' *precautionary principle*, must be safeguarded imperatively in order not to jeopardize the possibility of a real human existence in the future. Therefore, it follows from Jonas' precautionary principle and the ethical principle of a strong, in particular strong ecological sustainability that, within the bioeconomy, regenerative resources may only be used to the extent that they are capable of regenerating themselves naturally.[89]

This raises the question of whether and how non-regenerative resources should be dealt with in the bioeconomy: Should these resources no longer be used at all to protect the possibility of future real human existence? May they be used to the extent that they or their functions can be replaced by renewable resources?[90] Should this latter course of action turn out to be the safest option that poses the least threat to existence, then innovative, interdisciplinary research on nature in the context of the bioeconomy could provide a remedy for ensuring the satisfaction of specific needs, which hitherto have been based on the use of non-renewable resources, by showing ways of satisfying these needs on the basis of the demand-oriented use of renewable resources.

4.2.7 Condition 7: National and International Legal Anchoring

The seventh and last, less normative than structural condition, demands the necessity of local-national and global-international legal anchoring and implementation of principles of an ethically responsible and sustainable bioeconomy. This important connection between locality and globality is also referred to as the "Prinzip der Glokalität"[91] within the context of the sustainability debate.

This condition is necessary for the reason that if only one nation adheres to the normative conditions of a responsible and sustainable bioeconomy, which are based on the concepts of responsibility for the future and sustainability, this may achieve small steps towards a future in which genuine and dignified human existence is

[88] Markus Vogt emphasizes similar points, see Vogt (2016, p. 7). On the value of frugality in Jonas ethics, see Jonas (2016, pp. 175 ff.) and Jonas (2017d, pp. 67–70).

[89] Cf. Faucheux and Noël (2001, pp. 201 f., 455).

[90] Cf. ibid., 455; Meyer (2018, pp. 40 ff.).

[91] Pufé (2017, p. 116), "principle of glocality" (own translation).

possible, and the safeguarding of satisfaction of the needs of future generations in the sense of the underlying definition of sustainability is ensured. However, these steps will be too small and hardly sufficient to maintain the indispensable integrity of the biosphere. Accordingly, a more ethically responsible and sustainable bioeconomic approach to nature must be anchored nationally as well as internationally at the legal and political levels in order to make the bioeconomy appear as a practical way of securing the future real, genuine existence. This could require the establishment of an international system of sanctions to ensure compliance with these regulations.

References

Birnbacher, D., & Schicha, C. (2001). Vorsorge statt Nachhaltigkeit – ethische Grundlagen der Zukunftsverantwortung. In D. Birnbacher & G. Brudermüller (Eds.), *Zukunftsverantwortung und Generationensolidarität* (pp. 17–33).

Bostrom, N. (2018). Die Zukunft der Menschheit. In Id (Ed.), *Die Zukunft der Menschheit. Aufsätze*. Translated from English by Strasser, J.-E (pp. 7–47).

Buddeberg, E. (2011). *Verantwortung im Diskurs. Grundlinien einer rekonstruktiv-hermeneutischen Konzeption moralischer Verantwortung im Anschluss an Hans Jonas, Karl-Otto Apel und Emmanuel Lévinas*.

Claas, A. (2011). *Lässt sich das Prinzip Verantwortung noch verteidigen? Erneuter Versuch der Begründung des Prinzips Verantwortung von Hans Jonas auf der Basis der diskursiven und neo-pragmatistischen Ethik von Albrecht Wellmer*.

Ekardt, F. (2013). Nachhaltigkeit. In A. Grunwald (Ed.), *Handbuch Technikethik* (pp. 187–192).

Ekardt, F. (2014). *Das Prinzip Nachhaltigkeit. Generationengerechtigkeit und globale Gerechtigkeit* (3rd ed.).

Faucheux, S., & Noël, J. (2001). *Ökonomie natürlicher Ressourcen und der Umwelt*. Translated from French by S. Geisendorf, U. Hampicke.

Federal Ministry of Education and Research. (2010). *Nationale Forschungsstrategie BioÖkonomie 2030. Unser Weg zu einer bio-basierten Wirtschaft*. Retrieved March 29, 2021, from, https://www.bmbf.de/upload_filestore/pub/Nationale_Forschungsstrategie_Biooekonomie_2030.pdf

Federal Ministry of Education and Research. (2017). *Forschung für eine biobasierte Wirtschaft. Erfolge und Herausforderungen für die Bioökonomie in Deutschland*. Retrieved March 29, 2021, from, https://www.bmbf.de/upload_filestore/pub/Forschung_fuer_eine_biobasierte_Wirtschaft.pdf

Federal Ministry of Education and Research and Federal Ministry of Food and Agriculture. (2014). *Bioökonomie in Deutschland. Chancen für eine biobasierte und nachhaltige Zukunft*. Retrieved March 29, 2021, from, https://www.bmbf.de/upload_filestore/pub/Biooekonomie_in_Deutschland.pdf

Federal Ministry of Education and Research and Research for Sustainability – FONA. (n.d.). *Bioökonomie*. Retrieved March 31, 2021, from, https://www.fona.de/de/themen/biooekonomie.php

Global Foodprint Network. (2020). *Earth Overshoot Day is August 22, More Than Three Weeks Later Than Last Year*. Press Release from 05.06.2020. Retrieved March 29, 2021, from, https://www.overshootday.org/newsroom/press-release-june-2020-english/

Gottwald, F.-T., & Krätzer, A. (2014). *Irrweg Bioökonomie. Kritik an einem totalitären Ansatz*.

Grefe, C. (2016). *Global Gardening. Bioökonomie – Neuer Raubbau oder Wirtschaftsform der Zukunft?* Licensed edition for the German Federal Agency for Civic Education (Vol. 1737).

Grunwald, A., & Kopfmüller, J. (2006). *Nachhaltigkeit*.

Hamberger, J. (2013). Nachhaltigkeit – Entstehung und Formung eines Begriffs im Forstbereich und wie er zum Berufsethos einer Branche wurde. In M. Vogt, J. Ostheimer, & F. Uekötter (Eds.), *Wo steht die Umweltethik? Argumentationsmuster im Wandel* (pp. 427–445).
Hauff, V. (Ed.) (1987). *Unsere gemeinsame Zukunft. Der Brundtland-Bericht der Weltkommission für Umwelt und Entwicklung*. Unabridged edition with a new preface to the German edition.
Hirsch Hadorn, G. (2000). Verantwortungsbegriff und kategorischer Imperativ der Zukunftsethik von Hans Jonas. *Zeitschrift für philosophische Forschung* 54(2) (pp. 218–237).
Huber, W. (2018). Ehrfurcht des Heiligen. Zur Aktualität des Prinzip Verantwortung von Hans Jonas. *Pastoraltheologie* 107 (pp. 411–426).
Jischa, M. F. (2005). *Herausforderung Zukunft. Technischer Fortschritt und Globalisierung* (2nd ed.).
Jonas, H. (1993). Warum die Technik ein Gegenstand für die Ethik ist: Fünf Gründe. In H. Lenk & G. Ropohl (Eds.), *Technik und Ethik* (2nd revised and extended edition, pp. 81–91).
Jonas, H. (1994a). Zur ontologischen Grundlegung einer Zukunftsethik. In: Id (Ed.), *Philosophische Untersuchungen und metaphysische Vermutungen* (pp. 128–146).
Jonas, H. (1994b). Rassismus im Lichte der Menschheitsbedrohung. Ansprache bei der Entgegennahme des Premio Nonino am 30. Januar 1993 in Udine. In D. Böhler (Ed.), *Ethik für die Zukunft. Im Diskurs mit Hans Jonas* (pp. 21–29).
Jonas, H. (2016). Prinzip Verantwortung – Zur Grundlegung einer Zukunftsethik. In A. Krebs (Ed.), *Naturethik. Grundtexte der gegenwärtigen tier- und ökoethischen Diskussion* (8th ed., pp. 165–181).
Jonas, H. (2017a). *Das Prinzip Verantwortung. Versuch einer Ethik für die technologische Zivilisation* [1979] (6th ed.).
Jonas, H. (2017b). Podiumsgespräch (1981). "Möglichkeiten und Grenzen der technischen Kultur". In Id (Ed.), *Technik, Medizin und Ethik. Praxis des Prinzips Verantwortung* [1987] (9th ed., pp. 269–301).
Jonas, H. (2017c). Warum die moderne Technik ein Gegenstand für die Ethik ist. In Id (Ed.), *Technik, Medizin und Ethik. Praxis des Prinzips Verantwortung* [1987] (9th ed., pp. 42–52).
Jonas, H. (2017d). Auf der Schwelle der Zukunft: Werte von gestern und Werte für morgen. In Id (Ed.), *Technik, Medizin und Ethik. Praxis des Prinzips Verantwortung* [1987] (9th ed., pp. 53–75).
Jonas, H. (2017e). Warum die moderne Technik ein Gegenstand für die Philosophie ist. In Id (Ed.), *Technik, Medizin und Ethik. Praxis des Prinzips Verantwortung* [1987] (9th ed., pp. 15–41).
Jonas, H. (2017f). Vorwort. In Id (Ed.), *Technik, Medizin und Ethik. Praxis des Prinzips Verantwortung* [1987] (9th ed., pp. 9–13).
Jonas, H., & Gebhardt, E. (1994). Naturwissenschaft versus Natur-Verantwortung? In D. Böhler (Ed.), *Ethik für die Zukunft. Im Diskurs mit Hans Jonas* (pp. 197–212).
Jonas, H., & Löw, R. (1990). Das Prinzip Verantwortung in der technisch-industriellen Welt. In B. Engholm & W. Röhrich (Eds.), *Ethik und Politik heute. Karl-Otto Apel, Hans Jonas, Hans Küng im Gespräch* (pp. 17–35).
Kreß, H. (2017). Verantwortung in Religion und Kultur. In L. Heidbrink, C. Langbehn, & J. Loh (Eds.), *Handbuch Verantwortung* (pp. 645–665).
Mathis, K. (2017). *Nachhaltige Entwicklung und Generationengerechtigkeit. Eine interdisziplinäre Studie aus rechtlicher, ökonomischer und philosophischer* Sicht..
Meyer, K. (2018). *Was schulden wir künftigen Generationen? Herausforderung Zukunftsethik.*
Moser, S. (2016). Verantwortung im Spannungsfeld zwischen Machtentfaltung und Verletzlichkeit: Die Umkehr des Verantwortungsverständnisses bei Hans Jonas. *Labyrinth* 18(1) (pp. 58–78).
Müller, W. E. (1988). *Der Begriff der Verantwortung bei Hans Jonas.*
Ostheimer, J. (2013). Nachhaltigkeit. Zum moralischen Gehalt eines Naturnutzungskonzepts. In M. Vogt, J. Ostheimer, & F. Uekötter (Eds.), *Wo steht die Umweltethik? Argumentationsmuster im Wandel* (pp. 397–414).

Pinsdorf, C. (2022). Bioeconomy beneath and beyond: Persisting challenges from a philosophical and ethical perspective. In D. Lanzerath, U. Schurr, C. Pinsdorf & M. Stake (Eds.), *Bioeconomy and sustainability. Scientific, socio-economic and ethical dimensions* (pp. 343–377). Springer.

Pufé, I. (2017). *Nachhaltigkeit* (3rd revised and extended edition).

Ropohl, G. (1993). Neue Wege, die Technik zu verantworten. In Id & H. Lenk (Eds.), *Technik und Ethik* (2nd revised and extended edition, pp. 149–176).

Schmidt, J. C. (2013). Das Argument ‚Zukunftsverantwortung'. Versuch einer analytischen Rekonstruktion der naturphilosophischen Natur- und Technikethik von Hans Jonas. In Id, G. Hartung, C. Köchy, & G. Hofmeister (Eds.), *Naturphilosophie als Grundlage der Naturethik. Zur Aktualität von Hans Jonas* (pp. 155–186).

Sunstein, C. R. (2005). The precautionary principle as a basis for decision making. *The Economists' Voice* 2(2) (pp. 1–9).

United Nations Conference on Environment and Development. (1992). *Rio-Erklärung über Umwelt und Entwicklung*. Retrieved March 29, 2021, from, https://www.un.org/Depts/german/conf/agenda21/rio.pdf

Vogt, M. (2016). *Ethische Fragen der Bioökonomie*. Working Paper 5. Straubing: Bayerischer Forschungsverbund ForChange. Retrieved March 29, 2021, from, http://www.biooekonomierat-bayern.de/dateien/Publikationen/M.Vogt_Ethische_Fragen_der_Bio%C3%B6konomie_03.2016.pdf.

von Hauff, M., Schulz, R., & Wagner, R. (2018). *Deutschlands Nachhaltigkeitsstrategie*.

Werner, M. H. (2016). Hans Jonas' Prinzip Verantwortung. In M. Düwell & K. Steigleder (Eds.), *Bioethik. Eine Einführung* (4th ed., pp. 41–56).

Wesseler, J., & Kardung, M. (2022). Agriculture in the bioeconomy: Economics and policies. In D. Lanzerath, U. Schurr, C. Pinsdorf, & M. Stake (Eds.), *Bioeconomy and sustainability. Scientific, socio-economic and ethical dimensions* (pp. 225–245). Springer.

Wetz, F. J. (1994). *Hans Jonas zur Einführung*.

Bioeconomy as a Normative Concept of Resilience: Challenges and Opportunities

18

Sebastian Lenze

Abstract

For decades the concept of resilience has been used across different disciplines to describe features that make systems or entities robust towards various kinds of stresses. While there are multiple resilience concepts throughout the disciplines, the views on what resilience means also differ within fields of study. Notions vary between a conservative-reactive, an adaptive and a transformative understanding of resilience. This differentiation can lead to conflicting interpretations when attempting to analyse and foster resilience factors for a particular entity or system, as the outcomes of either variation would diverge significantly. For interdisciplinary concepts such as bioeconomy this is especially relevant. In this contribution, bioeconomy is presented as a resilience concept that can avoid conflicting interpretations if it is based on a transformational interpretation of resilience and includes the normative concept of sustainability. From the assessment of bioeconomy as a resilience-based concept, this article concludes that a system or entity has to fulfil three criteria in order to continue functioning in crisis situations: First, it must promote a second-order concept of resilience, that is, one that takes other contexts into account; second, the interpretation of the resilience concept used must be clarified—i.e. whether it holds a conservative-reactive, an adaptive or a transformational understanding; and, third, it has to contain a clearly defined normative dimension.

Keywords

Bioeconomy · Resilience · Sustainability

S. Lenze (✉)
Department of Philosophy, University of Bonn, Bonn, Germany

D. Lanzerath et al. (eds.), *Bioeconomy and Sustainability*,
https://doi.org/10.1007/978-3-030-87402-5_18

1 Introduction

In the years 2017 to 2019, the forest stand in Germany has suffered severe damage due to various extreme weather phenomena. In addition to storms in 2017 and 2018, the droughts and heat waves of 2018 and 2019, the associated forest fires and the spread of the bark beetle will leave an area of around 180,000 hectares in need of afforestation.[1] In addition, millions of trees show "sehr hohe Schadenssymptome".[2] Due to these circumstances, the Federal Ministry of Food and Agriculture published a key issues paper in September 2019 entitled "*Deutschlands Wald im Klimawandel*",[3] which presents measures for the creation of "klimaresilienter, naturnaher, leistungsfähiger, standortgerechter Mischwälder".[4] With the focus on resilience, the paper uses a concept that has been used in ecology since the 1960s.[5] In ecosystem theory, resilience describes "die Fähigkeit eines Ökosystems angesichts sozialer, sozio-ökologischer oder rein ökologischer Störungen seine grundlegende Organisationsweise zu erhalten".[6] In psychology, on the other hand, resilience is combined with the term crisis, in so far as it is understood as the sum of individual strategies of resistance to crises.[7] While further interpretations exist, among others, in the material sciences, social sciences, engineering and economics,[8] it is apparent that resilience is a strongly context-dependent concept. As such, it can have different meanings not only between different disciplines, but also within disciplines.[9]

The Federal Ministry of Food and Agriculture's key issues paper identifies three dimensions of sustainable forest use in Germany. In addition to the primary goal, the conservation of the forest, "die Leistungsfähigkeit und verschiedenen Funktionen des Waldes entlang der drei Säulen der Nachhaltigkeit (Ökologie, Ökonomie, Soziales)"[10] are to be safeguarded and promoted and the "verschiedenen gesellschaftlichen Interessen an den Wald im Ausgleich"[11] are to be accounted for. Accordingly, the conservation of the forest is linked to interests from different areas. From a social perspective, for example, it could serve as a local recreation area, fulfil economic factors as part of forestry, and, from an ecological point of

[1] Cf. Federal Ministry of Food and Agriculture (2019, p. 5).

[2] Ibid., "very high damage symptoms" (own translation).

[3] "Germany's Forests in Climate Change" (own translation).

[4] Ibid., "climate-resilient, near-natural, efficient, site-appropriate mixed forests" (own translation).

[5] Cf. Folke (2006, p. 254).

[6] Nida-Rümelin and Gutwald (2016, p. 252), "the ability of an ecosystem to maintain its basic mode of organization in the face of social, socio-ecological or purely ecological disturbances" (own translation).

[7] Cf. ibid.

[8] Cf. Weiß et al. (2018).

[9] Cf. Fletcher and Sarkar (2013).

[10] Federal Ministry of Food and Agriculture (2019, p. 4), "the performance capacity and various functions of the forest along the three pillars of sustainability (ecology, economy, social affairs)" (own translation).

[11] Ibid., "various social interests in the forest in balance" (own translation).

view, bind carbon dioxide. However, it is also possible that the different interests affecting the forest conflict. In addition to the pure conflicts of interest that can arise between the various claims, conflicting interpretations between the concepts of resilience of the individual disciplines are conceivable. Maintaining the status quo of one system may affect the maintenance of another system. One way to resolve this conflict in a sustainable way is to formulate a concept of resilience that takes all three pillars of sustainability into account and implements them. Therefore, in this paper bioeconomy is presented as a notion of resilience that can avoid conflicting interpretations if it is based on a transformational interpretation of resilience. A bioeconomy-based model offers a promising perspective, as it is located between different disciplines and includes the normative concept of sustainability.

To this end, the concept of resilience is examined, and possible conflicts are highlighted. Two lines of conflict are prominent here: First, the problem of conflicting concepts of resilience in the different disciplines is addressed. A further conflict arises from the different meanings of the concept of resilience. It is shown that bioeconomy itself is a concept of resilience that considers criteria that resolve the conflicts between the different concepts of resilience. Thus, the genesis of the concept of resilience is discussed under Sect. 2, before possible points of criticism of the concept are formulated under Sect. 3. This is followed by an interpretation of bioeconomy as a resilience concept, which removes the previous points of criticism (Sect. 4). Overall, it is argued that bioeconomy offers an analytical framework for identifying individual resilience factors that can have a positive effect across different contexts. Its perspective is thus particularly interesting for the transdisciplinary investigation of the concept of resilience.

2 Genesis and Use of the Concept of Resilience

2.1 Resilience Between Conservation, Adaptation and Transformation

The term resilience comes from the Latin verb *resilire*, which means, among other things, to *jump back* and *bounce off*. In its original use, the term thus refers to elements of conservation, i.e. of returning to a certain status, and to the aspect of defence—both of which also play a role in current concepts: "Der springende Punkt bei der Resilienz ist, dass sich die Entität (sei es ein Mensch, eine Institution oder ein System) zumindest zeitweise auch selbst verändert, 'verformt', nachgibt oder anpasst, also elastisch ist".[12] According to its conceptual origin, resilience is thus a conservative-reactive concept. When confronted with a disorder, a resilient system changes briefly and then regresses back to its original state. Besides this aspect,

[12] Vogt and Schneider (2016, p. 182), "The crucial point with resilience is that the entity (whether it is a person, an institution or a system) also changes, 'deforms', yields or adapts itself, i.e. is elastic, at least temporarily" (own translation).

adaptive and transformational interpretations of the concept of resilience are also being discussed in the current scientific debate. Carl Folke defines adaptability and transformability in relation to socio-ecological systems as follows:

> In resilience work adaptability is referred to as the capacity of people in a social-ecological system to build resilience through collective action whereas transformability is the capacity of people to create a fundamentally new social-ecological system when ecological, political, social, or economic conditions make the existing system untenable [. . .].[13]

Adaptation is thus the ability of such a system to work out resilience mechanisms for itself and to implement them, e.g. in form of policies that promote resilience. As in the case of persistence, which is aimed at "Gefahrenabwehr und Risikominimierung",[14] adaptation here is focused on the aspect of self-preservation "ohne weitergehende strukturelle Ursachen für die Verwundbarkeit zu ändern oder einen bestimmten Entwicklungspfad in Frage zu stellen".[15] Transformability, on the other hand, aims at the ability of system participants to create a completely new system when certain circumstances make it untenable. This, as Markus Vogt and Martin Schneider point out, is associated with a socio-cultural change that entails a "Umdenken in den Leitwerten und -zielen".[16] The distinction between a conservative-reactive, an adaptive and a transformational concept of resilience can be further illustrated by the following example. In the reaction to the anthropogenic climate change, resilience criteria of a coal-based energy industry are to be worked out. A conservative-reactive resilience concept could, for example, identify certain factors that help to survive this situation without making any fundamental changes. Imagine extended lobbying efforts[17] by electricity producers and an attempt at greenwashing as resilience strategies of this kind. An adaptive approach would focus on improving the existing infrastructure and, for example, making coal-fired power plants more efficient and potentially more climate-friendly. This has been attempted, for example, with the *Carbon Capture and Storage* (CCS) technology, which is designed to deposit waste carbon dioxide from coal-fired power plants in underground geological formations. A transformation-based resilience approach, on the other hand, would, in this scenario, advocate a complete switch to renewable energies and thus call for a system change, whereby even after a "coal exit" one could still speak of an energy industry, i.e. the identity of the system would be preserved. The concept of resilience creates an extreme tension between

[13] Folke (2006, p. 262).

[14] Vogt and Schneider (2016, p. 188), "averting danger and minimising risk" (own translation).

[15] Ibid., "without changing further structural causes of vulnerability or calling into question a particular developmental path" (own translation).

[16] Ibid., "rethinking of guiding values and goals" (own translation).

[17] In October 2019, various media reported that the five largest oil companies had spent more than 250 million euros on lobbying the European Union between 2010 and 2019 "to delay, mitigate or torpedo climate protection measures" (cf. Der Standard 2019, own translation), as one critic of these measures put it.

conservation/reaction, adaptation and transformation, which makes it a *terminus aequivocus*, i.e. an ambiguous term. The problem of equivocation is dealt with in more detail in Sect. 3.1.

2.2 Context Resilience

As mentioned in Sect. 1, another difficulty in defining resilience is the strong context-bound nature of the concept. Resilience can be formulated in a specific context of analysis or in context of a discipline. With regard to the scope of study, Vogt and Schneider distinguish between first-order resilience, which refers to a system, a specific entity, an institution or an actor, and second-order resilience, "die einer Entität nur zugesprochen wird, wenn sie in einem übergreifenden Zusammenhang anderen Entitäten keine unlösbaren Resilienzprobleme aufgibt".[18] This dimension, also called "contextual resilience", is explained in more detail later. Such a conceptualization makes it possible to examine the resilience of an entire system, but also that of an entity of a system in relation to other actors or to the overall functioning of the system in which it is located. In this context, the utilization of the reactive resilience concept in a subsystem can have a negative impact on the implementation of an adaptive or transformational resilience in the overall system—a line of conflict that is discussed in more detail in Sect. 3.3. Especially when different interests are overlapping, as is the case in the forest-example, it is necessary to balance and coordinate the different notions of resilience. Thus, in order to be able to present bioeconomy as a context-spanning concept of resilience that considers social, ecological and economic aspects, the varying resilience-ideas of the individual disciplines must be briefly outlined and related to each other.

2.3 Resilience in Ecology

In the 1960s–1970s, a radical change took place in ecological resilience research. The *stable equilibrium view*, a structurally conservative concept of resilience that considered ecosystems to be in equilibrium if no human stressors intervened and had dominated the scientific debate until then, was replaced by an adaptive concept of resilience.[19] The perspective shifted from a regulation of change with the aim of stability to an understanding of governance, which should strengthen the capacities of social-ecological systems to adapt to certain developments.[20] This *adaptive management approach* is also applied in the Federal Ministry of Food and Agriculture paper, as it allows for the adaptation of forests and forestry to the challenges of

---

[18] Vogt and Schneider (2016, p. 189), "which is only assigned to an entity if it does not cause unsolvable resilience problems for other entities in an overarching context" (own translation).

[19] Cf. Folke (2006, p. 253).

[20] Cf. ibid., p. 254.

climate change without making radical changes. The departure from equilibrium models in ecology stems from the view that life on earth has been strongly influenced by different catastrophes and that a mere fluctuation between different equilibrium states would have restricted decisive impulses for higher development and succession.[21] The focus of this thrust of resilience research is therefore clearly to be understood in distinction to a conservative concept of resilience. The question arises to what extent such a concept of resilience is not in fact a transformational one. Without claiming to make a final judgement, this question will be discussed in more detail in Sect. 3.1.

2.4 Resilience in the Economy

From an economic perspective, resilience can be seen as a counter-concept to efficiency, because "[e]s geht um das Vorhalten von Reserven und Spielräumen, um im Fall von Störungen weniger verletzlich zu sein und robuster reagieren zu können".[22] In this context, precaution (increasing resilience) is opposed to increasing efficiency, since it is about the allocation of resources that can only be used for one of the two processes at a time. This means that systems that have been optimized for efficiency are more susceptible to disturbances.[23] From a macroeconomic perspective, for example, the market economy can be described as a resilient system, since it is "in hohem Maße flexibel, anpassungs- und lernfähig sowie dynamisch, innovativ und wandlungsfreudig".[24] This capability of resistance and adaptability is based on the "Zerstörung von sich nicht bewährenden Teilsystemen".[25] In addition to this destruction, the externalization of costs is brought up as a criticism of the description of the market economy as a resilient system.[26] Because "[o]hne starke soziale und ökologische Rahmenbedingungen, die es bisher auf globaler Ebene nicht gibt, fördert der Markt ein Wettbewerbsverhalten auf Kosten der sozial Schwachen sowie der öffentlichen Güter".[27] The problem of externalization obliges economic resilience concepts to consider the environmental effects that actions in the economic sphere can have. One concept that takes this into account in a differentiated way is that of functional and normative resilience by Benedikt Gleich and Rebecca Gutwald, which is briefly described in the following section.

[21] Cf. Vogt and Schneider (2016, p. 184).

[22] Ibid., p. 186, "it is about maintaining reserves and margins in order to be less vulnerable in the event of disruptions and to be able to react more robustly" (own translation).

[23] Cf. ibid.

[24] Ibid., "highly flexible, adaptable and capable of learning, as well as dynamic, innovative and open to change" (own translation).

[25] Ibid., "destruction of subsystems that do not prove worthy" (own translation).

[26] Cf. ibid., p. 187.

[27] Ibid., "when there are no strong social and ecological framework conditions, which do not yet exist at the global level, the market promotes competitive behaviour at the expense of the socially weak and public goods" (own translation).

2.5 Functional and Normative Resilience

Functional resilience refers to the continuity of an entity or system through adaptability, stability and openness.[28] Normative resilience, on the other hand, refers to the ability of a person or system to achieve inherent goals despite possible crises.[29] The former is thus a purely descriptive concept that examines "die Eigenschaften eines Systems oder Wesens, die sein Überleben oder Weiterbestehen sichern".[30] Thus, functionally resilient markets are "reaktive Systeme, die Störungen unter Erhaltung ihrer Identität einfach aufnehmen".[31] Gleich and Gutwald divide functional resilience into intrinsic and contextual resilience. The intrinsic resilience of markets ensures the continued existence of core functions when disruptions occur. Normatively, this is not necessarily positive, since, as mentioned above, the effects of a resilient system on its contexts are not taken into account here (e.g. through consequence externalisation).[32] Context-resilience studies the effect of the market on specific resilience qualities of the periphery of markets, such as third parties, society or the environment. Only peripheral factors with a direct relation to market events are considered, such as educational systems, regulatory economic laws or welfare expectations.[33] The concept of resilience "bezieht sich sowohl auf die Stabilität und Widerstandsfähigkeit von Märkten als auch auf die von Umwelt und Gesellschaft".[34] In functional resilience, both factors are intertwined and "entfalten in dieser Dynamik ihre je besonderen Wirkungen, seien diese produktiv oder destruktiv".[35] Even a meaningful interlocking of functionally resilient markets, which is composed of intrinsic and contextual resilience, is therefore not necessarily accompanied by an orientation towards social goals.

This orientation and the associated orientation of resilient markets towards such a goal can only be made possible by a normative perspective, which is missing from the purely descriptive concept of resilience. Through this normative perspective a reflection of the target repertoire as well as a more specific analysis of the interactions between markets and their external effects can be carried out. Thus, "[wären] Kontextprobleme nicht von vornherein in die Logik von Markt und seinen

[28] Cf. Gleich and Gutwald (2017, p. 208).

[29] Cf. ibid.

[30] Ibid., "the characteristics of a system or entity that ensure its survival or continued existence" (own translation).

[31] Ibid., "reactive systems that easily absorb disruptions while maintaining their identity" (own translation).

[32] Cf. ibid., p. 207.

[33] Cf. ibid., p. 209.

[34] Ibid., p. 208, "refers both to the stability and resilience of markets and to that of the environment and society" (own translation).

[35] Ibid., p. 209, "unfold their specific effects, whether productive or destructive, within this dynamic" (own translation).

Externalitäten zu fassen",[36] but would "spezifischer den Gehalt der Resilienzfaktoren des Kontexts in ihrem Eigenwert [. . .] betonen".[37] A normatively resilient system is thus able to use a crisis constructively towards a normatively positive goal and thus also falls into the category of non-reactive systems. Gleich and Gutwald define normatively resilient markets as follows: "So verstanden sind normativ resiliente Märkte solche, bei denen Eigen- und Kontextresilienz produktiv verschränkt sind und darüber hinaus normative Anforderungen [. . .] erfüllen".[38] In Sect. 4, it is argued that bioeconomy is a concept of normative resilience, which—coupled with a departure from the growth paradigm—can resolve conflicts between the various concepts of resilience.

2.6 Attempt at a Transdisciplinary and Cross-Contextual Definition

Through a closer examination of the concept of resilience and its description from a conceptual and contextual perspective, divided into research context and subject area, it becomes clear that the concept of resilience allows for various possibilities of interpretation that may conflict with each other. From the concepts of resilience presented here, the following cross-context definition of a framework for analysis emerges:

The study of resilience describes a heuristic model of analysis that identifies *ex post* certain properties of a system or entity that have helped it to maintain itself (persistence) in response to a crisis (first-order/own resilience), to adapt (adaptation) or to change (transformation) without endangering the resilience factors of other entities/systems (second-order/contextual resilience), whereby no normative quantity is inherent in the concept of resilience itself apart from this functional one, but can be added to it (normative resilience).

Despite the possibility of an overarching definition, different patterns of interpretation remain, which can lead to conflicts. These lines of conflict are now to be examined in more detail in order to make clear that a concept of resilience that works in practice must be bound to certain criteria, such as normative assumptions. Furthermore, it will be shown that bioeconomy fulfils these criteria.

[36] Ibid., "contextual problems would not be a priori subsumed in the logic of the market and its externalities" (own translation).

[37] Ibid., "more specifically emphasize the intrinsic value of the resilience factors of the context" (own translation).

[38] Ibid., p. 207, "Thus, normatively resilient markets are those in which intrinsic and contextual resilience are productively intertwined and fulfil normative requirements" (own translation).

3 The Concept of Resilience Under Critical Review

3.1 Concept and Interpretation

As was made clear, there are various conceptual difficulties that the use of the concept of resilience can entail. These will first be briefly outlined and then discussed regarding the interpretation of bioeconomy in a specific way (Sect. 4). The first point of criticism that can be made against the concept of resilience has already been mentioned and relates to the problem of equivocation. The concept contains conflicting interpretations, ranging from conservative-reactive, to adaptive and transformational interpretations. The respective discipline in which the concept of resilience is used does not seem to play a role, since all interpretations are possible in all disciplines discussed here and can refer to individual entities and/or entire systems. It should be noted, however, that an investigation of resilience is always a procedure in which a system is analysed regarding certain aspects. Thus, the focus of the respective research is to be set individually. Charlotte Rungius, Elke Schneider and Christoph Weller nevertheless identify a conservative element in the fundament of the concept of resilience:

> Die Anwendung des Resilienzbegriffs auf etablierte und aktuell möglicherweise bedrohte oder in einer Krise befindliche soziale Systeme (Sicherheit, etc.) verzichtet nicht nur auf die Erörterung der zugrundeliegenden normativen Annahmen hinsichtlich der Existenzberechtigung des der Resilienzanalyse unterzogenen Systems und die Berechtigung der Infragestellung des Systems, sondern trägt auch noch zu seiner Legitimation bei: Die Prüfung der Überlebens- und Funktionsbedingungen eines sozialen Systems erkennt seine Existenz an, sichert seine Stabilität und ist damit per se ein konservierend-konservativer Beitrag zum gesellschaftspolitischen Diskurs.[39]

That the concept of resilience, whenever it is examined for a certain system, also assumes its right to exist can be justifiably argued for both the conservative-reactive and the adaptive concept of resilience. However, we find a divergent interpretation regarding the transformational concept of resilience. This concept examines a system for resilience factors which preserve its identity, i.e. which make it recognizable as a system of a certain type, but which focus on the change due to external circumstances. This is one of the most fundamental internal conflicts of the concept of resilience, since the concept is confronted with the question of up to what point a system maintains its identity in the face of major changes before it becomes another

[39] Rungius et al. (2018, p. 50), "The application of the concept of resilience to established and possibly (crisis-)threatened social systems (security, etc.) not only dispenses with the discussion of the underlying normative assumptions regarding the justification for the existence of the system subjected to the resilience analysis and the justification for questioning the system, but also contributes to its legitimacy: The examination of the survival and functional conditions of a social system acknowledges its existence, ensures its stability and is thus per se a conservative contribution to the socio-political discourse" (own translation).

system.[40] How long, then, can we speak of a system A that exhibits certain transformational resilience criteria before it becomes a system B based on these criteria? A resolution of this line of conflict lies in the consideration of the system level and the temporal aspect of the change. For example, if we look at the economic system of a country, we can assume that this system is made up of a sum of factors ($x + y + \ldots + n$). If we now examine the system for transformational resilience criteria and promote these (e.g. the decarbonization of the economy), we could examine the system for its factors again after some time and get a modified result ($x1 + y1 + \ldots + n1$). Although we still recognize the system as an economic system, it is based on different factors than the previous economic system.[41] In this respect, the above criticism would only be justified here in the form in which it raises the question of whether there must be an economic system at all.

3.2 The Missing Normative Dimension

A different point of criticism levelled against the concept of resilience is the lack of a normative dimension in the concept itself. The analysis and promotion, especially of conservative-reactive and adaptive, and possibly also of transformational resilience criteria, can be applied to any system or entity, even if it pursues normatively negative goals. In this context, North Korea is often cited as a negative example of a resilient state, whose resilience, however, can only be desirable for the fewest.[42] Julian Nida-Rümelin and Rebecca Gutwald attempt to resolve this conflict by linking the normative concept of authorship to that of resilience.[43] By assigning authorship, people are presented as responsible for their own actions, which implies the possibility of freedom of choice. Thus, there can be no concept of resilience that does not include the concept of authorship.[44] It makes sense to acknowledge the authorship of individuals or human responsibility in general for certain conditions. However, focusing on individual accountability may complicate the framework of the study and dilute the systemic perspective.

Another possibility to examine the resilience of systems or entities with respect to their normative goals is the model of Gleich's and Gutwald's normative resilience already described. Contextual resilience can be used to identify exactly how the resilience of a partial aspect of a system or of a complete system affects the resilience of another aspect or system. After the analysis, this should be considered when

[40] A similar question was already raised in antiquity by the Theseus paradox. In the paradox it is assumed that during the voyage of a galley parts of the ship are exchanged again and again until every single part has been changed once. This raises the question whether the ship at the time of arrival is still the same galley Theseus put to sea with. Cf. e.g. Rosenberg (2009).

[41] A vivid example of such a development is Nokia Corporation, which has transformed itself from a wood pulp producer to a telecommunications company.

[42] Cf. Nida-Rümelin and Gutwald (2016, p. 257).

[43] Cf. ibid.

[44] Cf. ibid.

implementing the promotion of certain mechanisms. These are two different ways of introducing a normative dimension into the resilience debate. Nida-Rümelin and Gutwald attempt to identify a normative dimension in the concept of resilience itself, while Gleich and Gutwald, like Vogt and Schneider, pursue the strategy of linking an external normative variable to the concept of resilience, because "[f]ür ein normativ gehaltvolles Konzept von Resilienz braucht man den Bezug auf zivilisatorische Leitwerte".[45] In the literature, the second strategy is more frequently used to resolve the conflict potential of the concept's lack of normativity. At this point, it is not yet possible to assess which strategy is the better one, but it will be necessary to show that a bioeconomy concept already entails certain normative assumptions. In this way, the problem of a lack of normativity in the term resilience can be avoided, since for the resilience concept "bioeconomy" the values of this specific concept apply (e.g. that of sustainability).

3.3 Conflicting Concepts of Resilience

The last challenge of the concept of resilience, which is the subject of this paper, is particularly relevant in the context of the debate on bioeconomy. As already indicated in Sect. 2, different conflicts of resilience between systems or entities are conceivable. Such conflicts can arise between two or more entities of a system, between an entity and a system, and between different systems. Folke describes this conflict potential as follows:

> A human society may show great ability to cope with change and adapt if analyzed only through the social dimension lens. But such an adaptation may be at the expense of changes in the capacity of ecosystems to sustain the adaptation, and may generate traps and breakpoints in the resilience of a social–ecological system. Similarly, focusing on the ecological side only as a basis for decision making for sustainability leads to too narrow and wrong conclusions. That is why work on resilience stress linked social–ecological systems.[46]

The concept of bioeconomy implies the claim to formulate a second-order concept of resilience that applies to all areas it affects. Since it is a cross-contextual concept, conflicts over resilience can become evident here. Vogt and Schneider illustrate such a conflict using the example of the German *Energiewende*, in which individual companies (entities) stand in the way of a systematic transformation towards an energy industry based on renewable energies.[47] In this example, a conservative-reactive interpretation of the resilience of an entity conflicts with the transformational concept of resilience of a system. Under certain circumstances, this

[45] Vogt and Schneider (2016, p. 189), "for a normatively substantive concept of resilience one needs to refer to civilizational guiding values" (own translation).

[46] Folke (2006, p. 260).

[47] Cf. Vogt and Schneider (2016, p. 189).

conflict can lead to a lock-in state that prevents progress. For this line of conflict, a reference to the core of the resilience concept seems necessary. A resilience analysis is a concept for examining an entity or system for certain factors. The focus of the research is on the respective work. Certain coal-fired power plant operators could also be examined for transformational aspects of their resilience. Besides research, the second important aspect is that of funding. As Folke points out, promoting certain aspects of resilience is a political decision:

> [R]esilience is an approach, a way of thinking, that presents a perspective for guiding and organizing thought and it is in this broader sense that it provides a valuable context for the analysis of social-ecological systems, an area of explorative research under rapid development with policy implications for sustainable development.[48]

From this point of view, it is a political decision to put certain resilience factors on the agenda and, depending on political style and value system, to promote them. The conflict described here thus arises from political decisions and can be resolved by them, through a conscious analysis and promotion of a certain, desired direction.

It has been shown that there are certain conflicts regarding the concept of resilience, that can, however, be resolved. By examining the problem areas, three criteria can be identified which, if they are met, prevent conflicts. First, when implementing policies based on a concept of resilience, a second-order concept should be chosen, i.e. one that considers the impairment of the resilience of other affected areas. In this way, resilience conflicts, e.g. between different systems, can be avoided. Secondly, in order to avoid a different interpretation of the concept of resilience, one of the three possibilities of interpretation must be chosen. Thus, for example, a conflict between the conservative-reactive concept of resilience of a subsystem and the transformational concept of the overall system could be avoided. Third, a normative dimension should be added to the respective concept of resilience. This could take the form of a value to be preserved or a socially relevant goal that can be promoted by the implementation of resilience. In the following, it will be shown that the concept of "bioeconomy" is such a resilience concept.

4 Bioeconomy as a Normative Concept of Resilience

Bioeconomy can be understood as a second-order concept of resilience, i.e. a normative resilience blueprint that contains transformational goals for different contexts and pursues the normative goal of sustainable development of all systems/entities concerned. The concept of bioeconomy is defined by Frank Simon and Hans-Christian Schaefer as follows:

> Die Bioökonomie beschreibt eine Wirtschaftsweise, bei der in allen Wirtschaftssektoren und der Gesellschaft biologisches Wissen zur Anwendung kommt und erneuerbare, biologische

[48] Folke (2006, p. 260).

> Ressourcen genutzt werden. [. . .] Zu ihrer weiteren Entwicklung werden grundlegend neue Technologien und systemische Innovationen notwendig sein, mit denen erhebliche Triebkräfte gesellschaftlichen Wandels und wirtschaftlicher Transformation verbunden werden. [. . .] Die Ziele der Bioökonomie sind [. . .] im Sinne einer nachhaltigen Entwicklung.[49]

As can be seen here, the concept of bioeconomy conceals several aspects that pursue a context-sensitive development based on normative aspects. The idea thus raises a global claim, since biological knowledge is to be applied in all sectors of the economy, but also in society. An application of knowledge for sustainable development in different sectors is an implementation of context-resilience. From a normative perspective, the concept of sustainability corresponds to an opposition to the economic dictum of increasing efficiency, which would lead to the "Umdenken in den Leitwerten und –zielen"[50] described by Schneider and Vogt. The aspects of societal change and economic transformation point to a transformational interpretation that can be applied in different contexts of resilience. Thus, the bioeconomy concept presented here meets the criteria of a positively entangled, functional and context-related resilience with normative elements that enable a crisis-proof development. How can the bioeconomy programme, which has been designated as a resilience concept, be conceived in the context of the example of comprehensive and sustainable forest use mentioned at the beginning of these considerations?

As described above, forests combine different social, ecological and economic interests, which can lead to various conflicts. The forest binds carbon dioxide, provides an important habitat for animals and plants, can serve touristic purposes and produces one of the most important raw materials for a functioning circular economy (a core aspect of bioeconomy): wood. In terms of a normative resilience, all these uses must be coordinated in such a way that the concept of resilience of one aspect does not undermine the resilience of another entity. A spruce monoculture, for example, can be profitable from an economic perspective, but is less able to withstand extreme weather phenomena such as droughts and storms than a near-natural mixed forest. Thus, from an ecological perspective, the functional resilience of a forest enterprise is harmful in the sense that an increase in yield results in a lower ecological resilience. This conflict has been largely resolved by bioeconomy's overall claim of and focus on sustainability, since the resilience of all affected areas is considered. Such a bioeconomy concept could, when applied, promote ecological resilience through sustainable use of forests and at the same time ensure the resilience of forest operations, since then, although the efficiency of timber production is not increased, a reliable and steady decomposition can take place

[49]Simon and Schaefer (2018, p. 4), "Bioeconomy describes an economic system in which biological knowledge is applied in all sectors of the economy and society and renewable biological resources are used. Its further development will require fundamentally new technologies and systemic innovations, which are associated with considerable driving forces of social change and economic transformation. The goals of bioeconomy are in the sense of sustainable development" (own translation).

[50]Vogt and Schneider (2016, p. 188), "rethinking of guiding values and goals" (own translation).

that is not influenced by extreme weather events and the associated fall in timber prices. In this way, the forest ecosystem and the steady income of a company would be equally secured. From this point of view, the resilience factors of one system are not only not harmful but can even be beneficial to another. In this case, the ecosystem and the system of corporate interests would be given equal consideration. Ecological, social and economic resiliencies are therefore not mutually exclusive, but can complement each other productively in the long term.

5 Conclusion

In this paper the concept of resilience was reflected from different perspectives. It has become clear that there are some justified concerns about the use of the term resilience, especially due to its equivocal meaning and the little differentiated normativity of the term. Furthermore, context-dependent applications of the term can be contradictory. Taking these problems into account, a framework of investigation, based on the concept of resilience, can be interpreted in such a way that certain resilience factors need to be identified that enable a system or entity to better deal with a crisis. For this purpose, three criteria must be met: (1) the promotion of a second-order resilience concept, (2) the decision for one of the three possible interpretations of resilience (conservative-reactive, adaptive or transformational) and (3) the addition of a normative dimension to the resilience concept. Following the analysis of the concept of resilience, bioeconomy can be understood as a conflict-adjusted concept of resilience, which, in the face of major crises, can facilitate a stable development of an economic system towards a more sustainable variant of it. The practical implementation of such an approach was illustrated using the example of the Federal Ministry of Food and Agriculture paper on forest use in times of climate change.

The resilience-related investigation of bioeconomy is still in its infancy. However, its interdisciplinary character and its application in different contexts offer special opportunities for the further development of an overarching, normative concept of resilience. Bioeconomy can be examined from a theoretical perspective to sharpen the concept and from an empirical perspective to identify the individual resilience factors. While it is presented as a sustainable, interdisciplinary and cross-contextual concept, there is still a lack of a precise investigation of the resilience factors of individual aspects of a bioeconomy. This includes an examination of the relationships between the factors in order to identify possible synergy effects. If the individual aspects are specifically promoted, a positive effect could be achieved for all application areas that fall under the overall bioeconomy system. In order to make such investigations possible and to identify individual resilience factors, the definition of the resilience investigation developed here can contribute.

References

Der Standard. (2019, October 24). *Fünf Ölkonzerne gaben seit 2010 250 Millionen Euro für EU-Lobbyarbeit aus*. Retrieved October 24, 2019, from, https://www.derstandard.at/story/2000110301675/fuenf-oilcompanies-gifts-since-2010-250-million-euros-for-eu

Federal Ministry of Food and Agriculture. (2019). *Deutschlands Wald im Klimawandel – Eckpunkte und Maßnahmen*. Retrieved October 10, 2019, from, https://www.bmel.de/SharedDocs/Downloads/Landwirtschaft/Wald-Jagd/Wald_Diskussionspapier.pdf

Fletcher, D., & Sarkar, M. (2013). Psychological resilience: A review and critique of definitions, concepts, and theory. *European Psychologist, 18*(1), 12–23.

Folke, C. (2006). Resilience: The emergence of a perspective for social–ecological systems analyses. *Global Environmental Change, 16*(3), 253–267.

Gleich, B., & Gutwald, R. (2017). Wie resilient sind Märkte? Eine kontextualisierte Analyse von Rohstoff- und Strommärkten. *Gaia, 26*(S1), 207–215.

Nida-Rümelin, J., & Gutwald, R. (2016). Der philosophische Gehalt des Resilienzbegriffs. *Münchener Theologische Zeitschrift, 67*(3), 250–262.

Rosenberg, J. F. (2009). *Philosophieren. Ein Handbuch für Anfänger*. Klostermann Verlag.

Rungius, C., Schneider, E., & Weller, C. (2018). Resilienz – Macht – Hoffnung. In M. Kardini, M. Schneider, & R. Gutwald (Eds.), *Resilienz. Interdisziplinäre Perspektiven zu Wandel und Transformation* (pp. 33–60). Springer.

Simon, F., & Schaefer, H.-C. (2018). Editorial. In *Jahresschrift des DNWE* 26 (special edition) (pp. 4–8).

Vogt, M., & Schneider, M. (2016). Zauberwort Resilienz. *Münchener Theologische Zeitschrift, 67*(3), 180–194.

Weiß, M., Hartmann, S., & Högl, M. (2018). Resilienz als Trendkonzept. In M. Kardini, M. Schneider, R. Gutwald (Eds.), *Resilienz. Interdisziplinäre Perspektiven zu Wandel und Transformation* (pp. 13–32). Springer.

Part VII

Conclusions and Outlook

Bioeconomy: Challenges and Conflicts from an Interdisciplinary Perspective 19

Mandy Stake

Abstract

In this chapter, the overlapping main challenges and conflicts in the bioeconomy debate that were recognized during the immersion workshop are outlined and discussed to gain a more comprehensive overview and understanding of their interdependency and severity. The challenges and conflicts are clustered around the key topics of sustainability, economical efficiency, and the human self-understanding and role in nature.

Keywords

Main challenges of bioeconomy · Bioeconomy and sustainability · Bioeconomy and ethics

1 Introduction

In this anthology, scientists from different disciplines examine modern biotechnological processes and a knowledge-based bioeconomy on the basis of their ecological, economic, legal, social, and philosophical-ethical aspects. Every chapter represents a valuable scientific contribution toward the latest research in the bioeconomical field, which is put in a context of critical ethical examination. The concept of bioeconomy requires us to engage in thorough ethical investigation and discussion that highlight the opportunities for bioeconomical approaches, but also to keep a critical and controlling eye on them.

Both the debates about the applications of biotechnological devices and the concepts of bioeconomy in general are a matter of interdisciplinary critical and

M. Stake (✉)
Forschungszentrum Jülich, Institute of Neuroscience and Medicine, Jülich, Germany
e-mail: m.stake@fz-juelich.de

D. Lanzerath et al. (eds.), *Bioeconomy and Sustainability*,
https://doi.org/10.1007/978-3-030-87402-5_19

diverse discussions. As is debated throughout this book, bioeconomy is a field with a wide range of topics, problem clusters, and solutions tackling some of the problems. One key result of the discussions is that we have to accept that there is not "the bioeconomy" and thus not "the solution" for "the problem." However, if we talk about "the concepts of bioeconomy," the idea of the interconnectedness of the different areas and topics we have pursued is supported.

The interdisciplinary discussions show a widespread connection of the topics that seem to impede clear suggestions for ethical guidelines toward action recommendations. As important as the development of such guidelines and recommendations would be, this cannot be done at a point where these interconnections have not yet been considered from a higher vantage point. Instead of creating new regulations, codes, or guidelines that could not yet encompass all topics or be applied universally, we propose to identify the overlapping main challenges and conflicts as a helpful first step toward that goal.

This chapter is based on the discussed and suggested main challenges and conflicts identified during the immersion workshop and in the written post-workshop remarks. In this light, this contribution is not a chronological retelling of the discussions during the workshop, but a systematic reconstruction of the discourse combined with a focus set by the author. The challenges and conflicts that were stressed the most are clustered around the key topics of sustainability, economical efficiency, and the human self-understanding and role in nature. Although these key topics are very interdependent, the specific conflicts and questions are outlined in this particular order to gain a more comprehensive overview and understanding. Of course, the contexts and subtopics mentioned here are not the only topics of interest in bioeconomy. In this light, this chapter is not an attempt to formulate a complete listing of the main challenges of bioeconomy in general, but rather to summarize what was accentuated the most during the immersion workshop. The attendants' works, opinions, and ideas are thus united in a new and comprehensive way and are ready to be proposed to a bigger audience.

2 Sustainability Context

Discussions concerning sustainability tackle questions of resource usage as well as the organization of economies, their processes, and products. The concept of sustainability is mostly connected to concepts of efficiency, which was critically discussed throughout the workshop. It has been suggested several times that the concept of sustainability should also encompass the concepts of sufficiency and adequacy. However, in connection to bioeconomy, the concept of sustainability seems to have become even more unclear, which is partly because of the scope of meaning of the concept of bioeconomy. The word "bioeconomy" is a composition of the terms "bio" and "economy." We will focus on both of these terms respectively and then connect the acquired understanding to the concept of sustainability. In the workshop, the word "bioeconomy" was considered to be an often used buzzword that emphasizes new and promising opportunities and technologies of an environmentally and climate-friendly energetic use of biomass. Although critical views about this understanding are already addressed in the literature, especially when it comes to the use of fossil fuels, the way how renewable biomass is used and measured to be more sustainable (for instance, by the high water footprint of green bioeconomies) or the connected risks that the expansion and intensive use of agricultural land will further increase the loss of biodiversity, is often disregarded in general discussions. Furthermore, the term "bio" in the compound word "bioeconomy" is first and foremost typically associated with economies using biomass as resources and for biotechnologies; it must not automatically be understood as an organic, good, or sustainable use of resources. These facts lead to the conclusion that, other than normally specified, advocates of bioeconomic concepts *cannot* yet oversee all the interconnections and that bioeconomies are by no means environmentally sustainable per se, as Marion Stahl, Florian Fiebelkorn, and Ulrich Schurr stated. As a result, the former impression of the positively connotated meaning of "bioeconomy" unexpectedly changed toward a more restrained understanding during the workshop: bioeconomic processes, products, or whole economies are not automatically bio, organic, good, or sustainable.

A similar thought was expressed concerning the concept of sustainability. In general, this concept is regarded to be of great importance for an environmentally friendly coexistence of the human life form with all other life forms. However, it can also be misused easily since it is still not politically and scientifically decided how to define what "sustainable" exactly means. If, for instance, ensuring more sustainability already means to use more fiber from woods to produce packaging instead of using plastics, and if we need fast growing trees or other plants growing in plantations for that, then the package could be thought to be more sustainable than plastic ones only because it is ecologically degradable. However, this production process is connected to other problem clusters concerning, for instance, the loss of diversity in plants and natural living spaces for animal species in former forests and landscapes that are then transformed into plantations. This type of "sustainable" production could only consider one-sided ways of planting, growing, and cutting down plants by "efficient" machines as much as possible in a shorter amount of time, which can have detrimental effects on the health of the soil, the diversity of plants and animals, and therefore the ecosystem. Such an unbalanced and short-sighted way of using the concept of sustainability as a label should be taken into consideration since it leads to widespread misconceptions in pursuit of more sustainability.

Some participants of the workshop, for instance Patrick Hohlwegler, warned about an increasing danger of greenwashing due to the labeling of bioeconomy as a "sustainable economy." As can be seen in the example above and the discussed politically promoted strategies, the economic paradigm of growth remains and outweighs new strategies like "degrowth." Strategies focusing on degrowth not only aim at the well-being of every person, but also at the preservation of the ecological basis of life. As such a strategy tackles the given paradigm of efficiency and constant economic growth, fundamental changes in and of the human living environment as well as substantial political, economical, and cultural changes are required.

This is why, apart from the "bio" in "bioeconomy," we should also be aware of the meaning of the word "economy" in the compound word "bioeconomy": bioeconomy is and was designed as a part of economy. As recent history shows, the most *efficient* way to use resources or the answer to the question what makes the most economical sense is usually not what is also best for keeping a balance and respectfully using natural resources and landscapes for our purposes. Implementing concepts of bioeconomy could thus also open new paths of exploitation, if we do not state more clearly what "efficient" in the context of sustainability actually means. We can see glimpses of the effects of such a fuzzy understanding of "efficient" in our everyday life but also by taking a look at definitions that characterize nature predominantly as a constant supply of biomass for resource usage. Also, new methods or biotechnological processes are called "efficient" in order to obtain those natural resources.

2.1 Sustainable Land Use

Concepts of bioeconomy offer solutions for a more sufficient and sustainable use of land. Similar to the concerns raised toward a risk of greenwashing, Hohlwegler also considered the danger of green grabbing (or land grabbing): A comprehensive switch to biomass as an energy source could increase conflicts about land use toward more bioenergy crops or crops for bio-industrial use. As Irwa Issa outlined, the land used for this purpose is about 10–15% of the worldwide arable land, whereas almost a third of global cropland, especially in industrialized and emerging countries, is used to produce food for animals in order to further produce meat and other animal food supplies like milk and eggs. In the future, the purpose of existing landscapes and areas that are now used for food production could change in favor of growing crops used for bioenergy. Dedicated land use for bioenergy purposes stands in direct competition to land use for growing food or animal feed. According to Issa, this needs to be seen as critical since world market prices for plant-based foods like rice, lentils, or beans are already increasing because of these circumstances, making the living conditions and social situation worse for poor people. Similarly, Issa emphasized that there is growing criticism against deforestation and especially against the conversion of virgin forests into agricultural land to produce palm oil, soybeans, or beef for export, as for instance in Southeast Asia and Latin America.

Land use is still the biggest factor for species extinction. The way that cultivated land changes the landscapes also changes the idea of what the land is for as well as the meanings and values of naturalness, nonhuman life forms, and the human life form's relationship to the natural environment, as Christina Pinsdorf pointed out. Ethical debates about principles, values, and their importance in everyday life play a major role in differentiating current, possible, and more desirable actions toward environmental surroundings: How should a forest be used or designed and for what purpose? Does everything need a specific purpose and how is it affecting adjacent variables? Can we even speak about variables as we would be able to measure and predict outcomes in the most "effective" or "sustainable" way? The usage of land plays the most important role in the production cycle, and one of the biggest is the production of food and feed.

2.2 Food and Feed Production

The discussions about food and feed production mainly focused on new technologies and solutions for better production rates with less environmental impact. It is clear that the current situation of mass productions and especially the global increase of meat consumption are the main and most problematic stress factors for the Earth's climate system: Deforestation of land used for food, feed, and animal farming comes with enormous agricultural emissions like greenhouse gas emissions, but also soil acidification, water pollution, and desertification. In vitro meat production and insect food as meat substitutes seem to be promising new ways to provide enough nutritious food for the increasing needs with less emissions and good protein availability for more people. In the lectures of Birgit Beck and Florian Fiebelkorn, it became clear that this could reduce environmental problems as well as ethical conflicts in connection with factory farming, if established as a mass product.

By the general definition, the agricultural sector has always been a bioeconomy. With newly introduced biotechnologies and digitization, further intensification and more efficiency in the production processes are possible, Katrin Beer stated. However, this fact alone does not change current distributive injustice, as for instance the provision of food, energy, and equal possibilities of lifestyle choices. There can be no fundamental change in the availability of food, resources, or energy supply through bioeconomy alone. As especially Stahl outlined, it also does not aim to change consumer behavior toward a more modest consumption but rather justifies current behaviors of mass consumption through the "bioeconomic label."

Food shortages must not be understood as the result of production methods, but as the consequence of the type of resource use and distribution processes. From a moral point of view, there was a wide agreement between the attendees that it would be necessary to fundamentally rethink our "Western" lifestyles and in particular our eating habits. According to Stahl, one way of doing this would be to not keep the standards of consumption but rather decrease our consumption of certain food groups or find alternative nutritious diets altogether. This alone could be an effective remedy against mass factory farming and the associated ethical and ecological problems that come with it: inefficient food sources like meat, food waste, and nutritional problems lead to basic predicaments that could be dealt with or be solved already today if the food sector was just organized differently, as Beer pointed out.

Another problem concerns the export and import of food supplies: Lioudmila Chatalova emphasized that industrialized countries, that are playing a pioneering role in bioeconomy, are getting most of their biomass for feed as well as new alternative "sustainable" products and material use from developing countries. As a result, a growing demand for biomass could shift the sustainability problems even further abroad, i.e., especially toward countries with less developed institutions, and exacerbate rather than solve them.

However, it is not clear which authority could make distribution decisions on a global level. So far there are some moral guidelines, such as the Sustainable Development Goals (SDGs) and the Universal Declaration of Human Rights. Nonetheless, there is no political authority that translates these guidelines, goals, and rights into binding rules for a fair distribution on a global level, so that all countries must adhere to them. Beer suggested that a global democratic process would be needed to regulate these global distribution issues: The liberal market economy mainly leads to exploitation and increasing injustice. Since global distribution and justice problems seem to be the result of insufficiently regulated global markets, instead of continuing these insufficient practices, a global social market economy with redistribution mechanisms and taxes was discussed as a possible solution to help tackling the problem. At a national, regional, or local level, this would lead to different dynamics in different areas, shifts in power, and related conflicts of interest with traditional commercial enterprises and political rulers, according to Beer.

2.3 Bioenergy

New technologies and processes enable more efficient uses of resources and could quickly be leveraged in a transformation phase toward a bioeconomy. However, too much waste is still produced and remains unused nowadays. The fact that residues and waste materials have not yet been used efficiently was a moral concern for the majority of the attendees since this practice would be one of the most obvious, feasible, efficient, and sustainable ones to save natural resources and energy, to expand recovery times for natural habitats, or to even shift the human mindset from a more or less grabby one toward a more humble and moderate one. Also, the goal of providing everyone with clean sustainable energy was considered to be a main critical issue. It was suggested that we need to define clear boundaries for resource usage as well as for the concept of efficiency. This could be done by establishing more coupling and cascade uses of resources in general in order to gain as much added value as possible from as little material as possible. This also includes a need to use the available resources in as many different ways as can be achieved. As emphasized by Beer, a circular economy with (almost) zero waste and the goal of creating each end product in such a way that it is a starting product for the same or a different product would be a convincing and pursuable path toward more sustainability. Beer also stressed that questions like how waste could be prevented, what waste cannot be currently recycled, what could be used energetically and what materially, or what waste hierarchy, waste policy, and the impact of economics, products, and earnings are better, must be considered. The attendees agreed that there is a need to set limits to sustainable and nonexploitative uses of resources. Bioenergy and biomass could be used energetically within these limits. Another important issue would then be which products, processes, techniques, and strategies will be accepted by society.

2.4 Social Acceptance

Social acceptance seems to play an important role for the introduction and implementation of new applications, technologies, processes, and products. This is why all mentioned aspects must be discussed in a larger social context and with a focus on everyday usage. However, the gap between research and everyday life practices for all people still seems to be too big. Since non-sustainable ways of living are considered widely "normal" at the moment, they are thus also widely accepted.

In the context of bioeconomy, people are often referred to as "consumers". One way of studying the gap between people's life practices and new developments in biotechnologies and bioeconomical concepts thus is to analyze consumer acceptance rates. In this manner, it can be investigated, as for instance Fiebelkorn did, to what extent consumer research can influence developments in bioeconomy, biotechnology, and everyday life practices, and vice versa.

People have an important role in understanding that the choices they make, the products they buy, and the waste they produce are all key factors changing our environment on an everyday basis. Awareness of the issues and environmental problems is already rising because a lot of information and knowledge about the key factors of environmental destruction, consumption, waste, exploitation, and pollution is given and communicated by scientific scholars and experts. Also, knowledge about possible solutions is already offered by various scientific writings, the media, and multiple aid organizations or other nongovernmental organizations. This information is in principle available to all. However, we can see that the everyday life choices of the people are only slowly, if at all, following the proposed advices given by experts.

Awareness about those topics and problems was considered as important but also as overrated by the attendees. Although knowledge is helpful, it does not do anything on its own: it must be applied. However, the successful application and implementation of new and more sustainable living habits is exactly the crux of the matter since this is going against already implemented habits that we are emotionally connected to. Applied perspectives always have strong cultural factors and values that are criticized or seen as endangered by new habits, processes, technologies, or foods. Thus, even if some already implemented or applied behavior is considered to be irrational, it seems hard to change it for a better one. Being aware of the problems and the solutions without being overwhelmed by all the information in our already complicated lifestyles seems not to be easy, as was highlighted by Dirk Lanzerath. However, this should not be seen as a reason to not trying to implement new and more sustainable habits.

In order to have a successful implementation of bioeconomic concepts and technologies, it was therefore suggested that we should not only focus on economic, technological, and ecological aspects, but also consider social and psychological aspects. This approach requires the willingness and enforcement of public communication to acquire understanding, discussion, and possible social acceptance of bioeconomic processes and projects. In this light, Stahl proposed that appropriate communication strategies should be created, but that they should also leave space for criticism and alternative suggestions. In order to address the new potentials, challenges, and risks of bioeconomical and biotechnological solutions, the attendees agreed that an equal debate between *all* of the groups of interest, i.e., civil society, sciences, industries, businesses, and politics is needed, in order to exchange, discuss, and weigh the different values and interests of individuals and groups. Such dialogue-oriented and deliberative procedures for consensus-oriented conflict situations are already recommended for the discourse about genome editing by the German Council of Bioeconomy, Robin Siebert accentuated.

This procedure creates collective responsibility for all sides, the consumers *and* the economies, industries, and businesses. Since the practice to generate more profits at lower costs is mostly associated with different types of exploitation, such as the depletion of natural resources, which negatively affects the natural balance and the ecosystem as well as the unequal access to these resources, and continues the exploitation of human labor, changes must be made: economies, industries, and other companies would also have to take responsibility and shift their focus from more profits with less costs to new applications that are agreeably and fully sustainable and fair for everyone.

All in all, it was suggested, as for instance by Beer, to think about the question whether sustainability and related topics like the resource usage or, more general, the value of nature should better be raised outside of or at least accompanied by discussions of the concept of bioeconomy. Since bioeconomy is designed as part of a circular economy to reach higher efficiency, we should carefully consider what "highly efficient" means for whom. Some ways of production only seem to be highly efficient in the short run, but are actually highly destructive and problematic in the long run. A good example of this claim is the way we plant monocultures around the world that are detrimental to diversity and especially the soil. Without better definitions and clearer concepts as well as ethical evaluations in every step, bioeconomy as an economy based concept cannot be the only comprehensive solution to counteract sustainability problems.

3 Economic Efficiency Context

Whether bioeconomy is able to contribute to encounter global environmental problems or rather stabilizes existing system structures of policies, market strategies, and the exploitative human-nature relationship by the systematic background context of following a policy of maximum efficiency and capital increase, is a very important question to be conscious of. If the problems are systematic, then we should tackle them at the core. This would mean thinking about new ways of creating economic businesses and power dynamics.

In the context of economic efficiency, these questions are approached with regard to economy and biotechnology. If we are able to rebuild economy in a more sustainable way and call it "bioeconomy," we should also be aware of the processes in this systematic transformation.

3.1 Economy: The Concepts of Efficiency and Sufficiency

Since bioeconomy is by definition a form of business, the limits for economic activity should be set politically and according to societal needs. In general, "economic use" means that capital is used to generate added value to achieve more capital. This value is attained by various types of resource usage and by producing goods for consumption. The consumers are on the other side and intersect with economy by various means. In this respect, the ethical issues that define margins of a concept of bioeconomy can be addressed with the following questions, as Beer concludes: Which resources can be used to what extent and in which way? How are the goods produced and distributed? Who gets what? Who is entitled to what and why? What are legitimate needs? What are fundamental rights? Where must the freedom of individuals be restricted to ensure the well-being of everyone?

From an economic point of view, at least some of these questions can be outlined and empirically surveyed. The underlying assumption is that we can measure and set objectives to the questions raised. However, the premise that all of the components in question are principally measurable and can be expressed by formulas leads to serious consequences. In some of the participant's contributions, for instance in Wesseler's and Kardung's, several conditions that we would think to be not measurable per se were proposed to be measurable goods, such as happiness. However, if a "felicity condition" is included in formulas that measure happiness, a definition of happiness and an explanation of how it can be measured must be given as well. It was suggested that happiness can be measured by the purchasing power of a person. Yet, as it was criticized by some attendees, this cannot be the only "normative" and objective condition that stands for "happiness" and it also leads to fallacies: it presumes, for example, that there is no happiness without consumption, it also disregards the fact that growing economies can lead to a growing social and economic inequality and then again could lead to a decline in happiness. Apart from these facts, there are clearly a lot of options to gain happiness or felicity out of nonconsumption ways—the easiest example would be fasting for health or religious reasons, or living according to the concept of asceticism, as Pinsdorf pointed out. In addition to that, Pinsdorf and Sebastian Lenze argued that a whole catalogue would have to be formulated that takes normative aspects into account and their argumentative weight would have to be discussed first. This is a clear line of conflict since it is not only unclear what happiness is, but also how we should measure and differentiate between happiness as a good for individuals and for the society or as for "the greater good."

If we go further and ask about the goals and the implementation of new production strategies and material usage, we have to acknowledge that even this is done in the "old way," namely by maintaining the economic fundamentals, production methods, and consumptions habits that are mainly not sustainable or even justify the creation of social costs, Chatalova emphasized. It is this fundamental economic condition of a constant pursuit of a growth-oriented approach that leads to today's main sociocultural and environmental challenges. Stahl stated that the very concepts of efficiency and outcome play a major role in the economic models, as they do in bioeconomic approaches.

Basically, it represents the ideas of "having your cake and eat it too," "produce more efficiently in terms of investing time and money to have the same goods," or "more new sustainable ways but only without cutting back financially or on personal wishes and preferences." Besides, there was a very simplistic, misleading, and outdated idea brought forward during the discussion, namely that one cannot manage what one cannot measure. However, this idea is flawed because it does not really make any claim about the goods or values that are measured. In this context, Pinsdorf stressed that this "price tag talk," as she called it, excludes important values and goods that are not, or at least not easily, measurable as for instance happiness, experiences in nature, well-being, or the value of life of the human and of nonhuman life forms.

If we do want to discuss outcomes and risks, it is not prima facie apparent why we should not use the precautionary principle for doing so. However, the interpretation of the precautionary principle as arguing along *unknown* risks seems to be not very helpful, according to Venghaus. Pinsdorf, Lanzerath, and Lenze proposed that we should rather focus on what we *do* know about risks and benefits, social values, and individual values, which include more than economic aspects. Especially Pinsdorf considered this to be a broader and more helpful approach than creating a one-sided and extremely narrow approach that only considers measurable economic aspects.

As was shown in particular by Chatalova, there is a need to introduce the concept of sufficiency in order to create sustainable results. Sustainability can be seen as a stability factor in bioeconomic systems over a longer period of time. This concept clashes, however, with the economic value of efficiency. The question that Lenze introduced in this context is: why do we still follow this paradigm of economic growth and an increase in efficiency, when it is shown very clearly that it is not ecologically and thus not economically sustainable in the long run? The answer depends, again, on what we mean with "efficient": we can introduce the term as it is commonly used in economics, but we can also use it in a more profound way of "sustainably efficient" that *in*cludes the concept of sufficiency rather than *ex*cluding it by definition.

If we, however, continue to reduce the environment to ecosystem services, then we would obtain the same results and economic "benefits" in terms of financial growth, which propels us to the same situation where the system of capitalistic economy has already brought us—this seems to be the root of the problem. The concept of a circular bioeconomy could be helpful and positively contribute to a change in the debate: If we ask how much benefits we get out of an action, then according to Pinsdorf the term "benefits" must not be understood only in economic terms, but rather be considered to relate to other goods as well—like understanding something new or having a broader knowledge about results and outcomes that we can transform again into better solutions and so on. This approach would engage with the basics of our understanding of capitalism and asks how it may be transformed.

Chatalovas' concept that stands in line with exactly this new way of understanding creates a new challenge to basic economic principles. Following her proposal, an overutilization of resources can be avoided. As sufficiency is shown to be indeed "more effective" than efficiency in the traditional understanding, the optimal solution can be to produce less in order to gain more.

By being so very interconnected, the concept of bioeconomy could be a new economic concept that also establishes a new openness which is helpful to tackle the outdated ideas of a profit-oriented economization of nature. As Schurr, Lanzerath, and Pinsdorf pointed out, instead of only operating with quantitative criteria, the discussion opens up to also include more qualitative criteria. Such criteria address various research fields in the natural, economic and social sciences and thus give rise to a new interdisciplinary challenge.

3.2 Biotechnology

Technology enables new and more efficient processes. Biotechnologies are means to use or modify living systems, organisms, or processes to develop products, improve production procedures, or enhance processes in living organisms. However, especially in the context of bioeconomy, the development and application of modern technologies and techniques must take into account their consequences.

To simply implement new techniques seems to be a rather short-sided thought and to disregard questions on how to assess technology and its risks. In this context, the precautionary principle was again referred to, which would have to be examined in various contexts and from the perspectives of different disciplines, as Jana Schoop proposed. The application of biotechnologies has strengths and weaknesses: As over-regulation by bio-based technologies and bioprospecting is a danger that could misbalance natural developments and processes, so could under-regulation be a detrimental cause of complying with social and environmental standards. Within this context, Patricia Osseweijer and Jan Börner emphasized that this would be a challenge for governance through legislation.

The belief in technological progress shows an overwhelming optimism regarding technical solutions to our current environmental problems. Nevertheless, one of the most pending questions was whether we can solve moral problems with technical solutions since they could have destructive late effects that cannot be considered in the present because of our lack of knowledge. There was an overall agreement between all participants that a complete trust in technical innovations alone is not sufficient to tackle the moral problems we are facing today. In the worst case, they could potentially lead to more or new moral problems, which we then would have to think about for further technological innovations and so on, Schoop argued.

The assumption that technology could solve problems that were caused by technology is the idea of a "technological fix" that has yet to be proven valid. Sustainability, social equity, and distribution are at the core of this problem cluster. Even if we use other resources, use existing resources differently or more efficiently, these procedures alone do not contribute to more justice, social equity, and sustainability. There are still rebound effects and distribution mechanisms that play a role in exploitative behaviors of the past, present, and future. Superordinate questions arise that lead to new challenges and conflicts: Nina Gmeiner, Max Mittenzwei, and Leonard Prochaska raised the question, to what extent bioeconomy is accompanied by a socially undesirable concentration of power through higher technology in individual sectors, and when this, as a social problem, outweighs generally desirable innovations. What could a post-capitalist or post-growth economic order be? How can we understand efficiency in a sense that goes hand in hand with sustainable movements?

Beer emphasized that such questions would have to be carefully considered and clarified in a social discourse. They do not only concern bioeconomy or biotechnological means, but our very everyday lifestyle choices, the organization of society, and world politics.

3.3 Process of Transformation to a Bioeconomy

The questions of the last paragraph are closely related to the challenges of a transformation process toward a bioeconomy. The interconnectivity of global and local problems requires solutions that observe local *and* global contexts. A central idea in the discussions of the immersion workshop was that a circular bioeconomy must be thought regionally *and* globally. In each closed economic cycle as well as coupled and cascade concepts, local cycles were favorably prioritized in order to enable and secure short transport routes and less emissions. A consequence of this would presumably be an increasing amount of more localized and decentralized markets that countertrends the globalization and centralization of industries that create the current power monopoles of industries, corporations, and states. Beer pointed out that, following this idea, resource usage would have to be planned, organized, and used differently, regional potentials would have to be analyzed and evaluated in the light of new technologies and the newest knowledge of the ecosystem of the region. However, other questions arise in that context: For instance, Gmeiner, Mittenzwei, and Prochaska questioned how differences in the development of bioeconomic concepts in the cities vs. the rural regions can be dealt with, or how difficulties between resource-rich versus resource-poor countries can be avoided: How local is local? Do we need any further distinctions? It could be challenging to promote a bioeconomic concept in the periphery, resource poor, and biotechnologically less developed areas. Yet, as we have already seen, it is also questionable whether the implementation and usage of the newest biotechnological means are at all criteria for a nation to be recognized as being well developed.

The process of the transformation to a bioeconomy is a highly political and social transformation. In this light, it is strange how much it is primarily recognized in elitist, academic discussions as well as by country leaders of so-called "Western countries," or heads of budget strong economies and industries, but less or not at all in public discussions or the daily news. There is a clear asymmetry between the stakeholders concerning the given information and knowledge about bioeconomy, biotechnologies, sustainability programs, and their implementations. If there are information events on bioeconomy, they are mainly product- and consumer-oriented. This has something to do with the fact that people as consumers have a key role in accepting the new products and solutions for them to be economically successful and create capital. Noticeably in these information events, there are already constructed solutions that are proposed and sold to the consumers. In these contexts, consumers mostly do not have a critical co-constructing, inclusive, and collaborating function. This is surprising because bioeconomy is marketed as a bottom-up innovative non-elitist concept that is constructed to provide for our needs while protecting the environment and the next generations to come, that drives product developments, ensures food security, proposes bio-based solutions, more sustainable alternatives, and promises economic growth and the creation of new jobs. However, the distribution of benefits and risks as well as the regulation and governance of a bioeconomy transformation are mostly in the hands of the few rich power possessors, whereas those people who are not in that power possession are primarily seen as consumers. This can easily be recognized in the bioeconomic debates about "consumer communication"—it is not called "people communication" or "social communication."

The successful integration of bioeconomic concepts in recent policy strategies can also be seen as generating a new and modern version of the same old basic principles making it only *sound* more innovative. We must ask: Where does the capital come from and why is so much money invested in these new technologies? Surely it is true that this happens partly out of an external pressure to act with regard to the increasing environmental problems that we are slowly getting more aware of over the last 10 to 20 years. The detrimental effects of decades of mistreatment and exploitation of nature start only now to increasingly affect us environmentally. However, the other part of the story is that these effects could have been seen much earlier. The so-called "developing countries" are suffering from it far more and for a much longer time, although they produce far less emissions and waste compared to the "western" world. It is mainly "western" factories and industries, the fat economies, with their monopolization and outsourcing strategies that are creating environmental challenges. The troubled lean economies have often been communicating openly about the environmental impacts, however, they are yet to receive due acknowledgment, interest, or contribution to tackling the problems created by the western countries, corporations, industries, or organizations. Why is it that especially western science gets funded by these very countries, corporations, industries, and organizations that recognize a market need for more "sustainability," "bioeconomy," or "more efficiency"? Why do they only now start to create strategies to "reverse" their doings of exploitation of people, land, and resources by saying that these new technologies can solve their self-created problems? Why are monocultures and forest clearing in "developing" countries (palm oil, soybeans, etc.) heavily criticized but in Western countries (wood, rapeseeds, etc.) not as much?

These difficult questions are only addressed to a certain degree by environmental and conservation organizations and cannot be answered easily. But it is one way of critically questioning the economic basics and underlying power structures that brought us where we are in terms of the transformation processes right now.

Power relations are real, leadership competitions happen, and a widespread overwhelming feeling of helplessness and powerlessness in a lot of people that are facing too much information and decisions daily are key factors that we have to think of. People have the freedom of choice and decisive control, but they also need the education and the awareness of the value of their decision-making and choices. That means we have to focus more on raising questions and giving easy access to information to everyone in order to have eye-to-eye discussions. This requires considerable work and time investment. First, we must really hear what the needs and the intentions of all the involved interest groups are and reach a well-informed understanding of those. Only then we can start to create solutions together in form of a real collaboration that is non-exclusive and not arrogant toward people in need. Since social transformation is based on individual transformation, only a way of open communication toward and informed collaboration with every interest group enables informed and responsible decisions. Thus it is too short-sided to view individuals only as consumers and to measure the value of their choices by the capital that they create. The formation of a well informed and critically thinking society that is eager to learn more and takes part in the political and economic discussions should be a priority. Every single person is part of that group. Since this seems to be the starting point of any further discussion in this context, the third and last main challenging context is that of human self-understanding and human's role in and toward nature.

4 Context of Human Self-Understanding and Human's Role in Nature

The role of the economy in society, questions about what and how resources may be used, the discussions about new bioeconomic concepts, or biotechnological innovations and solutions to environmental, social, and economic problems are rather surface talk. They are all part of the problem, but are not directly addressing the problem's roots.

Language and communication play a large and very important role. Words matter: as we have seen, the meanings of words vary and have effects on concepts and their outcomes. As stated by Gmeiner, Mittenzwei, and Prochaska, the understanding of well-being as a goal concept, for instance, varies between business and ethics. Also, our understandings of the words "efficient" or "sustainable" vary: they can be used in an economical and mechanizing way, but they can also hold great potential for a societal- and individual-oriented understanding that recognizes, respects, and appreciates the relationships between humans, humans and other life forms as well as allow a more confined understanding of the human being as a part of nature. Some of these connections are measurable, like the outcome of our concepts and the way we apply them, but how we experience and live with such an understanding, our connection toward it and our understanding itself cannot be measured, but is a matter of empathy, expression, and comprehension that is open to experience and may be analyzed by other scientific means of social and cultural sciences.

There seems to be a tension between the value of economy and the value of nature. A bioeconomic approach might be able to open up the understanding of nature and its life forms as intrinsically valuable instead of promoting an instrumental understanding of nature and its life forms as predominantly economically valuable, Schoop emphasized. The integration of the sustainability concept is one way to resolve this tension by changing the concepts of efficiency and economy and fit them into sustainable lifestyles and production forms that value nature for its own sake rather than for the sake of its usage for people.

The way we engage with each other and with nature is seen in our everyday practices which can thus be evaluated as good or bad practices. This evaluation is dependent on certain criteria that have to be discussed and listed. Two possible criteria the evaluation could be based on could be: the common or general good for all living beings and a clear understanding of the concept of sustainability. Although on the one hand, there is no one clear rule of what is to be called a good or a bad practice, on the other hand there are practices that can immediately and clearly be called as such based on the evidence and consequences they cause. Exploitative behavior, for instance, is a bad practice that can be evidently recognized through the consequences we are facing in the environment and social interactions right now. According to Dieter Sturma, this is reason enough to take responsibility and change our practices that do not only concern the way we use resources, but also how we value them, how we value nature, and how we value ourselves and our role toward and within nature and society. These values are not only economic values that are part of an economic formula. Pinsdorf, Schoop, and Stahl addressed that we rather need to consider the value of the human life and arguably also the intrinsic value of all life forms independently of an exclusively human-centered evaluation and find ways to integrate those into our ethical concepts.

The precautionary principle should also be assessed in an interpretative way that considers risks of actions beyond the economic losses. The evaluations related to economic growth and the monetization of nature are only one way of talking about values. It seems to be clear that this kind of evaluation is part of a bad and exclusive practice toward people, other living beings as well as natural living spaces and that we would need to reconsider our roles and choices of lifestyles, our consumption decisions, the economic system, global trade, market designs, prices, and costs. This procedure questions the whole capitalist system upon which also bioeconomic concepts are based.

5 Conclusion

We explored the main challenges and conflicts that were addressed in the immersion workshop. These were summed up according to the main contexts of sustainability, economic efficiency, and human self-understanding and humanity's role in nature. We have to keep in mind that all the contexts are interconnected. However, as we are able to conclude now, some contexts are more fundamental than others. The discussions started with questions concerning sustainability which is the main goal promoted by bioeconomy concepts as well as biotechnological means. The focus in this context lies more upon the land and resource use as well as the products and means that we gain from that. Although an anthropocentric view on environmental and social development toward more sustainability was reasonably questioned, we have to acknowledge that it is also appropriate to promote sustainability and offer information in a way that people will accept more willingly, encompassing also values that are not quantifiable. Social acceptance thus does not only mean "consumer acceptance" or "producer and investor acceptance."

The second context of economic efficiency comes into play because of the need to have other perspectives than the one that only calculates with economical risks and efficiency rates. This presupposes to question the meanings of the very concepts of efficiency, value, and sustainability in bioeconomy. Any recent concept of bioeconomy relies on basic economic and capitalist principles. This is also the main reason why any bioeconomy aims toward more efficiency. The current main goal is to use less biomass to produce more but better versions of the products by more efficient, i.e., cheaper or less destructive means. This is what is then called "more sustainable," and partly rightly so: Compared to the former practices, such a new production line is indeed more sustainable in terms of the amount of biomass that is used, the way it is produced, or the ecological footprint that it has. Nevertheless, the concepts these new products are relying on fail to recognize more forward-looking societal and economic changes that are suspected to be at the roots of the environmental problems. In any system of bioeconomy where the value system is based on an anthropocentric perspective, nature conservancy is egoistical since it is not created to protect nature for its own sake or because it is inherently valuable, but because it is necessary and better for human survival and life.

The current developments of bioeconomy are surely very exciting and full of new inventions and sensational insights into natural processes and interconnections that stimulate further discussions and technologies. We can acknowledge that technology itself is not good or bad, but that our actions make it so. Who we are as persons, as a people, and as a society depends on the way we use technology and shape our economic processes. We thus have to be very mindful about the choices that we make. With the dangers of the system that even concepts of bioeconomy are relying on to keep the capitalist system and thus speculation, distributional asymmetries, and strong competition going, we also have to be mindful about our expectations, needs, and choices. This brings us to the third context of the human self-understanding, which seems to be at the heart of the matter.

What matters is what kind of human beings we are and want to be. It is the remarkable power of the human life form that we are able to determine what we want to be, how we want to act, and in what system we want to live in. The fact that more countries are starting joint discussions about new ways of implementing bioeconomy on a global and more collaborative level, is a tremendous step toward more humble practices and the protection of the environment. However, if we want to develop a more sustainable and just society, we also have to address the given structural and political problems that are hindering this goal to be achieved. Together, we have to figure out ways to avoid all forms of exploitation and consider how we are able to decide in the most responsible and humble way. This indicates that we would have to reconsider and think in new progressive ways to take active steps towards a truly balanced, cooperative, and reciprocal give and take of natural resources, ideas, and knowledge that lies in the interest of all of us. Is this possible? To answer this question, the biggest challenge of all would be to define new frameworks as well as political and economic regulations that are truly in line with the ideas of equality, equity, economic degrowth, and sufficiency, which are, however, not yet central in the bioeconomy debates.

Bioeconomy Beneath and Beyond: Persisting Challenges from a Philosophical and Ethical Perspective

20

Christina Pinsdorf

Abstract

The concept of bioeconomy is currently discussed worldwide as an attempt to solve global problems relating to climate change, ecological crisis, and global population growth. Bioeconomic applications are of enormous range and affect key sectors of society, such as the food and feed sector, the energy, transportation and construction sector, the chemical sector as well as the textile and clothing industry. Social and environmental justice are meant to be central aims of the concept of bioeconomy just like sustainable economic growth and prosperity. But as promising as the concept of bioeconomy may sound, it still faces various challenges, both from a more theory-driven philosophical perspective and from a rather application-oriented ethical point of view. The present study analyzes persisting philosophical challenges underlying the concept of bioeconomy in view of tensions concerning the relations between economy and man as well as between economy and nature and reveals bioeconomic promises and disillusions. Persisting ethical challenges are scrutinized on the basis of the Precautionary Principle (PP), the principle of Responsible Research and Innovation (RRI) as well as the differentiation of a technological and a behavioral fix. Eventually, it is argued that bioeconomy is no panacea. What is needed rather is a great sustainable transformation to globally address the urgent ecological, social and economic problems of the Anthropocene.

Keywords

Bioeconomy and philosophy · Bioeconomy and ethics · Great sustainable transformation

C. Pinsdorf (✉)
Institute of Science and Ethics (IWE), University of Bonn, Bonn, Germany
e-mail: pinsdorf@uni-bonn.de

D. Lanzerath et al. (eds.), *Bioeconomy and Sustainability*,
https://doi.org/10.1007/978-3-030-87402-5_20

1 Introduction

Among the most challenging threats for humanity living in the Anthropocene are climate change, ecological crisis—along with the destruction of nature and biodiversity loss—and food security for a growing human population. At this point in time, the most prominently discussed concept worldwide to address these challenges is the concept of bioeconomy. It is of enormous range and affects key sectors of society, such as the food and feed sector, the energy, transportation and construction sector, the chemical sector as well as the textile and clothing industry. Within the concept of bioeconomy very important issues such as the allocation of renewable, climate-friendly energy- and resource-supplies as well as the rejection of the current dependency on fossil resources like coal, oil, and gas are addressed just as the global food situation. Furthermore, social and environmental justice as well as sustainable economic growth and prosperity are listed as the most prominent aims of the concept of bioeconomy.

However, as promising as the concept of bioeconomy sounds, there are various persisting challenges discernible both from a more theory-driven philosophical perspective and from a rather application-oriented ethical point of view.

In what follows, general philosophical challenges still underlying the concept of bioeconomy shall be examined and major ethical challenges still evoked by it shall be ventilated. For a clearer structure, the philosophical challenges are differentiated under the subtitles "Economy and Nature," "Promises and Disillusions," and "Economy and Man." The first subtitle covers the ongoing debates between neoclassical and ecological economics, the controversy over the interpretation of bioeconomy as economization of nature or ecologization of economics and the dispute over the adequate approach to nature as natural capital or intrinsic value. The second subtitle discusses two main promises of the bioeconomic concept. By means of the ideas of decoupling and a circular bioeconomy their related disillusions are briefly outlined. Under the third subtitle "Economy and Man," the relation between economic growth and human flourishing is scrutinized just as the connection between economic growth and capitalism.

The following part focuses on remaining ethical challenges evoked by the concept of bioeconomy that are mainly tackled by means of two internationally qualified moral principles for the ethical accompaniment and review of modern biotechnologies, especially concerned with their ecological, social, and economic record: the Precautionary Principle (PP) together with the principle of Responsible Research and Innovation (RRI). Last but not least, the ethically relevant question of technical solutions to moral problems is introduced. This question is raised in the juxtaposition of a technological and a behavioral fix as proposed solutions for environmental, social, and economic injustices prevalent in the Anthropocene.

The upshot holds that the concept of bioeconomy is not a panacea for the urgent challenges of our time, but only one piece in the puzzle of possible solutions that needs to be handled with care; especially in view of economic—whatsoever green—growth. Ecological, social, and economic justice on a finite planet may only be achieved via a *Great Sustainable Transformation* that establishes new ways of human flourishing within ethical and planetary boundaries.

2 Philosophical Challenges Underlying the Concept of Bioeconomy

Denominating an outstanding transdisciplinary project, the term "bioeconomy" lacks conceptual clarity. Oftentimes it remains unclear, who in an interdisciplinary context means what when talking about the concept of bioeconomy. Furthermore, the term has undergone various conceptual changes in the course of its history[1]: Starting off in the 1980s—formulated by the economist Nicholas Georgescu-Roegen (as well as the Club of Rome as an idea of economic modesty located within biophysical limits)[2]—its focus became very much constricted to genetic engineering during the 1990s and was accompanied by high, yet still unfulfilled expectations. At the end of this decade, geneticist Juan Enríquez-Cabot was considered to having delivered the definition of "bioeconomy" as a tool to use new biological knowledge for commercial and industrial purposes.[3] Since 2000 the focus shifted again, this time involving Artificial Intelligence (AI), Big Data and their fusion with biotechnological developments. More recently, the original connotations of modesty and economic limitation sometimes sneak back into the concept of bioeconomy, when economic growth—however green it may be—is not thought of as the highest-ranking goal, but ideas of a—not necessarily growing—circular economy return to mind instead.

A glimpse at the historical variation of meanings shows that bioeconomy is a politically-scientifically informed concept of economic transformation,[4] and it becomes at least more comprehensible why there are so many opposing views and assessments of its concept.

Moreover, there are further semantic uncertainties concerning the delimitation of other concepts such as green economy, green growth, green deal, etc.

2.1 Economy and Nature

The main reason why the concept of bioeconomy causes confusion is the ambiguous relationship between the two terms it is composed of: the term "bio" on the one hand—stemming from the ancient Greek term "βίος (bios)" meaning "life" and relating to nature, the habitat and lifespan of species as well as their livelihood—and the term "economy" on the other hand—etymologically linked to the ancient Greek

[1]Cf. Grefe (2018, 21 f.) and Vogt (2018, p. 32).

[2]Cf. Georgescu-Roegen (1971) and Meadows et al. (1972).

[3]Cf. Enríquez-Cabot (1998, 925 f.) and Birner (2018, p. 19).

[4]Cf. von Braun (2018, p. 11).

terms "οἶκος (oikos)" meaning "household" and "νόμος (nomos)" meaning "law" and "custom" and relating to husbandry and market but also to enterprise, commerce, business, industry and trade.[5] There are various dimensions in which the relation between "bio" and "economy" is assessed differently or even contradictory. Three of these dimensions will be discussed in the next three subsections: neoclassical vs. ecological economics, economization vs. ecologization, and natural capital vs. intrinsic value of nature.

2.1.1 Neoclassical or Ecological Economics?

Between the two main streams of neoclassical and ecological economics, views on the role of nature within economy resp. economy as a part of nature differ widely.

In *neoclassical* economics, nature is seen as an *object* whose value is to be judged exclusively by means of its impact on human well-being. Hence, neoclassical economics may be characterized as an anthropocentric utilitarian approach, according to which human needs can be met by goods derived from nature. Mineral oil is an example for a natural product satisfying the human need for heat energy. Environment in general—such as a relatively pollution-free atmosphere enabling human breathing—as well as plants and animals in particular—such as plant based or animal source foods enabling human nourishment—serves existential human needs. However, from the neoclassical economist's point of view, nature is thought to be subordinated to the economic system and natural goods are considered to be principally substitutable by man-made goods.[6] This assessment of the substitutability of natural by man-made goods is a key aspect that differentiates different positions of sustainability.[7] Economist Volker Radke summarizes the core message of neoclassical economics as follows: if natural assets decline in a period, social welfare can still be maintained if this decline is offset, for example, by sufficient investments in machinery or in people's education. In line with this economic perspective, sustainability is achievable by substituting natural assets with other goods. According to Radke, sustainability in this context is defined as non-declining per capita human well-being over time and the central prerequisite for achieving well-being is seen in the overall stock of capital. In this light, marketable natural resources are valuated monetarily: for instance, mineral oil is quantified in barrel or logging is quantified in solid cubic meter. According to the neoclassical understanding of sustainability—all the different components of the total assets are substitutable against one another without decrease in total prosperity—it thus principally is irrelevant in what kind of assets one invests in.

Inter alia, Radke criticizes the underlying conception of aggregating individual utility on an abstract level by means of a certain social welfare function, as this presupposes the cardinality of individual utility. Only if scores can be assigned to

[5]Cf. for instance Leshem (2016), Müller (2017). Cf. also Lanzerath and Schurr (2022) in this volume.

[6]Cf. Radke (2004, pp. 147–155).

[7]Cf. for instance Schoop (2022) in this volume.

individual utility, the margin between two levels of use may be calculated. In comparison with *ordinal* utility functions, this is a very strong demand on *individual* utility functions in Radke's view.

In *ecological* economics, nature is regarded as a *subject*. The most fundamental critique of neoclassical environment- and resource-economy, as well as its ethical basis, is voiced by advocates of a so-called *ecocentric position*, under the terms of which nature is seen as a natural or legal person provided with inherent rights. Nature in itself is not substitutable and, not least due to their highly complex network of interdependencies, neither are natural goods substitutable by man-made ones. Eventually, every capital good is nothing but natural matter formed by man.[8] Under the ecological economics' term of sustainability, it is then not man, but nature making up for its focal point, and sustainability is equivalent to a long-term preservation of the viability of ecosystems. Ecological economics may be characterized as an ecocentric position as its main focus is not on human or economic good, but on the well-being of ecosystems. With this idea in mind, governmental institutions shall try to influence individual preferences in the best interest of societies and ecosystems as a whole. In particular, individuals shall be induced to foster those kinds of need satisfaction that are not accompanied by materialistic consumption. In this paternalistic manner, environmental pollution shall be prevented without decline in individual well-being. Yet, nature is thought to be superordinated to the economic system, which may not be considered in isolation, but only embedded in the natural environment.[9] Radke's critique of ecological economics refers to its elitist valuation of nature and the associated paternalistic preference order that might impinge on liberal-democratic principles and could lead toward an expertocracy or even an ecological dictatorship.[10]

2.1.2 Economization or Ecologization?

One of the most prominently debated juxtapositions in the bioeconomic context is the understanding of bioeconomy as an *economization of nature,* on the one hand, and as an *ecologization of economy,* on the other hand. One can thus read the meaning of economic transformation ascribed to bioeconomy in two directions: with an accentuation of *economy* in the sense of a program pursuing the goal of further economically exploiting the profitable resource 'nature', or with an accentuation of *bio* in the sense of a program pursuing the goal of further protecting nature and the environment as a value on its own. However, the far more widespread notion and application of bioeconomy leads into the first direction.[11]

In the year 2000, it has been the European Council's announced "way forward" to "*become the most competitive and dynamic, knowledge-based economy in the*

[8] Cf. Bonaiuti (2015).

[9] Cf. Radke (2004, pp. 157–162).

[10] Cf. ibid., 163.

[11] Cf. Gottwald and Krätzer (2014, p. 12) and Vogt (2018, 31 f).

world",[12] especially by means of innovation and a digitalized information society. Adversaries mostly share the view that today's bioeconomy is a worsening program of the most competitive and dynamic, knowledge-based economization of nature in general or of life in particular.[13]

In Germany, a much-noticed critique of the concept of bioeconomy as a wrong track and a totalitarian approach has been put forth by Theo Gottwald and Anita Krätzer in 2014. On the basis of an analysis of the bioeconomy programs of the European Union (EU), publications of the German Bioeconomy Council and different spheres of bioeconomic activity, Gottwald and Krätzer conclude that bioeconomy not only represents a new, but an absolutizing dimension of economic thinking. This thinking is accompanied by a reassessment of every living entity as a discretionary tradable and negotiable commodity of biomass within ever new paths of exploitation.[14] More moderate positions at least agree with regard to the explicit orientation toward growth, which somehow appears to be disguised behind the "bio" of "bioeconomy" in order to promise a win-win business for economy and nature. But as a matter of fact, the focus is almost entirely on "green growth," i.e. an "economization of ecology" rendering the gap between promise and reality concerning climate and environment policies allegeable.[15]

Furthermore, within the concept of bioeconomy, the term "sustainability"[16] is reinterpreted in so far as it is not the precautious conservation of nature and environment but instead its enduring exploitation, which is considered to be sustainable only.[17] Pursuant to the opinion of Gottwald and Krätzer, the leading ethical principle guiding sustainable biotechnological innovation—the Precautionary Principle (PP)—is undermined. According to the PP, ecological damage must be avoided instead of being addressed ex post facto. Thus, in line with PP, a successful reorientation toward an *e*fficient and especially a *su*fficient way of doing business that relies on self-limitation in view of growth and consumption is proposed. In fact, due to rebound effects and distribution mechanisms, bioeconomy's focus on

[12] European Parliament (2000). At the conference "New Perspectives on the Knowledge-Based Bio-Economy" of the European Commission in 2005, the European Commissioner for Science and Research, Janez Potočnik, held a talk entitled "Transforming Life Sciences Knowledge into New, Sustainable, Eco-Efficient and Competitive Products" which is meant to be a definition of the knowledge-based bioeconomy (cf. Birner, 2018, 20).

[13] Cf. Birch (2006), Gottwald and Krätzer (2014), Fatheuer et al. (2015) and Grefe (2018).

[14] Cf. Gottwald and Krätzer (2014, 8 f).

[15] Cf. Fatheuer et al. (2015, pp. 137–167), Vogt (2018, p. 33) and Pies et al. (2018, p. 107).

[16] In its relevant report "Our Common Future" (also known as "Brundtland Report"), the United Nations World Commission on Environment and Development (WCED) (also known as Brundtland Commission) defines "sustainable development" as "development that meets the needs of the present without compromising the ability of future generations to meet their own needs" (World Commission on Environment and Development, 1987, par. 27). Furthermore, at the 1992 United Nations Conference on Environment & Development in Rio de Janeiro the global action program "Agenda 21" has been worked out, which determined three dimensions of sustainable development: environmental, social, and economic (cf. United Nations, 1992).

[17] Cf. Gottwald and Krätzer (2014, p. 19).

efficiency does not seem to qualify for the paramount idea to sustainability and justice if mechanisms of exploitation remain in place. Until now, however, bioeconomy neither fosters sufficiency nor a real alternative policy of sustainability, that is to say, a consistency approach which calls for an adjustment of innovation along the cycles of nature and not vice versa.[18]

This fundamental critique is based on the general opinion that the concept of bioeconomy pushes "a neoliberal regime in which market values are installed as the overriding ethic in society and the market rule is imposed on all aspects of life".[19] Associated with this assessment is the claim that bioeconomic strategies are promoted in the interest of large-scale industries whose utmost goal is the commercialization of innovations in the life sciences for profit, which oftentimes happens at the expense of small scale enterprises and of the majority of citizens who rather reject technological applications such as genetic engineering or synthetic biology.[20]

Besides this main critique, doubts are raised concerning the integrity and execution of the concept of bioeconomy under the accusation of greenwashing. As part of this critique, the general potential of bioeconomy to contribute to a more sustainable way of economic activity is acknowledged, yet the realization of this potential by diverse allegedly sustainable approaches is questioned.[21] Indeed, putting to use biotechnological innovations fed by bio-based, renewable materials and energy is not sustainable per se—not even environmentally benign. This challenge is fueled by the concern that the prefix "bio" is misused and becomes a fraudulent label behind which actually unsustainable practices are hidden and even fostered.[22]

Beyond that, the language-game of bioeconomy itself already unleashes an influence that affects the human-nature-relationship in a potentially worrisome way, as the question of economizing nature or ecologizing economy is also linked to the conception of nature as capital or as a value in itself.

2.1.3 Natural Capital or Intrinsic Value of Nature?

Some criticize the bioeconomic terminology for mechanizing and/or economizing nature. Nature supplies bioeconomy with useful energy and materials and stores or assimilates its waste. As if that was not enough, nature provides a biosphere and an enormous number of further offerings, which are indirectly as well as directly beneficial to humans, let alone their economic activity.[23] These offerings are commonly referred to as *ecosystem services*. Ecosystem services make up the core of the

[18] Cf. ibid., 154. Cf. also Schleissing (2018, p. 72).

[19] Birch (2006, p. 4).

[20] Cf. Birner (2018, p. 24).

[21] Indeed, bioeconomic applications can be energy-intensive, have negative water-footprints and/or negative biodiversity records (cf. for instance Fritsche and Rösch, 2017; Heimann, 2018; Lago et al. 2019).

[22] Cf. Birner (2018, 24 f).

[23] Cf. Victor (2019, p. 49).

German project Naturkapital Deutschland—TEEB DE.[24] Those services of nature consist not only in the most basal processes such as soil formation, photosynthesis or nutrient circulation which build the prerequisite for life on earth (basic services), but also in the production of drinking water, food, feed, and raw materials (providing services) or climate regulation, flood control, pollination and filtration effects (regulating services) or even in the contribution to cultural aspects of human life such as recreation, aesthetics, spirituality, education, and personal identity (cultural services).[25] Therefore, the concept of ecosystem services stretches far beyond that of bioeconomy, yet clearly overlaps with it—especially with regard to the category of providing services. All kinds of ecosystem services are subsumed under the so-called *natural capital.*[26]

Especially the terms "ecosystem services" and "natural capital" are bound to the utility of the biosphere for man.[27] And this is not accidentally or unwittingly so, but on purpose. Firstly, advocates of the ecosystem services approach suppose that the worthiness of nature consists in its usefulness to humans. Secondly and beyond that, they espouse an economic understanding and monetary valuation of nature's utility for man. That is because, according to TEEB DE, it is oftentimes overlooked that unimpaired ecosystems provide important and gratuitous capacities and services, which would otherwise require cost-intensive technical solutions (e.g., concerning climate protection, flood control and the cleaning of air and waters) and cause high social costs (e.g., concerning health and recreation).[28]

Further theses of TEEB DE claim that the worth of nature frequently remains hidden, because its goods and services appear to be at unlimited disposal and free of charge. But, in the economic sense, nature in fact constitutes a *capital* and its performance may be conceived of as a *dividend* accruing to society. Pursuant to TEEB DE, an economic perspective thus helps to uncover nature's worth and its diverse goods and services while providing economic arguments for the preservation of natural capital at once.[29]

Although these arguments are meant to complement ethical and ecological reasons, they assume the ideas of ecosystem services and natural capital to be inevitably given, instead of grasping that those arguments emanate from a thoroughly human-centered, if not utterly capitalistic, mindset.[30] The good being worthy

[24] TEEB DE links to the international study "The Economics of Ecosystems and Biodiversity (TEEB)" and exhibits that through the use of natural resources, valuable biospheres get lost also in Germany (cf. Naturkapital Deutschland – TEEB DE 2012).

[25] Cf. Naturkapital Deutschland – TEEB DE (2012, p. 23).

[26] After a typical definition, natural capital is "the world's stocks of natural assets which include geology, soil, air, water and all living things. It is from this Natural Capital that humans derive a wide range of goods and services, often called ecological goods and services, which make human life possible" (World Forum on Natural Capital, 2017).

[27] Cf. Naturkapital Deutschland – TEEB DE (2012, p. 10).

[28] Cf. ibid., 6.

[29] Cf. ibid., 9, 15.

[30] Cf. Pinsdorf (2020).

of protection here is not nature as such, but only that kind of nature suited for useful capital. Although TEEB DE reaffirms that the motive to preserve nature as a value in itself shall not be neglected, let alone replaced,[31] there is the factual risk of ethical and ecological arguments to become overlaid or even displaced by economic arguments, whose powerfulness seems to be overriding in enduring times of economic supremacy.

For Markus Vogt, catholic theologist and professor for Christian Social Ethics, ethical and economic perspectives are not mutually exclusive, but may complement each other as TEEB DE in his view impressively shows. Here, the protection of nature and biodiversity in particular is motivated by the quantification of its economic worth. Only the numbering of costs caused by climate protection deferral (up to 20% of Gross World Product (GWP) in non-action, approx. 3% of GWP in rapid action) has startled up the world's public.[32] Vogt says, one could now lament on the circumstance that the most sensitive organ of human perception is the modern citizen's wallet, or one can adapt to this actuality and foster an economic valorization (*In-Wert-Setzung*) of nature. As incentives for the protection of nature in market economies only emerge, insofar resource scarcity is prized in resp. insofar external costs are internalized, economization of ecology is for Vogt neither theoretically nor practically unethical, but sometimes just necessary. This is not a seldomly uttered argument of pragmatism. It is considered to be uncontroversial that the internalization of external costs via a systematic juxtaposition of all costs and benefits, i.e. including, e.g., restricted or even lost ecosystem services, gives impetus for the implementation of more sustainability.[33] A prominent example of an economic valorization of nature in the realm of climate change is certificate trading, i.e. carbon trade or carbon tax to reduce CO_2 emissions. This shows that pricing can induce a change of behavior in people and organizations and can thus be supportive to meet environmental objectives.

Pragmatist arguments such as "nature protection is worthwhile because it is less expensive than attempts to restore destroyed ecosystems or to substitute natural resources"[34] are sound from an economic perspective but misleading from an ethical point of view.[35] Although their content is true, they promote a problematic motivation for the protection of nature and, what is more, reduce nature to an object of human disposability. Although in economics the maximum willingness to pay is equivalent to individual utility, hence value of a good,[36] the monetary measurement is not only inappropriate for pivotal interpersonal relations such as love and friendship, but also for primal relations between man and nature. If nature is perceived only in terms of its real or potential market values, it amounts to nothing more than

[31] Cf. Naturkapital Deutschland – TEEB DE (2012, p. 14).

[32] Cf. Vogt (2018, 34 f).

[33] Cf. for instance Naturkapital Deutschland – TEEB DE (2012, 46) and Jackson (2009, p. 174).

[34] Naturkapital Deutschland – TEEB DE (2012, p. 11).

[35] Cf. Pinsdorf (2020).

[36] Cf. ibid., 79.

an exchangeable value without any consideration for its intrinsic value. Analyses of willingness to pay for nature's goods and services or even living beings require, for instance, an inventory of nature, which may seem to be an inadequate as well as impossible endeavor.

Furthermore, the valorization of nature may become subject to a fallacy of composition: Global environmental challenges like climate change, biodiversity loss, or deforestation are phenomena of which critical thresholds—the so-called *tipping points*—have either been crossed already or are likely to be crossed at any time soon. Fueled by financial tradeoffs, one might be fooled into asking "What harm can one more ton of greenhouse gas emissions do? What is the value of losing just one more hectare of old-growth forest?".[37] Since ecological thresholds are inherently uncertain, the mindset of monetary governance becomes more and more hazardous.

A possible commercialization or commodification of nature in terms of its pricing and subsequent marketing is sometimes even feared to lead to a sellout of nature and is oftentimes at least judged to be inadequate from an ethical point of view. The reverse argument, i.e. economic valorization of nature's goods and services is conducive for the development of mindfulness toward and appreciation of nature as well as for exercising one's moral responsibility toward nature,[38] is less than convincing. As soon as something is priced, it is given an exchange value and becomes financially negotiable. In this context, nature as an arbitrary commodity is discretionary tradable. Conceptionalizing nature as capital paves the way forward to an over-exploitative attitude of man toward nature, its ecosystems and living creatures since in this mode of thinking *homo sapiens sapiens* is the only entity of relevance. If, instead, nature and other life forms also matter, and we strive for an attitude of respect for nature and its ecosystems as well as for the recognition of other life forms, it might be more reasonable not to conceive of them as just another form of capital whose purpose is to serve mankind. Mindfulness and appreciation are not achieved via market integration, but rather with reference to a monetary non-negotiable (intrinsic) value, that mirrors the moral status and ethical standing of non-human nature.[39]

All in all, the pragmatist critique of the pragmatist argument contends that it is shortsighted, only works short term and essentially encourages a mindset which has led to and will eventually worsen the ecological crisis altogether. The fundamental critique of the pragmatist argument refers to the epistemological mistake it rests upon—"if [. . .] no commodities available through markets are adequate substitutes for the unpriced ecosystem services, then it makes no sense to estimate a monetary

[37] Victor (2019, p. 89).

[38] Cf. Naturkapital Deutschland – TEEB DE (2012, pp. 12, 21, 47, 62). Peter A. Victor explains: "Commodification [. . .] refers to the conversion of something outside the economy into a commodity for purchase and sale. [. . .] The success of capitalism owes much to this process through which the market takes over aspects of society that were previously outside the economy" (Victor, 2019, 53).

[39] Cf. Pinsdorf (2016, 143 ff.) and Pinsdorf (2020).

value for them"[40]—as well as its ethical inadequacy. Concerning ecosystem services, the fundamental critique accuses the concept by and in itself to put forward an awry understanding of the human-nature-relationship on different levels: for one thing, nature does not produce goods and services in order to place them at the disposal of humans. Nature is an entire and complicated complex, evolved over a period of millions of years, whose center is not mankind and even less mankind's mental constructions of "services," "economics," or "monetary units." In reality, mankind does not know what the focal point of nature is—it might well be that it has none. Be that as it may, for mankind to think and act as if it was the focal point of nature is factually false. For not too few ethicists it is also morally wrong. In economic approaches to nature, such as TEEB DE, it is almost exclusively human well-being what matters. In this perspective nature proves itself to be valuable if and only if it delivers goods and services of direct or indirect utility for human wants and needs.[41] In philosophical ethics such an approach is assigned to normative anthropocentrism, which has been thoroughly criticized for its bias and ethical unjustifiability, i.e. with reference to the discriminatory ideas of *speciesism* or *human chauvinism*.[42]

TEEB DE commendably intends to make people aware of the connection between nature, economic net product, and human well-being. It surely initiates a visualization of the so-called *ecosystem services* and their economic worth and lays the foundations for integrating natural capital in private, entrepreneurial, and political processes of decision-making in order to maintain the basis for human existence. The valorization of natural capital facilitates the incorporation of nature's goods and services as an integral part of commercial calculus from the outset.[43] This, however, rather protects the economy instead of nature. Ecological economist and professor emeritus of Environmental Studies, Peter A. Victor, also comes to a sobering conclusion concerning the monetary valuation of nature and ecosystem services:

> In a culture in which monetary values have such a dominating presence, assigning large monetary values to nature can have considerable rhetorical power, which is important, given the precarious state of nature and the overriding importance of attracting attention to possible solutions [...]. But that does not make it good economics nor does it make it ethically sound.[44]

Summing up, bioeconomy's valorization of nature is at least problematic and only tolerable if it conceptually includes the intrinsic value of nature and the living

---

[40] Victor (2019, p. 78). For further critique of the conceptual framework underlying the monetarization of ecosystem services or ecological damage, cf. Victor (2019, 77, 80 ff).

[41] Cf. Naturkapital Deutschland—TEEB DE (2012, 10 f). In TEEB DE, for instance, human well-being and usefulness for humans is emphasized throughout (cf. ibid., 9, 10, 15, 18, 23, 49, 80).

[42] Cf. for instance Singer (1977), Bradie (2011, 567 f.), Breitenbach (2009), Sturma (2013), Pinsdorf (2016), Thompson (2017, 85 ff.) and Kopnina et al. (2018).

[43] Cf. Naturkapital Deutschland —TEEB DE (2012, p. 64).

[44] Victor (2019, p. 91).

being's own good and flourishing in equal measure—however, the tension between the two conceptions—natural capital and intrinsic value of nature—might in reality be irreconcilable.

2.2 Promises and Disillusions

2.2.1 The Promise of Decoupling

Societies all over the world are facing a profound dilemma: economic growth (at least in its current form) is unsustainable whereas economic degrowth (at least for capitalistic societies) is unstable.[45] The concept of bioeconomy is meant to induce the solution to this fiddly dilemma inasmuch as it is designed to secure the dynamic of growth that keeps (capitalistic) societies going, but on sustainable, hence ecologically, socially and economically safe grounds.

A feature of bioeconomy to achieve this end is the idea of decoupling: "Decoupling refers to the proposition that economies can grow and yet reduce their use of materials and energy through a combination of technological change and a switch from goods to services."[46] Through increased efficiency and innovation, interlocked cascades of resource utilization as well as the substitution of fossil commodities and energy, bioeconomy is due to separate resp. decouple economic growth from resource usage and environmental impact:

> Production processes are reconfigured. Goods and services are redesigned. Economic output becomes progressively less dependent on material throughput. In this way, it is hoped, the economy can continue to grow without breaching ecological limits – or running out of resources.[47]

Decoupling is about efficiency enhancement, about doing more with less, i.e. more economic activity and productivity with fewer resource inputs, fewer waste outputs, and less environmental damage. A relevant difference is given between relative and absolute decoupling:

> Relative decoupling of materials from GDP [Gross Domestic Product (C.P.)] occurs when, over time, material use per dollar of GDP declines (that is, material intensity) but total material use does not. Absolute decoupling occurs when material intensity declines faster than GDP growth, so that total material use also declines.[48]

Relative decoupling is easier to accomplish than absolute decoupling, but only the latter leads to a potentially significant reduction of environmental burdens.

---

[45] Cf. Sukhdev (2009, p. xix) and Jackson (2009, p. 65).

[46] Victor (2019, p. 107).

[47] Jackson (2009, p. 67). Cf. also Hamm (2018, p. 138) and Victor (2019, p. xiii).

[48] Victor (2019, p. 108). Cf. also ibid., 38; Jackson (2009, pp. 67–76).

While the German Bioeconomy Council and other proponents of the bioeconomic transformation are enthusiastic in view of the possibilities of decoupling, many critical voices remain. For instance, some already question the possibility to generate steadily increasing incomes for a growing world population without pushing ecological boundaries too far.[49] Others recognize that the concept of decoupling is not unreasonable in itself, as, for instance, energy expenditure per commodity unit has significantly decreased in Germany since 1970. However, despite this development, no ecological release ensued. In fact, energy consumption in total did not drop, but augmented instead.[50] This non-appearance of ecological release is again attributed to the phenomenon of the so-called *rebound effect*: e.g., the costs saved from energy conservation are deployed to expand the production of goods; or, money saved from, e.g., energy efficiency is spent on other goods and services. As the production and consumption of these other goods and services have energy costs on their own, savings achieved through efficiency can either be offset or even be outreached: "In short, *relative* decoupling sometimes has the perverse potential to decrease the chances of *absolute* decoupling."[51]

Next to the rebound effect, there are two further important factors that disrupt the aim of decoupling, namely population growth and augmented per person consumption. Taken together, they may cause an increase of material and energy use in total and over time, even though initially higher efficiencies cause less intense material and energy requirements.[52]

Concluding on the idea of decoupling, ecological economist and professor of Sustainable Development, Tim Jackson, differentiates:

> It's clear [...] that history provides little support for the plausibility of decoupling as a sufficient solution to the dilemma of growth. But neither does it rule out the possibility entirely. A massive technological shift; a significant policy effort; wholesale changes in patterns of consumer demand; a huge international drive for technology transfer to bring about substantial reductions in resource intensity right across the world: these changes are the least that will be needed to have a chance of remaining within environmental limits and avoiding an inevitable collapse in the resource base at some point in the (not too distant) future.[53]

2.2.2 The Promise of a Circular Bioeconomy

In more recent times, a directional change in the configuration of the concept of bioeconomy is observable: away from the growth-minded course for acceleration of

---

[49] Cf. Jackson (2009, p. 68).

[50] Cf. Herrmann (2015, p. 3), United Nations Environment Programme (2016, p. 16) and Hamm (2018, 138 ff). As Victor notes, "the twenty-first century has witnessed an unprecedented period of relative and absolute *re-coupling* of material extraction and global GDP" (Victor, 2019, p. 109).

[51] Jackson (2009, p. 95).

[52] Cf. Victor (2019, p. 108).

[53] Jackson (2009, p. 75).

(green) growth and toward a renewable resources-based circular flow economy.[54] The Communiqué of the Global Bioeconomy Summit, for instance, considers "it an important task to align the principles of a sustainable bioeconomy with those of a circular economy. This would involve systemic approaches across sectors (i.e., nexus thinking), particularly innovation policy measures that aim at optimizing Bioeconomy value networks and minimizing waste and losses."[55] One of the main drivers of making bioeconomy circular is the so-called *cascade utilization*, a process of using biomass initially materially, then chemically and only last of all energetically. In doing so, as much added value output as possible shall be gained out of as little material input as possible producing as little waste as possible. However, ultimately all materials being used stem from the ecosphere and all of them wind up as waste being disposed back into the ecosphere; this phenomenon applies to all materials being used and is referred to as "materials balance principle".[56] In a similar way, the energy balance principle—which encompasses nothing other than the *first* and *second law of thermodynamics*[57]—applies to all uses of energy. According to the first law of thermodynamics, in any process the quantity of energy is maintained and only its form changes: "An example is the conversion of the chemical energy on gasoline to mechanical energy and heat when used to power an automobile."[58] According to the second law of thermodynamics, energy's capacity to perform necessarily declines each time it is used:

> For example, in a conventional electric power station, energy from coal combustion is used to boil water. The steam drives a turbine that produces electricity. Some energy is released to the environment as waste heat, which is unavailable for further work. Only about 35 percent of the chemical energy in the coal leaves the power station as electricity and then there are further losses during transmission and use.[59]

Sustaining a constant level of any (economic) activity requires a constantly new energy supply.[60] Hence, the theoretical possibility of a circular bioeconomy with 100% of reuse (material recycling and energy efficiency) is precluded by the energy balance principle. Economic activity depending on nature's materials and energy will thus have to keep going back for more and will never produce zero-waste.

But even if an imperfect ecological circular economic system would be possible—and numerous doubts on its feasibility remain—two important questions have

[54] Cf. for instance World Wide Fund for Nature (WWF) (2009, 5 ff). For a further exploration of possible futures of a wood-based circular bioeconomy in Germany, see, for instance, Hagemann et al. (2016).

[55] Global Bioeconomy Summit (2015, p. 5).

[56] Cf. Victor (2019, p. 46). Some, however, still believe a fully closed loop economic system producing no waste to be possible (cf. World Wide Fund for Nature (WWF), 2009, 5, 15 f).

[57] Cf. Georgescu-Roegen (1971, p. 4–7, 17, 129, 197, 280).

[58] Victor (2019, p. 46).

[59] Ibid.

[60] Cf. ibid., 117.

not yet been properly considered: what is the bridge leading from a capitalistic to a post-growth economy and how should the process of transformation look like?[61]

2.3 Economy and Man

2.3.1 Economic Growth and Human Flourishing

Another conceptual lack of clarity applies to the idea of human flourishing or human well-being, which is strongly connected to bioeconomy as a concept for economic growth and prosperity.

In the logic of (at least capitalist) economies, the conventional thesis of economics suggests that economic growth is essential for maintaining economic and social stability, whereas economic degrowth is tantamount to economic collapse and social adversities.[62] In 1987, the Brundtland Report ascertained that global economy and global ecology are intertwined in new ways. Whereas in the past the main concern has been about environmental impacts of economic growth, now the impacts of environmental degradation on future economic prospects come in addition.[63]

Possible biophysical limits to growth can be divided into the four categories: "sources, sinks, services, and synthesis".[64] While "sources" refer to the supply of materials and energy, "sinks" refer to their disposal, and "services" relate to the way nature is anthropogenically transformed with the upshot of decreasing essential ecological functions. Last but not least, "synthesis" refers to the interrelation of the three categories before, thus setting up an even more complex biospherical limit on top.

[61] Herrmann, for instance, is convinced that due to purely economic reasons, this transition is either impossible or extremely difficult (cf. Herrmann, 2015, p. 3). In the second edition of his forward-thinking book *Managing without Growth*, Victor actually raises related fundamental questions: "How might an advanced economy function in the absence of growth? Would it collapse or is there a configuration of production, consumption, employment and other aspects of importance that is both feasible and attractive without relying on economic growth?" (Victor, 2019, p. 31). And, by the meaningful subtitle of his book, *Slower by Design, not Disaster*, Victor furthermore points to the most probable, if not certain vision that growth is coming to an end and the only freedom of choice left to us is either making it end (sooner) accompanied by well-informed decisions and knowledgeable measures or watching it end (later) disordered and tragically.

[62] Cf. for instance Gordon and Rosenthal (2003), Binswanger (2009b), Jackson (2009, 61 ff.), Smith (2010) and Dörre (2013).

[63] For the added dimension of alarm, see, for instance, the following statement in the Strategy Paper of the German Bioeconomy Council: "Originally, the concept of a bio-based economy was promoted in the light of expected rapidly depleting petrol, gas and coal reserves. However, the move into bioeconomy is no longer driven predominantly by expectations of rising prices of fossil fuels. In view of the exploitation of new fossil reserves and due to energy efficiency improvements, this argument has become less pressing but it nevertheless remains strategically essential. Without major adjustments, the continued emission of greenhouse gases and the related changes in climate conditions will irreversibly damage the global ecosystem and will involve incalculable economic risks" (German Bioeconomy Council, 2014, p. 1). Cf. also Victor (2019, 95 ff., 116, 135).

[64] Victor (2019, p. 100).

What is more, the increasing economic interdependence among nations is accompanied by an accelerating ecological interdependence on local, regional, national, and global scales.[65] Not least in consequence of the links between poverty, inequality and environmental destruction, the Brundtland Report claims: "What is needed now is a new era of economic growth—growth that is forceful and at the same time socially and environmentally sustainable."[66]

But what precisely is that supposed to mean? Or, what exactly is meant to be growing?

As Victor explains,

> economic growth is usually measured by the pace of change of gross domestic product (GDP) after adjustment for inflation also known as real GDP' [...]. This conventional definition of economic growth is not accompanied by a separate explicit definition of the economy, that is, that which grows.[67]

This scanty differentiation is, among other things, insufficient to grasp economy's embeddedness in and dependency on nature, as well as it is insufficient for establishing alternative approaches.[68] Concerning prosperity, GDP—the total value of all goods and services that have been produced by a national economy within one year for the purpose of consumption—has been criticized as an insufficient measure for quite a while. Critics argue that various elements of national wealth and well-being—such as accounting for social costs or unequal distribution of income, qualitative aspects of health and education or depletion of natural resources—cannot be captured on the basis of GDP growth.[69] Under headings

[65] Cf. World Commission on Environment and Development (1987).

[66] Ibid., 7.

[67] Victor (2019, 44 f). Victor ascertains further: "It is also difficult to find official definitions of economic growth even from organizations such as the OECD, the IMF and the World Bank that are dedicated to promoting it. We are simply told that economic growth is measured by changes in real GDP or real GDP per capita. What is being measured has become synonymous with its measurement" (Victor, 2019, 42 f.).

[68] Cf. also the following statement of ecological economist Herman E. Daly: "Exactly what is growing? One thing is GDP, the annual marketed flow of final goods and services. But there is also the *throughput*— the metabolic flow of useful matter and energy from environmental sources, through the economic subsystem (production and consumption), and back to environmental sinks as waste. Economists have focused on GDP and, until recently, neglected throughput. But throughput is the relevant magnitude for answering the question about how big the economy is—namely how big is the economy's metabolic flow relative to the natural cycles that regenerate the economy's resource depletion and absorb its waste emissions, as well as providing countless other natural services? The answer is that the economic subsystem is now very large relative to the ecosystem that sustains it" (Daly, 2009, xi f.).

[69] Cf. Sukhdev (2009, p. xvii) and Jackson (2009, p. 179). Cf. also Jackson (2009), Chap. 4 which analyzes data concerning life expectancy, health and educational participation in relation to GDP collected by the United Nations Development Programme (UNDP) over several decades.

such as "qualitative growth," "eco-social-product," or "beyond GDP," a supplementary measure of the quality of life, well-being and sustainability is claimed.[70]

The term "prosperity" itself seems to be disputable in view of a growing world population confronted with the threats of climate change and resource scarcity. At least it seems to be clear that prosperity under the current prognostic symptoms cannot mean the same as at the time of industrialization. For Mary Robinson, former president of Ireland, former United Nations High Commissioner for Human Rights and founder of the nongovernmental organization *Realizing Rights: The Ethical Globalization Initiative*, these days prosperity "cannot mean business as usual. It cannot mean more of the same".[71] What then can prosperity mean nowadays?

Outstanding ecological economist Herman E. Daly distinguishes between *quantitative* growth and *qualitative* development. Growth in that sense is based on an increased use of materials, whereas development in that sense means an achievement of more desirable goals—such as sustainability—with the same or even less use of materials. Along these lines, economies can simultaneously grow and develop, grow without developing or develop without growing.[72] The crux seems to be that *su*fficiency and ecologically oriented *e*fficiency seem to be complementary concepts consistent with core elements of sustainability, yet incompatible with economic growth. But is this necessarily bad news for prosperity and human well-being? Daly suggests it is not, because sustainable development without growth would lead to an economy that is not bigger, but better.

One key determinant in the semantic field of economic growth, prosperity, and human flourishing is an economic narrowing in the understanding of the term "felicity". In some economic theories, the underlying concept of felicity focuses solely on utility as the consumption of the economy's stock of capital assets, including manufactured goods, services provided by nature, health services, and others. From that economic point of view, felicity is only based on consumption of capital assets derived by diverse sources, such as marketed consumption goods, leisure, various health services, and consumption services supplied by nature. In comparison with former approaches in economics, this might already be judged to be quite a holistic approach. However, it excludes major aspects of felicity that are explicitly independent of the idea of consumption, such as non-material, spiritual or idealistic values, or idle time, which all seem to be unquantifiable in themselves, but are at the same time crucial aspects for the concept of felicity and also decisive for capturing comprehensive wealth.

While financial income provides access to vital as well as comforting goods and so-called *status goods* establishing social standing, some studies have shown that a

---

[70] Cf. Naturkapital Deutschland – TEEB DE (2012, 46 f.) and Victor (2019, 43 f).

[71] Robinson (2009, p. xvi). Also the OECD itself resumes that positive developments in environmental respect are still only at the margin and far from appropriate (cf. Organisation for Economic Co-operation and Development, 2015, p. 7).

[72] Cf. Daly (1996, 166 f). It is also interesting how Daly translates the meaning of consumption as destruction (cf. ibid., 62) and growth—at least in the global North—as some impediment to sustainable development (cf. ibid., 8, 13 ff.)

growing income beyond a certain threshold does not add or only marginally adds value to the well-being of individuals.[73] In the poorest countries of the Global South, people suffer extraordinary deprivations connected to infant mortality, general life expectancy, nutritional supply, clothing and shelter, or educational participation—here, economic growth and increased financial income are required to achieve urgently needed betterments. But in richer countries of the Global North this is hardly the case.[74] Concerning the global threat of poverty and hunger, customary economic development has not proven to be a solution, but rather to be a reproducer or even reinforcer of problems.[75] Bioeconomy in particular is criticized for globally reinforcing social injustice insofar as it is a capital-intensive endeavor, primarily framed within the industrial paradigm and geared toward international marketing. At least so far, bioeconomy cannot be evaluated as a facilitator of smallholder agriculture and food sovereignty of the poor.[76]

At the same time and at least in ecological respect, it is not feasible to turn the Global South into a Global North. It hence stands to reason that degrowth-strategies and a locally oriented sufficiency economy pave the way for a more socially just future. As the program of sufficiency for all living people may indeed involve further loads for the overburdened planetary ecosystems, there is another argument for the limitation of economic growth and the constant rise of material living standards of the world's most affluent societies. As the overall gain of economic growth is only significant in poor countries, and economic growth found in rich countries is—due to biophysical constraints—not applicable worldwide, it is economic growth in affluent countries that needs to be addressed: "So, in a world where economic growth is constrained by biophysical limits it makes sense for rich countries to manage without growth so as to leave room for growth in poorer economies."[77] This statement retains its validity even more as slow growth or even degrowth should not affect the real prosperity, hence happiness, well-being or felicity of people living in affluent societies, in a negative way. Diverse studies have shown that "higher incomes do make people happier but only up to a point".[78] The realignment of what it means to lead a good life can help people to live more fulfilled and contented lives without continuous raise in consumption necessary for economic growth. Degrowth on the basis of sufficientarianism and a subsistence economy could simultaneously render

---

[73] Cf. for instance Jackson (2009, p. 52, 59).

[74] "Economic growth has made it possible for people to live longer, healthier lives at a level of comfort that even the wealthy in pre-industrial societies could scarcely imagine. [. . .] But economic growth has its costs. These can be categorized as environmental costs and social costs. [. . .] Social costs include the breakdown of communities, alienation, crowding and crime" (Victor, 2019, p. 241).

[75] Cf. Read and Alexander (2020, p. 52).

[76] Cf. Vogt (2018, p. 39).

[77] Victor (2019, p. 216). Cf. also Jackson (2009, 180 f.) and Grefe (2018, p. 29).

[78] Victor (2019, p. 209). For further arguments on why economic growth does not or is at least not necessary to promote happiness and well-being cf. Victor (2019), Chap. 9.

economies more resilient against catastrophes and people more satisfied, once they have internalized an attitude of contentment.[79]

Swiss economist Mathias Binswanger ascribes the discrepancy between higher incomes and happiness to four so-called *treadmills* guiding people's consumption-related needs modulation[80]:

1. Positional Treadmill: People possess and consume goods and services because of their search for status and standing in society.
2. Hedonic Treadmill: People's aspirations adjust relative to their income, the more they earn, the higher their material living standard.
3. Multi-Option Treadmill: The increasing range of possibilities and choices accompanying higher income lead to overload and frustration.
4. Time-Saving Treadmill: Time-saving devices render leisure time jam-packed and more stressful instead of taking time pressure off.

Especially the status-related positional treadmill frequently undermines people's happiness and well-being, as the most widespread pursuit of status through consumption is all too often self-defeating.[81] This is partly because the extent of happiness someone draws from a certain level of possession and consumption depends on the possession and consumption level of others. As long as the level of possession and consumption increases for everybody under economic growth, nobody is better off at the end of the day. Decisive for the happiness factor resulting from the consumption of positional goods and services is the relative rather than the absolute consumption.[82] Similar considerations also apply to the aspirations-related hedonic treadmill: if someone's happiness hinges on the relationship between demands and their satisfaction, it will not enhance with an increase of demand satisfaction as demands will grow in turn.[83]

2.3.2 Economic Growth and Capitalism

Capitalism requires economic growth—so the widespread belief.[84] Capitalism is not a stable system, prone to balance or to producing reliable income that may well be cut. To the contrary, as soon as growth stops, chaotic shrinkage may impend, and a dwindling production may lead to frenetic attempts of maintaining jobs. The global financial crisis from 2007/2008 and the acute global coronavirus pandemic are instructive examples. Unfortunately, nature and environment are not at all inevitable profiteers of a faltering world economy, but oftentimes its first casualty. According

[79] Cf. Kasser (2002) and Read and Alexander (2020, p. 55).

[80] Cf. Binswanger (2006).

[81] Cf. Sen (1998) and Victor (2019, p. 209).

[82] Cf. Easterlin (1974, 113 ff.) and Victor (2019, 212 f).

[83] Cf. Easterlin (1974, 111 ff).

[84] Cf. for instance Gordon and Rosenthal (2003), Binswanger (2009b), Jackson (2009, 61 ff.), Smith (2010) and Dörre (2013).

to Ulrike Herrmann, an established German economic journalist, no-growth would end capitalism, but the upshot would not be an ecological circular economy—as wished for by many environmentalists—but an economy in free fall, a panic-creating event.[85] Distinctive Swiss economist Hans Christoph Binswanger was driven by the question, whether capitalism could in principle forgo (destructive) economic growth. He concluded that without growth, investment chains would collapse since companies only invest in anticipation of profits, which in turn are macroeconomically identical with growth.[86] No-growth makes companies fear financial losses, absent profits lead to investment freeze and no-investments make the economy collapse. Eventually, an uncontrollable downward spiral of recession is thought to set in: jobs would get lost, demands would decrease, the overall production would shrink, and unemployment would rise.

It thus appears to be imperative to keep the dynamic of growth running. This is normally achieved by at least two interrelated factors that Jackson calls "the 'iron cage' of consumerism"[87]: first, the motive of making (more) profit that provokes continued innovation and the so-called "creative destruction",[88] which in turn causes production and leads to an endless supply and flooding of the market with new products and services; second, the demand of consumers for (more) goods and services which is perpetuated by a complex social logic relating to the aforementioned treadmills.

At the same time, the dynamic of growth imperative has led to and further on leads to ecological crises, climate change, population increase, social injustice, etc. Thus, the world's (especially capitalistic) societies are facing the more than challenging dilemma already mentioned: without economic growth the whole system becomes dysfunctional and collapses, but with ongoing economic growth the whole system inescapably leads into ecological and social catastrophe. Ecological catastrophe in view of anthropogenic climate change, biodiversity loss and nature destruction, social catastrophe in view of the related global imbalance of suffering from ecological catastrophe and the resulting intra- as well as intergenerational injustice. Capitalism, which has brought wealth and technological progress, is now about to bring ruin as it is an oxymoron to have infinite economic growth in a finite world.[89]

As we have seen, however, green growth in the form of bioeconomy is not an overly promising concept, on the basis of which humanity will be able to dissolve the dilemma. It is a concept still implying the economic dynamic that capital is invested to generate added value and more capital, mostly via the production of consumable products and services. Bioeconomy thus banks on *new* raw materials and production

[85] Cf. Herrmann (2015, p. 3).

[86] Cf. Binswanger (2009a). Cf. also Jackson (2009, p. 65), Binswanger (2009b), Herrmann (2015, p. 3) and Binswanger (2019).

[87] Jackson (2009, p. 88).

[88] Cf. Schumpeter (1994 [1942/43], 81 ff.), Jackson (2009, p. 97) and Victor (2019, 50 f).

[89] Cf. Daly (1996, 33 ff.), Herrmann (2015, p. 3), Read and Alexander (2020, p. 33). The according demand for degrowth is not new (cf. Meadows et al., 1972).

mechanisms, but *old* targets for growth and modes of consumption. Hence the question arises whether bioeconomy in fact is no *progress*,[90] but stabilizes established modes of overexploitation and overconsumption as well as unsustainable standards of living and lifestyle by justifying them via reference to only allegedly sustainable modes of production? At least concerning bioeconomic growth, Jackson arrives at the disillusioning view "that there is as yet no credible, socially just, ecologically sustainable scenario of continually growing incomes for a world of 9 billion people" and "it is entirely fanciful to suppose that "deep" emission and resource cuts can be achieved without confronting the structure of market economies".[91] The answer to the former question thus seems to be affirmative, because what is needed to get out of the dilemma of growth is a system change, which the concept of bioeconomy certainly is not.

3 Ethical Challenges Evoked by the Concept of Bioeconomy

A thorough ethical evaluation of a new biotechnology or its application requires to depict as comprehensively as possible, in which respects the said technology or its applications are assessed to be ethically untenable, problematic, acceptable, or required by different ethical theories. An encompassing ethical evaluation of the concept of bioeconomy and its diverse applications cannot be accomplished here. However, two internationally accepted moral principles for the ethical accompaniment and review of modern biotechnologies, especially concerned with their ecological, social, and economic record, shall be presented briefly in the following passages.

3.1 Precautionary Principle (PP)

The EU has taken on the leadership role in fostering the precautionary principle (PP), specifically applied to decision-making processes in the realm of environmental policy.[92] But also the United Nations are pushing the precautionary approach in order to protect the environment.[93] However, in political guidelines and declarations, the understanding of the PP or its methodologies for assessing risks is either

[90]Insofar 'progress' is understood as a normative term which is oriented towards an improved way of life (cf. Schleissing, 2018, p. 75).

[91]Jackson (2009, p. 86).

[92]Cf. e.g. Commission of the European Communities (2000), Sunstein (2005, p. 1), Consolidated Version of the Treaty on the Functioning of the European Union 2016, Art. 191; European Commission Directorate-General for Environment (2018).

[93]Cf. e.g. Principle 15 of the Rio Declaration on Environment and Development (United Nations Conference on Environment and Development, 1992).

controversial or hardly spelled out. This is not the least reason for the existing academic debate on the correct application and interpretation of the principle.[94]

In his most important work, *Das Prinzip Verantwortung. Versuch einer Ethik für die technologische Zivilisation* (*The Imperative of Responsibility. In Search of an Ethics for the Technological Age*), philosopher Hans Jonas develops an ethics for the future of man and nature according to which man—under the conditions of technological progress and the massively extended range of his actions accompanying it—must take on his responsibility for life on planet earth.[95] One key element for the condition of possibility to take on responsibility under the prevailing circumstances is the question whether man is able or unable to generate sufficient foreknowledge to evaluate his new and evermore powerful influence, or whether the increasingly complex, but only to a limited extent foreseeable and controllable side effects of his actions can or cannot be met by his inventing technical solutions. Against this backdrop Jonas formulates his so-called *heuristics of fear* (*Heuristik der Furcht*) as a consequence of which, in case of doubt, the bad prognosis is to be given priority over the good prognosis and the PP must be guiding man's actions.[96]

For the most part however, Jonas' heuristics of fear is nowadays criticized as being too defensive and—given the open dynamics of modern technological research and development—ultimately paralyzing.[97] Contemporary authors like Ortwin Renn, Cass Sunstein, and Ingo Pies et al. have engaged in formulating a more balanced version of the PP by focusing equally on possible risks of action *and* non-action as well as by taking the plurality of risk perceptions into account.[98] The well-established approach of judging innovations on the basis of the PP—the authors claim—should be applied to the principle itself and hence to the evaluation of possible outcomes resulting from the omission of innovative methods, as risks "can arise from action as well as from inaction".[99] Precautionary risk assessment still requires a conservative assessment of risks in the sense of *one should rather err on the side of caution than on the side of daring*. Yet, one of the major challenges for the PP is its referentiality to uncertainty in risk-assessment. Whereas dealing with uncertainties whose probability of occurrence can be mentioned within a certain calculable probability amplitude seems to be feasible, how shall risks be regulated whose probability of occurrence is just unknown? In this context, the determination of reasonable assumptions for cautious procedures is not by itself scientifically predefined, but always requires a value judgment in the sense of balancing against

[94] Cf. for instance Bogner and Torgersen (2018), Boldt (2018, p. 82) and Pies et al. (2018, p. 115).

[95] Cf. Jonas (2017). For an intensive discussion of Jonas' ethics and its implications for the ethical evaluation of the concept of bioeconomy cf. Schoop (2022) in this volume.

[96] Cf. Jonas (2017, 7, 36, 63 ff., 70 ff., 81 ff).

[97] Cf. Sunstein (2005, p. 5).

[98] Cf. Renn (2002), Sunstein (2005, 2 ff.), Renn (2014, 246–285, 533 ff.), Bogner and Torgersen (2018), Pies et al. (2018, p. 115) and Vogt (2018, p. 46).

[99] Sunstein (2005, p. 2).

too much and too little caution.[100] Commonly accepted here is the formula, the more far-reaching and irreversible the consequences of a risky activity are, the more precaution is due.

Concerning the impact area, controllability and irreversibility of bioeconomic applications, disagreement prevails. Unintended side effects are mostly discussed under the keyword "biosafety," potential for misuse under the heading "biosecurity."[101] The assessment of harm an benefit also depends on how much risk—consisting of the probability of occurrence and the magnitude of damage of an adverse event—one is generally willing to take for what advantage.[102] Although one might think that the general willingness to take risks depends first and foremost on a social consensus, in fact the enormous competitive pressure in research and science as well as the forces of the market lead to the acceptance of ever increasing risks in order to remain competitive. In the course of this development, profits including those generated by means of high-risk technologies benefit companies, while systematically co-produced risks oftentimes cannot be limited locally and group-specifically. Thus, profits generated by risk technologies flow into private coffers, whereas society and nature as a whole suffer the consequences of the eventual costs.[103] This privatization of profits coupled with the socialization of costs is called *externalities* and has led to the claim that applying the PP is ever so important for internalizing external effects for the purpose of social and environmental justice.

Opponents of the PP, however, apprehend a status quo bias: "Advocates of the principle might [...] say that *new* risks are unacceptable, but *existing* risks are fine. [...] How does one account for tradeoffs between present and future risks? [...] Does one value a life today more than one tomorrow?"[104] In contrast, proponents of the precautionary approach claim that applying the principle in the first place concedes intrinsic value to all living entities.[105] They consider it to be imperative for man to come to a humbler, more precautionary attitude toward his place in the natural order. Moreover, they argue that man cannot guarantee the availability and applicability of complex technologies with certainty in the long term, so that all things considered, man must learn to be less dependent on complex technologies in order to become more resilient for survival in the long run.[106] Current defenders of the PP consider it to be an ethically broadly legitimized and morally significant action-guiding principle that should be applied whenever there is an uncertain or

[100] Cf. Renn (2002, p. 44) and Rippe and Willemsen (2018).

[101] For an assessment of biosafety and biosecurity in the field of synthetic biology, cf. e.g. Boldt (2018, 79 f.) and Lanzerath et al. (2020).

[102] Cf. Boldt (2018, p. 82).

[103] Cf. Kuttruff and Then (2018, 88 f., 97).

[104] Sunstein (2005, 5 f).

[105] Cf. Kuttruff and Then (2018, p. 98) and Read and Alexander (2020, p. 19).

[106] Cf. Read and Alexander (2020, 24 f).

vague possibility of new technologies causing serious damage to (depending on the underlying ethical theory) humans, animals, plants, and/or the environment.[107]

3.2 Responsible Research Innovation (RRI)

With its emphasis on risk, the PP is frequently criticized to impede technology implementation and innovation.[108] Hence, the European Commission came up with a new, so to say, counterweight principle to focus on the opportunities instead of the risks of a new technology to better promote the implementation of biotechnologies in society. Rather than eliminating risks, the focus of the Responsible Research and Innovation (RRI) principle is to reconcile technological developments with societal values and expectations. The EU has thus developed a value-based understanding of innovation with the aim of orienting technical innovations not only toward commercial interests and profits, but also toward ethical requirements and social needs. The official website of EU's biggest research and innovation program Horizon 2020 lists public engagement, open access, gender, ethics and science education as the thematic elements of RRI.[109] Here, RRI is further classified as key action of the "Science with and for Society"-objective, which is oftentimes accompanied by an RRI-definition of the European Commission's Directorate General for Research, René von Schomberg:

> Responsible Research and Innovation is a transparent, interactive process by which societal actors and innovators become mutually responsive to each other with a view to the (ethical) acceptability, sustainability and societal desirability of the innovation process and its marketable products (in order to allow a proper embedding of scientific and technological advances in our society).[110]

In the wake of RRI, innovations are assessed to be the answer to major challenges of humankind, such as climate change or world nutrition. It postulates a knowledge-based and reflexive technology policy that does not leave innovations solely to the steering power of markets and supposed constraints, but reflects, correlates, and promotes them in relation to the common good.[111] Both principles taken together—the PP as a risk management tool and the RRI as a process shaping tool—are meant to "adequately represent the double face of technological innovation".[112]

However, just as PP generates its specific problems, also RRI faces problems of its own kind. One important part of RRI is the worldwide development paradigm of

---

[107] Cf. for instance Rippe and Willemsen (2018), Kuttruff and Then (2018, p. 98) and Read and Alexander (2020).

[108] Cf. for instance Sunstein (2005, p. 5) and Bogner and Torgersen (2018).

[109] Cf. European Commission (2014) and European Commission (2020).

[110] Von Schomberg (2013, p. 63).

[111] Cf. Vogt (2018, p. 45).

[112] Bogner and Torgersen (2018, p. 4).

digitalization with its very own (ethical) challenges such as data security, data privacy or knowledge and participation.[113] In the following, only some problematic and general aspects of participation may be discussed.

In the context of bioeconomy, RRI has not yet succeeded to establish a clear ethical definition of objectives, a binding framework or a process-oriented formalization of decision-making procedures and participation rights.[114] Some critics say that even if RRI would have succeeded, it could ever hardly be more than a public awareness tool only.[115] But even its focus on continuous involvement of a heterogeneity of relevant actors, i.e. participation of various stakeholders, policy and administration as well as academia and the broader public, which surely adds value in form of alternative perspectives and rationalities that widen the decision-making horizon, faces some significant challenges. Sociologist Alexander Bogner and biologist Helge Torgersen from the Institute for Technology Assessment (ITA) enumerate among other challenges, social difficulties, issue-framing, and problems of timing and of definition. By social difficulties they refer, for example, to the unskilled trait of tolerating opinion pluralism or the unknown obligation of reasonable argumentation that lead to conflict. With the problem of issue-framing, they point to the fact that leading voices of the participation processes are mostly the same experts and institutions formulating the same standard arguments and questions, thus marginalizing alternative modes of thought from the outset. The problem of timing consists in the circumstance that at an early stage, new technologies or innovations do not interfere with the everyday life of people yet and are not yet broadly discussed in the media and do therefore not yet arouse people's interest. People tend to become motivated to critically engage with new technologies or innovations not before their trajectories have already become quite immutable or at least hardly influenceable by RRI efforts. Last but not least, the definition and deliberation of problems in the participatory, expert-led process runs the risk of remaining either too concrete and narrow, or too abstract and little committed.

In their comparison of PP and RRI Bogner and Torgersen come to the following conclusion:

> Although the PP and RRI have little in common content-wise, [...] they shared a political function, albeit using different strategies: they both should prevent or bring down controversies over particular applications among stakeholders and the public. These controversies were seen as the major obstacles to the implementation of biotechnology. [...] [Yet (C.P.)] both tools with their respective reference to risk or ethical principles and

---

[113] In the realm of agricultural bioeconomy, it is, for instance applications such as smart farming or precision agriculture that represent the predominant practices of digitalization. In general, the significance of AI for bioeconomic applications and the sustainability context is increasing rapidly. For a general conception of AI *for* sustainability and the sustainability *of* AI see van Wynsberghe (2021).

[114] Cf. Vogt (2018, p. 46).

[115] Cf. Bogner and Torgersen (2018, p. 1).

societal values could not sustainably cope with the recalcitrant problems of 'making biotechnology happen' [. . .].[116]

Besides the attested failure of RRI to have significantly overcome obstacles of technology implementation, it may further be criticized for its fixation on technical solutions.

3.3 Technological Fix Versus Behavioral Fix

Bioeconomy answers to the global ecological, social, and economic challenges with technological innovations. By some, it is even considered to be a game changer, for instance in its contribution to technologically based defossilization, decarbonization, and climate protection.[117]

Despite all the successes that have already been achieved and all the supplementary hopes and expectations for the future, two profound questions remain to be addressed: (1) Is technology eventually able to solve technologically induced problems, or does it perpetuate a never-ending spiral that creates necessity for new technologies to fix the shortcomings or deficiencies of the old ones? (2) Is it reasonable to deploy technical solutions for in actual fact moral and psychological problems?[118]

Concerning the first question, the insight about the need for developing amendments and alternatives to technological problem-solving strategies is starting to develop in some people's minds. On the one hand, people still seem to press for technological solutions, on the other hand, the message too gladly heard, *Don't you worry, technology will protect us from ourselves*, is questioned more and more.[119]

In spite of its seemingly green and sustainable appearance, the concept of bioeconomy is called into question concerning its inherent potential to prolong an inadequate mindset that has led humanity to the current situation of ecological crisis and global injustice in the first place.[120] Furthermore, it is reflected whether the installation of some bioeconomic applications necessarily always already requires innovative technical compensation for their destructive after-effects hence fostering a vicious cycle of a technological arms race: "What seems to be taking shape is a race between the advancing exhaustion of nature on the one hand and technological innovation on the other."[121] Although, it can never be entirely ruled out that the most

[116] Bogner and Torgersen (2018, p. 2).

[117] Cf. for instance Organisation for Economic Co-operation and Development (2011), Aguilar et al. (2018), von Braun (2018), European Economic and Social Committee (2018). Cf. also bioökonomie.de (2018), an initiative of the German Federal Ministry of Education and Research (BMBF).

[118] Cf. also Beck (2022) in this volume.

[119] Cf. Boldt (2018, p. 83) and Read and Alexander (2020, p. 17, 21).

[120] Cf. World Wide Fund for Nature (WWF) (2009, 2 ff.); Gottwald (2018, 100 f).

[121] Streeck (2016, p. 62).

helpful technological invention might be just within arm's reach, it would still be foolish to rely on that possibility.[122]

Thus, no matter how much a comfortable technological solution that reconciles excessive consumption patterns and business as usual with the Anthropocene's challenges is wished for, the concept of bioeconomy is not able to offer the single solution.[123] On the contrary, what is needed to combat ecological crises and mitigate climate change as well as to rectify the human-nature-relationship is a combination of biotechnological and predominantly socio-economic-ecological solutions, i.e. considerable changes in personal behaviors as well as, e.g., altered value and economic systems, and lifestyle and mobility concepts.

The second question scrutinizes whether technology optimism and faith in progress represent the advisable approach for dealing with nature, and whether it is able to adequately consider human's place in nature, the flourishing of human and non-human lifeforms as well as the planetary boundaries.

Among the reasons why a predominantly technological and bioeconomic way of dealing with living nature is conceptually misleading, Gottwald enumerates its irreducibility (beings are holistic entities which cannot be reduced to mere bricks of life), inalienability (if at all, beings may only be commodifiable and monetizable by strictest rules), unavailability (beings are equipped with intrinsic value), and unconditionality (beings are for their own sake worthy of protection).[124]

Among the steps required to bring about the overdue sustainable transformation, degrowth, post-growth, and voluntary abstinence from consumption are listed. To achieve this, capitalistic growth, materialistic individualism, and the logic of consumption need to be abolished, which of course is anything but easy and would require huge efforts for change in various areas.

One decisive area for inducing change in the logic of consumption is human psychology. The logic of consumption relies to a significant degree on status thinking. Status is a social phenomenon that is determined by and for groups and creates consumption standards and habits for group membership.[125] In this respect, status is a competitive concept that relies on social inequality. For the purpose of keeping the capitalistic economy going, structural incentives for the consumption of ever new goods that promise to further enhance status are marketed. Novelty keeps people buying more goods, which in turn keeps the economy going and the chase for status through consumption running.[126] Next to novelty *per se*, there are two further features impelling a throw-away attitude of consumers, namely creative destruction

---

[122]Cf. Jackson (2009, p. 83).

[123]Cf. Hagemann et al. (2016, p. 18) and Read and Alexander (2020, p. 9).

[124]Cf. Gottwald (2018, p. 103). I do neither subscribe to Gottwald's further conception of creatures having dignity and a right to freedom, nor to his theological viewpoint that creatures are intended by the Creator as they are. Instead, I argue for asymmetrical relations of recognition within which the morally relevant intrinsic good of all non-human lifeforms may be considered adequately (cf. Pinsdorf, 2016, 233 ff.).

[125]Cf. Victor (2019, p. 237).

[126]Cf. Sen (1998), Jackson (2009, p. 161) and Victor (2019, p. 235).

and planned obsolescence. Already in the early 1940s, influential Austrian national economist and politician Joseph Schumpeter coined the phrase *process of creative destruction* to describe the essence of capitalism, according to which old technologies are replaced by new ones and old companies are driven out of the market by the dominance of new ones in an endless cycle.[127] In combination with the feature of planned obsolescence, contemporary ecological economists observe an increasing intensification and acceleration of these structurally embedded cycles of creative destruction and novelty:

> Product lifetimes plummet as durability is designed out of consumer goods and obsolescence is designed in. Quality is sacrificed relentlessly to volume throughput. The throw-away society is not so much a consequence of consumer greed as a structural prerequisite for survival. Novelty has become a conscript to the drive for economic expansion.[128]

Now, the avoidance of status competition through consumption could already establish new ways to lessen harmful destructive practices toward the environment and the people. The never-ending spiral of producing, distributing, buying, consuming, and disposing of goods not only expands pressure through the increased material throughput and waste on the environment, but it also reinforces social inequality and creates distress, anxiety, and a fear of missing out on the people.[129]

Independently of each other, Jackson and Victor hint at ways out of this moribund vicious cycle. In a first step, structural incentives for consumption-based status competition would have to be revealed and criticized for promoting an unsustainable, sickening, and ethically unjustifiable practice. In a second step, these practices would have to be dismantled and replaced by new structures that foster the people's capabilities to flourish in much less consumptive ways and to fully participate in social life without materialistic status goods.[130] Over the course of the second step, people would have to be willing to change their value orientation and way of life. Because a human attitude of sufficiency and humility appears to be without any alternative in saving life on planet earth.[131] A general rethinking, accompanied by behavioral changes, is necessary, not least to avoid the aforementioned technologically induced rebound effects.

Even if such a development requires colossal changes and efforts on the part of society as a whole, it no longer seems to be pure utopia—for consumer culture spreads some kind of spiritual malaise, e.g. an apathetic sadness of the soul, as more

---

[127] Cf. Schumpeter (1994 [1942/43], pp. 81–86, 104) and Victor (2019, 50 f).

[128] Jackson (2009, p. 97). On obsolescence cf. also Daly (1996, p. 102).

[129] Cf. Jackson (2009, p. 154) and Victor (2019, p. 236).

[130] Cf. Kasser (2002), Jackson (2009, 153 ff., 180 ff). For the differentiation between status goods, useful goods and public goods cf. Victor (2019, 220 ff).

[131] Cf. Herrmann (2015, p. 3), Vogt (2018, p. 36) and Read and Alexander (2020, p. 19). On the huge impact of changed consumption patterns such as a less meat-based diet see, for instance, the pilot report on the monitoring of German bioeconomy by the Center for Environmental Systems Research (2020).

and more people discover that material things are not able to satisfy the human need for a meaningful life.[132] Moreover, ancient philosophical and social virtues such as temperance, appropriateness, and frugality cease to appear outdated, but are on the rise to be perceived as ever so fashionable.[133] These budding feelings, together with grassroots movements, such as Fridays for Future, claim a system change for environmental and social justice and open the door to a better future on planet earth for at least a little bit more.

To sum up, what is needed is a *Great Sustainable Transformation*[134] that encompasses both a technological *and* a behavioral fix. The first fix connects with socially acceptable technological innovations that support, among others, a new and stable economic framework which is not structurally dependent on ceaseless consumption but operates within ecological limits.[135] The second fix connects with a change of the social logic of consumerism that promotes socially meaningful and ecologically sustainable ways of human flourishing which are not structurally dependent on material accumulation and unproductive status competition, but instead enable people to fully participate in social life on ecologically sound grounds.[136]

4 Conclusion

Bioeconomy is neither a panacea for urgent challenges of the diverse crises in the Anthropocene nor is it sustainable *per se*. Real sustainability on a finite planet can only be achieved via a *Great Sustainable Transformation*. As the threefold understanding of the term sustainability—ecological, social, and economic—elucidates, economies, environments, and the socio-cultural sphere are interdependent. Economic growth effects not only the natural basis it is built upon, but also the social systems in which it is embedded. Today, economic growth runs the risk of undermining and damaging both the ecological and the social sphere. As such, the bioeconomic understanding of human flourishing or human well-being, which is still strongly connected to the concept of economic growth and prosperity, needs to be realigned in a manner that supports humanity to establish ways of flourishing meaningfully and within ethical and ecological boundaries.

[132] Cf. Read and Alexander (2020, 87 f).

[133] There are, for instance, diverse trends countering self-indulgence, such as downshifting, minimalism, vegetarianism and veganism, etc. Besides, there are more and more consumers who want to buy fewer and fewer products from companies "that do not pay attention to ecological and social aspects in their business policy" (Naturkapital Deutschland – TEEB DE, 2012, p. 66).

[134] Here I am borrowing and at the same time sharply distancing from *The Great Transformation* described by Karl Polanyi in 1944 (cf. Polanyi, 1973 [1944]).

[135] Cf. for instance the model of Contraction and Convergence (C&C) promoted by the Global Commons Institute (http://www.gci.org.uk/ [17.03.2021]).

[136] Cf. Jackson (2009, 157 f.); German Advisory Council on Global Change (2011, p. 1).

Solving the profound dilemma of growth requires rectifications on the technological and even more so on the behavioral level. It demands human society to change

> its economics, its accounts, its implicit biases against natural capital (versus man-made capital), against public wealth (versus private wealth) and against logical and less consumption (versus manic and more). And perhaps above all, human society needs to re-examine and change its relationship with nature to one of harmony and co-existence.[137]

Indian environmental economist Pavan Sukhdev, former head of the Green Economy Initiative of the United Nations Environment Programme (UNEP), Study Leader of TEEB and current president of the World Wildlife Fund (WWF), here summarizes the way in which economic reasoning has to change in order to aim for ecological as well as social justice and a sustainable economic system.

Aside from its persisting and problematic orientation toward (albeit green) growth, several semantic ambiguities of the concept of bioeconomy remain: Not least because of the various dimensions in which the relation between "bio" and "economy" is assessed contradictorily—as it is, e.g., the case concerning neoclassical versus ecological economics, the understanding of bioeconomy as economization of nature resp. ecologization of economy, or the conceptualization of nature and living beings as mere capital providing resources and ecosystem services to humans versus the conceptualization of nature and living beings, first and foremost, as entities of intrinsic value in and for themselves.

Over-optimistic promises and expectations concerning phenomena or ideas like decoupling and a zero-waste resp. renewable resources-based circular flow economy are further aspects still in need of being critically evaluated. On that front, PP and RRI are able to facilitate the process of judgment formation and critical public discourse, but are quickly stretched to their limits: for a profound and comprehensive ethical evaluation of the concept of bioeconomy prompts serious questions of relevance for philosophy of nature, anthropology, political philosophy, social philosophy, philosophy of technology, nature and environmental ethics, social ethics, animal ethics, business ethics and others.

Concerning the global questions and problems of environmental, social, and economic justice, there only exist moral guidelines such as the UN Paris Agreement or the UN Sustainable Development Goals (SDGs). There is, however, no political authority to translate those guidelines into binding and enforceable regulations so that perpetrators of globally relevant crimes against nature or mankind are really held accountable. To achieve this, a globally legitimized world government or world court would need to be set up in order to foster humanity's way out of the environmental, social, economic and—once more to our way of dealing with nature related—pandemic crisis via a truly concerted effort.

[137] Sukhdev (2009, p. xix).

References

Aguilar, A., Wohlgemuth, R., & Twardowski, T. (2018). Perspectives on bioeconomy. *New Biotechnology, 40*(A), 181–184.

Beck, B. (2022). Food as a moral problem. In D. Lanzerath, U. Schurr, C. Pinsdorf, & M. Stake (Eds.), *Bioeconomy and sustainability. Scientific, socio-economic and ethical dimensions* (pp. 33–60). Springer.

Binswanger, H. C. (2009a). *Die Wachstumsspirale. Geld, Energie und Imagination in der Dynamik des Marktprozesses*. Metropolis.

Binswanger, M. (2006). Why does income growth fail to make us happier? Searching for the treadmills behind the paradox of happiness. *The Journal of Socio-Economics, 35*, 366–381.

Binswanger, M. (2009b). Is there a growth imperative in capitalist economies? A circular flow perspective. *Journal of Post Keynesian Economics, 31*(4), 707–727.

Binswanger, M. (2019). *Der Wachstumszwang. Warum die Volkswirtschaft immer weiterwachsen muss, selbst wenn wir genug haben*. Wiley-VCH.

bioökonomie.de (2018, December 6). *Fighting climate change with bioeconomy*. Retrieved March 8, 2021, from, https://biooekonomie.de/en/news/fighting-climate-change-bioeconomy

Birch, K. (2006). The neoliberal underpinnings of the bioeconomy: The ideological discourses and practices of economic competitiveness. *Genomics, Society and Policy, 2*(3), 1–15.

Birner, R. (2018). Bioeconomy concepts. In I. Lewandowski (Ed.), *Bioeconomy* (pp. 17–38). Springer.

Bogner, A., & Torgersen, H. (2018). Precaution, responsible innovation and beyond – In search of a sustainable agricultural biotechnology policy. *Frontiers in Plant Sciences, 9*, 1884. Retrieved February 15, 2021, from, https://www.frontiersin.org/articles/10.3389/fpls.2018.01884/full

Boldt, J. (2018). Ethik in der Bioökonomie: Wishful Thinking? In: *Forum Wirtschaftsethik*, Jahresschrift des Deutschen Netzwerks Wirtschaftsethik (DNWE), 26. Jahrgang, Sonderausgabe Bioökonomie und Ethik, 78–86. Retrieved October 14, 2020, from, https://www.forum-wirtschaftsethik.de/files/biooekonomie/DNWE-Biooekonomie_und_Ethik_180913.pdf

Bonaiuti, M. (2015). Bio-economics. In G. D'Alisa, F. Demaria, & G. Kallis (Eds.), *Degrowth: A vocabulary for a new era* (pp. 25–28). Routledge.

Bradie, M. (2011). The moral life of animals. In T. L. Beauchamp & R. G. Frey (Eds.), *The Oxford handbook of animal ethics*. (pp. 547–576). Oxford University Press.

Breitenbach, A. (2009). *Die Analogie von Natur und Vernunft. Eine Umweltphilosophie nach Kant*. De Gruyter.

Center for Environmental Systems Research (CESR). (2020). *Pilotbericht zum Monitoring der deutschen Bioökonomie*. Retrieved March 8, 2021, from, https://kobra.uni-kassel.de/themes/Mirage2/scripts/mozilla-pdf.js/web/viewer.html?file=/bitstream/handle/123456789/11591/PilotberichtMonitoringBiooekonomie2020.pdf?sequence=6&isAllowed=y

Consolidated Version of the Treaty on the Functioning of the European Union. (2016). Part Three – Union Policies and Internal Actions. Title XX – Environment. Article 191 (ex Article 174 TEC). In: *Official Journal*, C 202, 07.06.2016. Retrieved February 11, 2021, from, https://eur-lex.europa.eu/eli/treaty/tfeu_2016/art_191/oj

Commission of the European Communities (COM). (2000). *Communication from the Commission on the Precautionary Principle*. Brussels, 02.02.2000. Retrieved February 11, 2021, from, https://eur-lex.europa.eu/legal-content/EN/TXT/PDF/?uri=CELEX:52000DC0001&from=DE

Daly, H. E. (1996). *Beyond growth. The economics of sustainable development*. Beacon Press.

Daly, H. E. (2009). A foreword by Herman E. Daly. In T. Jackson (Ed.), *Prosperity without growth. Economics for a finite planet* (pp. xi–xii). Earthscan.

Dörre, K. (2013). Kapitalismus im Wachstumsdilemma. Die Verdrängung der ökologischen Krisendimension und ihre Folgen. *WSI-Mitteilungen, 66*(2), 149–151.

Easterlin, R. A. (1974). Does economic growth improve the human lot? Some empirical evidence. In P. A. David & M. W. Reder (Eds.), *Nations and households in economic growth. Essays in honor of moses abramovitz* (pp. 89–125). Academic Press.

Enríquez-Cabot, J. (1998). Genomics and the World's economy. *Science Magazine, 281*(5379), 925–926.

European Commission (EC). (2014). *Responsible research and innovation. Europe's ability to respond to societal challenges*. Retrieved February 15, 2021, from, https://ec.europa.eu/research/swafs/pdf/pub_rri/KI0214595ENC.pdf

European Commission (EC). (2020). *Responsible research & innovation. Horizon 2020*. Retrieved March 17, 2021, from, https://ec.europa.eu/programmes/horizon2020/en/h2020-section/responsible-research-innovation

European Commission Directorate-General for Environment. (2018). *Study on the precautionary principle in EU environmental policies*. Final Report. Retrieved February 11, 2021, from, https://op.europa.eu/en/publication-detail/-/publication/18091262-f4f2-11e7-be11-01aa75ed71a1/language-en

European Economic and Social Committee. (2018). *Bioeconomy – Contributing to achieving the EU's climate and energy goals and the UN's sustainable development goals*. Retrieved March 8, 2021, from, https://www.eesc.europa.eu/en/our-work/opinions-information-reports/opinions/bioeconomy-contributing-achieving-eus-climate-and-energy-goals-and-uns-sustainable-development-goals-exploratory-opinion

European Parliament. (2000). *Lisbon European Council 23 and 24 March 2000. Presidency conclusions*. Retrieved October 20, 2020, from, https://www.europarl.europa.eu/summits/lis1_en.htm

Fatheuer, T., Fuhr, L., & Unmüßig, B. (2015). *Kritik der Grünen Ökonomie*. Oekonom.

Fritsche, U., & Rösch, C. (2017). Die Bedingung einer nachhaltigen Bioökonomie. In J. Pietzsch (Ed.), *Bioökonomie für Einsteiger* (pp. 177–203). Springer.

Georgescu-Roegen, N. (1971). *The entropy law and the economic process*. Harvard University Press.

German Advisory Council on Global Change (WBGU). (2011). *World in transition. A social contract for sustainability*. Flagship Report. Retrieved October 14, 2020, from, https://www.wbgu.de/fileadmin/user_upload/wbgu/publikationen/hauptgutachten/hg2011/pdf/wbgu_jg2011_kurz_en.pdf

German Bioeconomy Council. (2014). *Positions and strategies of the German bioeconomy*. Retrieved October 13, 2020, from, https://biooekonomierat.de/fileadmin/Publikationen/Englisch/Strategy_paper.pdf

Global Bioeconomy Summit. (2015). Communiqué of the Global Bioeconomy Summit 2015. *Making Bioeconomy Work for Sustainable Development*. Retrieved October 14, 2020, from, https://gbs2015.com/fileadmin/gbs2015/Downloads/Communique_final.pdf

Gordon, M. J., & Rosenthal, J. S. (2003). Capitalism's Growth Imperative. *Cambridge Journal of Economics, 27*(1), 25–48.

Gottwald, F.-T. (2018). Auf der Suche nach Regeln für eine nachhaltige Bioökonomie – Sechs Thesen zur Regulierung aus ethischer Sicht. In: *Forum Wirtschaftsethik*, Jahresschrift des Deutschen Netzwerks Wirtschaftsethik (DNWE), 26. Jahrgang, Sonderausgabe Bioökonomie und Ethik, 100–105. Retrieved October 14, 2020, from, https://www.forum-wirtschaftsethik.de/files/biooekonomie/DNWE-Biooekonomie_und_Ethik_180913.pdf.

Gottwald, F.-T., & Krätzer, A. (2014). *Irrweg Bioökonmie. Kritik an einem totalitären Ansatz*. Suhrkamp.

Grefe, C. (2018). Dasselbe in Grün? – Konfliktfelder, Konfliktlinien und Alternativen der Bioökonomie. In *Forum Wirtschaftsethik*, Jahresschrift des Deutschen Netzwerks Wirtschaftsethik (DNWE), 26. Jahrgang, Sonderausgabe Bioökonomie und Ethik, 20–30. Retrieved October 14, 2020, from, https://www.forum-wirtschaftsethik.de/files/biooekonomie/DNWE-Biooekonomie_und_Ethik_180913.pdf

Hagemann, N., Gawel, E., Purkus, A., Pannicke, N., & Hauck, J. (2016). Possible futures towards a wood-based bioeconomy: A scenario analysis for Germany. *Sustainability,* 8(2), 98. Retrieved March 22, 2021, from, https://www.mdpi.com/2071-1050/8/1/98

Hamm, H. (2018). Kann Biotechnologie etwas dazu beitragen, unsere Klimaziele im Verkehr zu erreichen? In *Forum Wirtschaftsethik*, Jahresschrift des Deutschen Netzwerks Wirtschaftsethik (DNWE), 26. Jahrgang, Sonderausgabe Bioökonomie und Ethik, 138–145. Retrieved October 14, 2020, from, https://www.forum-wirtschaftsethik.de/files/biooekonomie/DNWE-Biooekonomie_und_Ethik_180913.pdf

Heimann, T. (2018). Bioeconomy and SDGs: Does the bioeconomy support the achievement of the SDGs? *Earth's Future, 7*, 43–57. Retrieved March 18, 2021, from, https://doi.org/10.1029/2018EF001014

Herrmann, U. (2015). Über das Ende des Kapitalismus. In *die tageszeitung* (taz), 10.04.2015, Sektion: Le Monde Diplomatique, 3.

Jackson, T. (2009). *Prosperity without growth. Economics for a finite planet.* Earthscan.

Jonas, H. (2017). *Das Prinzip Verantwortung. Versuch einer Ethik für die technologische Zivilisation* (6th ed.). Suhrkamp.

Kasser, T. (2002). *The high price of materialism.* MIT Press.

Kopnina, H., Washington, H., Taylor, B., & Piccolo, J. (2018). Anthropocentrism: more than just a misunderstood problem. *Journal of Agricultural and Environmental Ethics, 31*, 109–127.

Kuttruff, M., & Then, C. (2018). Bioökonomie: Vom Waren-Wert des Lebens. In *Forum Wirtschaftsethik*, Jahresschrift des Deutschen Netzwerks Wirtschaftsethik (DNWE), 26. Jahrgang, Sonderausgabe Bioökonomie und Ethik (pp. 87–99). Retrieved October 14, 2020, from, https://www.forum-wirtschaftsethik.de/files/biooekonomie/DNWE-Biooekonomie_und_Ethik_180913.pdf

Lago, C., Caldés, N., & Lechón, Y. (2019). *The role of bioenergy in the emerging bioeconomy. resources, technologies, sustainability and policy.* Academic Press.

Lanzerath, D., Giese, B., & Jaeckel, L. (2020). *Synthetische Biologie. Naturwissenschaftliche, rechtliche und ethisch Aspekte.* Karl Alber.

Lanzerath, D., & Schurr, U. (2022). Introduction. In D. Lanzerath, U. Schurr, C. Pinsdorf, & M. Stake (Eds.), *Bioeconomy and sustainability. Scientific, socio-economic and ethical dimensions* (pp. 3–10). Springer.

Leshem, D. (2016). Retrospectives. What Did the Ancient Greeks Mean by *Oikonomia*? *The Journal of Economic Perspectives, 30*(1), 225–238.

Meadows, D. H., Meadows, D. L., Randers, J., & Behrens III, W. W. (1972). *The limits to growth. A report for the club of Rome's project on the predicament of mankind.* Universe Books.

Müller, A. (2017). Bios (Leben, Lebensform). In J. Ritter, K. Gründer, & G. Gabriel (Eds.), *Historisches Wörterbuch der Philosophie online*. Retrieved March 15, 2021, from, https://www.schwabeonline.ch/schwabe-xaveropp/elibrary/start.xav#__elibrary__%2F%2F*%5B%40attr_id%3D%27verw.bios.leben.lebensform%27%5D__1615807467119

Naturkapital Deutschland – TEEB DE. (2012). *Der Wert der Natur für Wirtschaft und Gesellschaft – Eine Einführung*. Ifuplan.

Organisation for Economic Co-operation and Development (OECD). (2011). *Industrial biotechnology and climate change. Opportunities and challenges*. Retrieved March 11, 2021, from, https://www.oecd.org/sti/emerging-tech/49024032.pdf

Organisation for Economic Co-operation and Development (OECD). (2015). *Environment at a glance: OECD indicators*. Retrieved October 19, 2020, from, https://www.oecd-ilibrary.org/docserver/9789264235199-en.pdf?expires=1603099391&id=id&accname=guest&checksum=84081DF52B6692F7F0369CE8FFDD9293

Pies, I., Hielscher, S., Valentinov, V., & Everding, S. (2018). Gesellschaftliche Lernprozesse zur Förderung der Bioökonomie – eine ordonomische Argumentationsskizze. In *Forum Wirtschaftsethik*, Jahresschrift des Deutschen Netzwerks Wirtschaftsethik (DNWE), 26. Jahrgang, Sonderausgabe Bioökonomie und Ethik (pp. 106–116). Retrieved October

14, 2020, from, https://www.forum-wirtschaftsethik.de/files/biooekonomie/DNWE-Biooekonomie_und_Ethik_180913.pdf

Pinsdorf, C. (2020). Haben Flora und Fauna einen Preis oder eine Würde? In Deutsche Wildtier Stiftung (Hg.), *Fauna, Flora und Finanzen – Welchen Wert hat die Natur?* (pp. 18–27). Deutsche Wildtier Stiftung.

Pinsdorf, C. (2016). *Lebensformen und Anerkennungsverhältnisse. Zur Ethik der belebten Natur.* De Gruyter.

Polanyi, K. (1973). *The Great Transformation. Politische und ökonomische Ursprünge von Gesellschaften und Wirtschaftssystemen* [1944] (6th ed.). Suhrkamp.

Radke, V. (2004). Naturverständnisse in der ökonomischen Nachhaltigkeitsforschung. In D. Rink & M. Wächter (Eds.), *Naturverständnisse in der Nachhaltigkeitsforschung* (pp. 141–172). Campus.

Read, R., & Alexander, S. (2020). *Diese Zivilisation ist gescheitert. Gespräche über die Klimakrise und die Chance eines Neuanfangs.* Felix Meiner.

Renn, O. (2002). Vorsorge als Prinzip: Besser in der Vorsicht irren als im Wagemut. *GAIA – Ecological Perspectives for Science and Society, 11*(1), 44–45.

Renn, O. (2014). *Das Risikoparadox. Warum wir uns vor dem Falschen fürchten* (2nd ed.). Fischer.

Rippe, K. P., & Willemsen, A. (2018). The idea of precaution: Ethical requirements for the regulation of new biotechnologies in the environmental field. *Frontiers in Plant Sciences, 9*, 1868. Retrieved February 15, 2021, from, https://www.frontiersin.org/articles/10.3389/fpls.2018.01868/full

Robinson, M. (2009). A foreword by Mary Robinson. In T. Jackson (Ed.), *Prosperity without growth. Economics for a finite planet* (pp. xv–xvi). Earthscan.

Schleissing, S. (2018). Bioökonomie als gesellschaftlicher Fortschritt? – Ethische Überlegungen zur Politikstrategie Bioökonomie. In: *Forum Wirtschaftsethik*, Jahresschrift des Deutschen Netzwerks Wirtschaftsethik (DNWE), 26. Jahrgang, Sonderausgabe Bioökonomie und Ethik (pp. 70–77). Retrieved October 14, 2020, from, https://www.forum-wirtschaftsethik.de/files/biooekonomie/DNWE-Biooekonomie_und_Ethik_180913.pdf

Schoop, J. F. (2022). Conditions for an ethically responsible and sustainable bioeconomy based on Hans Jonas' ethics of responsibility. In D. Lanzerath, U. Schurr, C. Pinsdorf, & M. Stake (Eds.), *Bioeconomy and sustainability. Scientific, socio-economic and ethical dimensions* (pp. 281–306). Springer.

Schumpeter, J. A. (1994). *Capitalism, socialism and democracy* [1942/43]. Revised Edition with a New Introduction by Richard Swedberg. Routledge.

Sen, A. (1998). The living standard. In D. Crocker, & T. Linden (Eds.), *The ethics of consumption* (pp. 287–311). Rowman & Littlefield.

Singer, P. (1977). *Animal liberation: a new ethics for our treatment of animals.* Random House.

Smith, R. (2010). Beyond growth or beyond capitalism? *Real-World Economics Review, 53*(2), 28–42. Retrieved March 16, 2021, from, http://www.paecon.net/PAEReview/issue53/Smith53.pdf

Streeck, W. (2016). *How will capitalism end? Essays on a failing system.* Verso

Sturma, D. (2013). Naturethik und Biodiversität. In L. Honnefelder & D. Sturma (Eds.), *Jahrbuch für Wissenschaft und Ethik* (Vol. 17, pp. 141–155). De Gruyter.

Sukhdev, P. (2009). A foreword by Pavan Sukhdev. In T. Jackson (Ed.), *Prosperity without growth. Economics for a finite planet* (pp. xvii–xx). Earthscan.

Sunstein, C. R. (2005). The Precautionary Principle as a Basis for Decision Making. *The Economists' Voice, 2*(2), 1–9.

Thompson, A. (2017). Anthropocentrism. Humanity as peril and promise. In S. M. Gardiner, & A. Thompson (Eds.), *The Oxford handbook of environmental ethics* (pp. 77–90). Oxford University Press.

United Nations (UN). (1992). *Agenda 21*. United Nations Conference on Environment and Development. Rio de Janeiro, Brazil, 3–14 June 1992. Retrieved October 14, 2020, from, https://sustainabledevelopment.un.org/content/documents/Agenda21.pdf

United Nations Conference on Environment and Development. (1992). *Rio declaration on environment and development*. Rio de Janeiro, Brazil, 3–14 June 1992. Retrieved February 5, 2021, from, https://www.un.org/en/development/desa/population/migration/generalassembly/docs/globalcompact/A_CONF.151_26_Vol.I_Declaration.pdf

United Nations Environment Programme (UNEP). (2016). *Global material flows and resource productivity: Assessment report for the UNEP international resource panel*. Retrieved October 16, 2020, from, https://www.researchgate.net/publication/305597138_Global_Material_Flows_and_Resource_Productivity_An_Assessment_Study_of_the_UNEP_International_Resource_Panel

van Wynsberghe, A. (2021). Sustainable AI: AI *for* sustainability and the sustainability *of* AI. *AI and Ethics*. Retrieved April 14, 2021, from, https://doi.org/10.1007/s43681-021-00043-6

Victor, P. A. (2019). *Managing without growth. Slower by design, not disaster* (2nd ed.). Edward Elgar.

Vogt, M. (2018). Bedingungen ethisch verantwortbarer Bioökonomie. In *Forum Wirtschaftsethik*, Jahresschrift des Deutschen Netzwerks Wirtschaftsethik (DNWE), 26. Jahrgang, Sonderausgabe Bioökonomie und Ethik (pp. 31–51). Retrieved October 14, 2020, from, https://www.forum-wirtschaftsethik.de/files/biooekonomie/DNWE-Biooekonomie_und_Ethik_180913.pdf

von Braun, J. (2018). Lösungsansätze der Bioökonomie zur Begegnung der großen globalen Herausforderungen. In *Forum Wirtschaftsethik*, Jahresschrift des Deutschen Netzwerks Wirtschaftsethik (DNWE), 26. Jahrgang, Sonderausgabe Bioökonomie und Ethik (pp. 9–19). Retrieved October 14, 2020, from, https://www.forum-wirtschaftsethik.de/files/biooekonomie/DNWE-Biooekonomie_und_Ethik_180913.pdf

von Schomberg, R. (2013). A vision of responsible research and innovation. In R. Owen, J. Bessant, & M. Heintz (Eds.), *Responsible innovation. Managing the responsible emergence of science and innovation in society* (pp. 51–74). Wiley-VCH.

World Commission on Environment and Development (WCED). (1987). *Report – Our common future*. Retrieved October 14, 2020, from, https://sustainabledevelopment.un.org/content/documents/5987our-common-future.pdf

World Forum on Natural Capital. (2017). *What is natural capital?* Retrieved October 15, 2020, from, https://naturalcapitalforum.com/about/

World Wide Fund for Nature (WWF). (2009). *Industrial biotechnology – More than green fuel in a dirty economy? Exploring the transformational potential of industrial biotechnology on the way to a green economy*. Retrieved October 14, 2020, from, http://industrialbiotech-europe.eu/wp-content/uploads/2015/04/wwf_biotech.pdf

Correction to: Problem Structures of Bioenergy Policy in the Power and Heat Sector in Germany

Katrin Beer

Correction to:
Chapter 9 in: D. Lanzerath et al. (eds.),
***Bioeconomy and Sustainability*,**
https://doi.org/10.1007/978-3-030-87402-5_9

The book was inadvertently published with the incorrect figures for Fig. 9.2, Fig. 9.3, and Fig. 9.4 in Chap. 9, which have been updated as below:

Fig. 9.2:

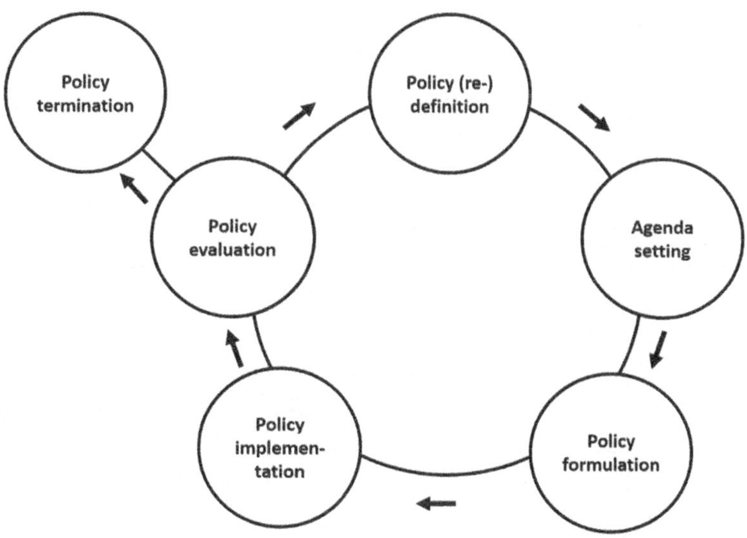

The updated version of this chapter can be found at https://doi.org/10.1007/978-3-030-87402-5_9

D. Lanzerath et al. (eds.), *Bioeconomy and Sustainability*,
https://doi.org/10.1007/978-3-030-87402-5_21

Fig. 9.3:

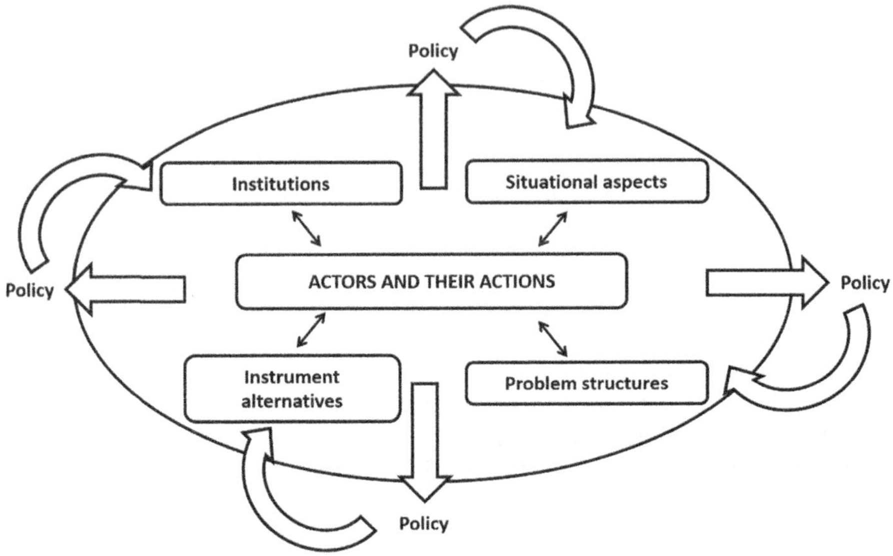

Fig. 9.4:

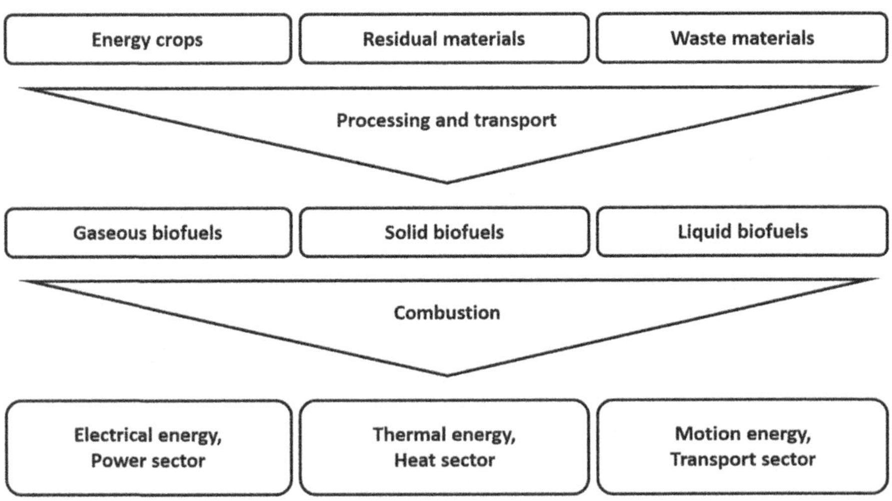

The corrections have been carried out in the chapter and the updated chapter has been approved by the author.

The manufacturer's authorised representative in the EU is Springer Nature Customer Service Centre GmbH, Europaplatz 3, 69115 Heidelberg, Germany. If you have any concerns regarding our products, please contact ProductSafety@springernature.com

Printed and bound by CPI Group (UK) Ltd, Croydon, CR0 4YY
22/09/2026
02220956-0011